THE CHEMISTRY OF IMIDOYL HALIDES

THE CHEMISTRY
of
IMIDOYL HALIDES

HENRI ULRICH

Donald S. Gilmore Research Laboratories
The Upjohn Company
North Haven, Connecticut

PLENUM PRESS • NEW YORK • 1968

Library of Congress Catalog Card Number 68-26773

ISBN 978-1-4684-8949-1   ISBN 978-1-4684-8947-7 (eBook)
DOI 10.1007/978-1-4684-8947-7

PREFACE

Primarily, the aim of this book is to provide a reference work for senior students and research workers engaged in the synthetic aspects of chemistry. The various classes of compounds under discussion provide useful intermediates for the synthesis of numerous nitrogen-containing derivatives. Imidoyl halides are also intermediates in several classical name reactions, such as the Gattermann, Houben–Hoesch, and Vilsmeier–Haack syntheses of aldehydes and ketones, the Beckmann rearrangement, and the v. Braun degradation. Some imidoyl halides have shown interesting agricultural activities, and the generation of highly reactive species (ketenimines, nitrile oxides, nitrile imides, carbodiimides, etc.) from imidoyl halides has contributed to the study of polar cycloaddition reactions.

To enable researchers to utilize this chemistry without consulting the original references, I have included a number of selected working examples. This procedure will facilitate the transformation of written information into well designed experiments, especially since part of the cited literature is not readily available.

The book is organized around classes of imidoyl halides, with synthesis and chemistry discussed in an orderly fashion. The physical properties of the known imidoyl halides are listed in tables, and I have attempted to draw attention to the more recent literature. The one or two references provided for each compound represent those which best describe its physical constants and synthesis.

Of course, the selection and emphasis devoted to the various topics are undoubtedly influenced by my own personal interest. Since this volume is designed to provide a comprehensive review of syntheses and reactions of imidoyl halides, mechanistic interpretations are only incorporated in order to stimulate discussion and to provide impetus for more detailed studies which are clearly needed.

I am greatly indebted to my colleagues who have contributed in this field, and I am thankful to Nancy Erardi, who has taken part in preparing the manuscript. In particular, I would like to thank my wife, whose encouragement and kind support made this volume possible.

Northford, Connecticut   Henri Ulrich
June, 1968

CONTENTS

Chapter 1

INTRODUCTION

I. SCOPE

Imidoyl halides, by definition, are a group of reactive organic compounds characterized by the presence of a halo group attached to the carbon atom of a carbon–nitrogen double bond. Although a great deal of the chemistry of this class of compounds has been done on the chloro derivatives, similar chemical behavior is expected for both the bromo and the iodo compounds. The chemistry of the fluoro derivatives differs somewhat, as will be shown in the text.

By the above definition two additional groups have to be specified, namely, the second group R attached to the carbon atom and the substituent R' on the nitrogen.

$$R-\underset{\underset{X}{|}}{C}{=}NR' \qquad X = F, Cl, Br, I$$

Furthermore, the free electron pair on the nitrogen confers to the imidoyl halides basic character, which can be utilized for salt formation. The iminium salts in their most general form can be written in the polar form (I) and in the nonpolar geminate dihalo form (II).

$$R-\underset{\underset{X}{|}}{C}{=}\overset{\oplus}{N}HR']X^{\ominus} \longleftrightarrow R-\underset{\underset{X}{|}}{\overset{\overset{X}{|}}{C}}-NHR'$$

$$\text{I} \qquad\qquad\qquad \text{II}$$

In the case of X = Cl, Br, and I, the iminium salts exist predominantly in the polar form (I), as evidenced by spectral data, whereas in the case of X = F, the compounds are in the geminate halide form (II).

The organization of this book is centered around several major groups of imidoyl halides, which are shown in Table I. Of course, this classification is arbitrary, but it provides a framework for useful discussion. A variety of

1

TABLE I
Imidoyl Halides

Chapter	Class of compounds	Name
2	X—C=NR | X	Carbonimidoyl dihalides
3	R—C=NR′ | X	Imidoyl halides
4	R_2N—C=NR′ | X	Haloformamidines
5	RO—C=NR′ | X	Haloformimidates
5	RS—C=NR′ | X	Halothioformimidates
6	R—C=NOH | X	Hydroxamoyl halides
7	R—C=NNR′$_2$ | X	Hydrazidoyl halides
8	⌐(CH$_2$)$_n$⌐ ⌐C=N⌐ | X	Heterocyclic imidoyl halides

other substituents, such as the pseudohalo groups (CN, NCO, OCN, NCS, SCN) and carboxyl groups, can be attached to the —C(X)=N— system, and compounds of this type are discussed within the outlined scheme. Likewise, O-substituted derivatives of hydroxamoyl halides are included in Chapter 6. Since the imidoyl halides can be part of a heterocyclic system, the cyclic homologs are treated separately in the last chapter. The fully aromatic compounds, such as halopyridines, halopyrimidines, and halo-1,3,5-triazines, are discussed only briefly in Chapter 8, as their chemistry is somewhat different.

The nomenclature of the imidoyl halides is not clearly established; I have therefore attempted to select names which identify the compounds as imidoyl halides, and which furthermore, relate to their chemistry. For example, *Chemical Abstracts* lists the hydroxamoyl chlorides as oximes of the corresponding carboxylic acid chlorides. Although this name is formally correct, it does not reflect the chemistry of this class of compounds, which is closely related to hydroxamic acids, its hydrolysis products.

II. REACTIVITY

A halo group attached to a carbon which has a heteroatom attached to it has increased reactivity. This concept is amply explained by the electronic assistance of the heteroatom. For example, the halogen groups in α-haloethers and α-chloroamines are considerably more reactive than halo groups attached to aliphatic carbon atoms. The reactivity of halo groups is further enhanced if the halo group is directly attached to a C=O double bond. For example, acid chlorides (III) are considerably more reactive than alkyl or aryl halides. Likewise, halo groups directly attached to C=N double bonds have an increased reactivity, quite comparable to that of acid chlorides. Thus imidoyl chlorides (IV) resemble acid chlorides in their chemical reactions.

$$\begin{array}{cc} R-C=O & R-C=NR' \\ | & | \\ Cl & Cl \\ III & IV \end{array}$$

Since imino groups contain an additional substituent, R', modification of R' will most certainly influence the mobility of the halo group. It is indeed surprising that rate studies related to the solvolysis of imidoyl halides have not been conducted extensively. This is perhaps related to the enormous complexities encountered in the reaction pattern of this type of molecule. For example, Hudson and his co-workers ([4]) noted the ambident behavior of acid halides in their solvolysis reactions. Likewise, in imidoyl halides the nucleophilic displacement reaction can not be explained exclusively by the S_{N2} mechanism. Imidoyl halides, having α-hydrogen atoms attached to the alkyl substituent (R), can undergo a bimolecular elimination reaction. Thus a merged mechanism involving both elimination and substitution may operate in these cases.

Only one pertinent rate study can be found in the recent literature, namely that of Ugi and his co-workers ([19]), who measured the rate of solvolysis of a variety of imidoyl chlorides, in acetone/water. Since these reactions did not obey first-order rate law, a two-step mechanism, leading to a nitrilium chloride ion-pair intermediate, was postulated. However, the exceedingly fast rates observed in imidoyl chlorides, having α-hydrogens, may indicate elimination, with formation of a ketenimine intermediate. In the absence of α-hydrogens (R = aryl) the reaction is considerably slower, and electron-withdrawing substituents attached to the aromatic ring further retard the rate of hydrolysis ([19]). If the substituent attached to the carbon atom is an acyl group (R = RCO) the rate of solvolysis is even slower.

The effect of the substituents attached to the nitrogen (R') on the rate of solvolysis of imidoyl chlorides in acetone/water is similar, i.e., R' = alkyl > R' = aryl, and electron-withdrawing substituents attached to the aryl moiety retard the rate, whereas electron-donating groups enhance it.

Cyclic systems, such as 2-chloro-Δ^1-pyrrolin, react considerably slower, because elimination can not readily occur in the cyclic configuration.

The hydrolysis of hydroxamoyl chlorides, observed by Souchay *et al.*([16]), also occurs *via* an elimination sequence, with formation of the corresponding nitrile oxide intermediate. Likewise, solvolysis of N,N'-disubstituted chloroformamidines may involve the carbodiimide as the reactive intermediate.

Studies related to the nucleophilic displacement in halo N-heterocycles indicate that the "imidoyl character" determines the mobility of the halo group ([2,25]). For example, the rate of reaction of 2-bromoquinoline (V) with methoxide ion is about 80 times faster than that of 2-bromopyridine (VI). Apparently, in the benzo derivatives, the aromatic character of the ring with the heteroatom is reduced, thereby increasing the imidoyl character and thus enhancing the rate.

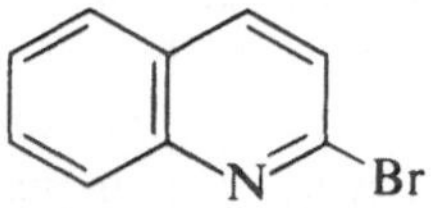

V 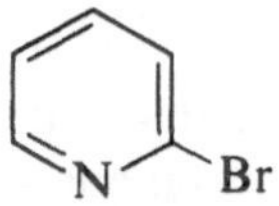　　　　　　　　　　　　VI

Likewise, 2-chlorobenzthiazole (VII) reacts considerably faster than 2-chlorothiazole ([2]).

The rate of reaction of 2-halobenzthiazole with nucleophiles has been studied by a group at the University of Bologna, and the results are as expected, i.e., F $\gg$ Cl > Br > I ([13]). The effect of substituents on the aromatic ring was investigated also; electron-withdrawing groups in 4,5- and 6,7-substituted 2-chlorobenzthiazoles increase the rate of reaction with sodium methoxide, whereas electron-donating groups lower the rate ([14,17,18]).

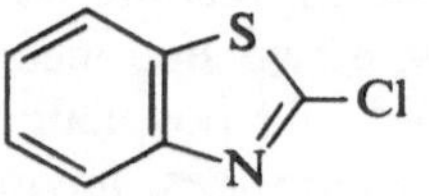

VII 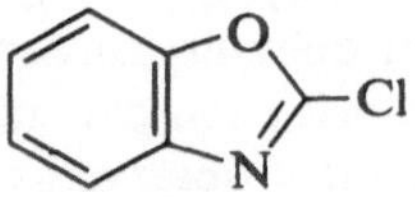　　　　　　　　　　　　VIII

In 2-chlorobenzoxazole (VIII) a similar behavior has been observed ([3]).

Likewise, 2-chloroquinoline reacts considerably faster than 4-chloroquinoline in nucleophilic substitution reactions, and Illuminati and his colleagues ([7]) refer to this phenomenon as the "α-aza" effect.

Further work is certainly needed to ascertain the magnitude of the influences operating here.

III. GENERAL REACTIONS

Imidoyl halides are readily synthesized from the corresponding carboxylic and thiocarboxylic nitrogen derivatives. The best reagents for the generation of imidoyl halides are carbonyl halides, such as phosgene, because the by-products are gases, and therefore can be easily removed with nitrogen. Although imidoyl halides have been known since 1876, they have been used extensively only by v. Braun and his students; by now their utility in organic synthesis is almost forgotten. It is the purpose of this book to focus attention to this class of compounds, and to demonstrate its usefulness for the synthesis of numerous linear and cyclic nitrogen compounds.

The chemistry of imidoyl halides is centered around the mobile halo group which can either be displaced or eliminated. Thus, imidoyl halides resemble very much carboxylic acid halides, except that the scope of their reactions is enlarged by the possibility of substitution on the nitrogen.

The displacement reactions are both nucleophilic and electrophilic and the nucleophilic substitution reactions especially dominate the chemistry of this class of compounds. In fact, imidoyl halides are so reactive that they have to be protected from atmospheric moisture, which is known to react readily with many imidoyl halides. In this reaction the $-CONH-$ group is regenerated.

The solvolysis can be autocatalytic because the generated acid could catalyze further reaction. In addition to water, a great variety of nucleophiles, such as $OR^{\ominus}$, $SH^{\ominus}$, $SR^{\ominus}$, $NR_2^{\ominus}$, $CN^{\ominus}$, $COO^{\ominus}$, $N_3^{\ominus}$, etc., have been reacted with the imidoyl halide system and the results obtained from several main groups of imidoyl halides are summarized in Table II.

$$
\underset{\substack{|\\ Cl}}{RC}=NR' + :B^{\ominus} \longrightarrow \underset{\substack{|\\ B}}{R-C}=NR' + Cl^{\ominus}
$$

TABLE II
Nucleophilic Substitution Reactions of Imidoyl Halides

$$RCX=NR' + :B^{\ominus} \longrightarrow RCB=NR' + X^{\ominus}$$

Starting material	Products
$Cl-C(Cl)=NR'$	Isocyanates, isothiocyanates and their O- and S-diacetals, guanidines, halo- and azidotetrazoles
$R-C(Cl)=NR'$	Amides, thioamides, imidates, thioimidates, amidines, tetrazoles, acylamides
$R_2N-C(Cl)=NR'$	Ureas, thioureas, pseudoureas, pseudothioureas, guanidines, aminotetrazoles
$R-C(Cl)=NOH$	Hydroxamic acids, thiohydroxamic acids, amide oximes, tetrazoles

The resultant classes of compounds are often difficult to synthesize by other procedures.

A complication can occur when an aliphatic carboxylic acid amide is converted to the corresponding imidoyl halide, because isomerization to the corresponding enamine (IX) occurs. The generated enamine undergoes rapid reaction with the imidoyl halide to form the condensation product X.

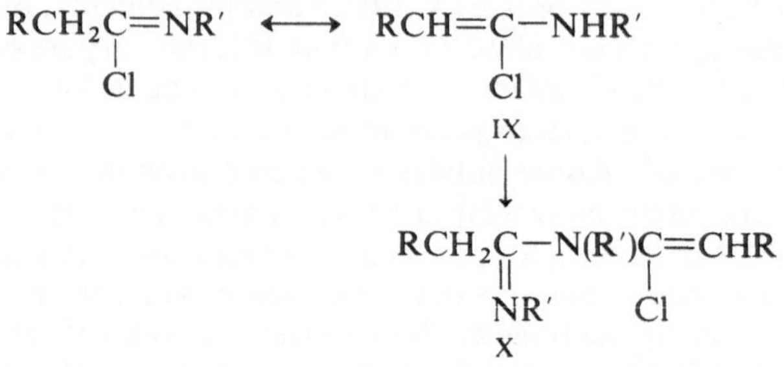

$$\underset{\underset{Cl}{|}}{RCH_2C}=NR' \longleftrightarrow \underset{\underset{Cl}{|}}{RCH}=\underset{}{C}-NHR'$$

$$IX$$

$$\downarrow$$

$$\underset{\underset{NR'}{||}}{RCH_2C}-N(R')\underset{\underset{Cl}{|}}{C}=CHR$$

$$X$$

This condensation reaction can be prevented by either introducing steric hindrance, or by lowering the basicity on the nitrogen by suitable substitution (for example, when the R' group is an arylsulfonyl group).

In the reaction of monosubstituted aliphatic carboxylic acid amides with phosgene, the enamine intermediate can be preferentially reacted with phosgene to form the carbamoyl chlorides XI, a valuable group of chemical intermediates ([10]).

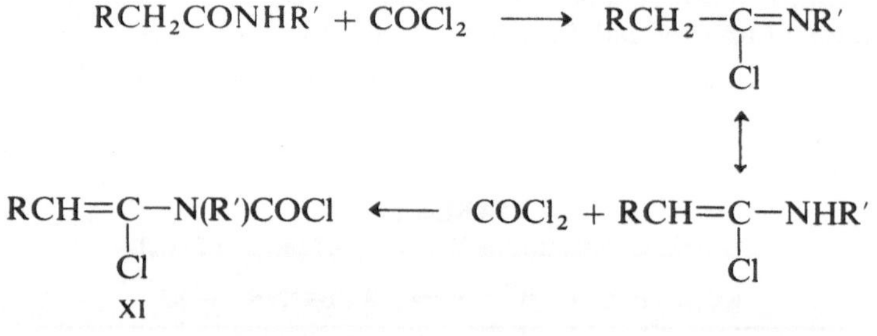

$$RCH_2CONHR' + COCl_2 \longrightarrow RCH_2-\underset{\underset{Cl}{|}}{C}=NR'$$

$$\updownarrow$$

$$\underset{\underset{Cl}{|}}{RCH}=C-N(R')COCl \longleftarrow COCl_2 + \underset{\underset{Cl}{|}}{RCH}=C-NHR'$$

$$XI$$

This reaction is similar to the conversion of chloroformamidines (XII) with phosgene to the corresponding N-carbonyl chlorides (XIII) ([20]).

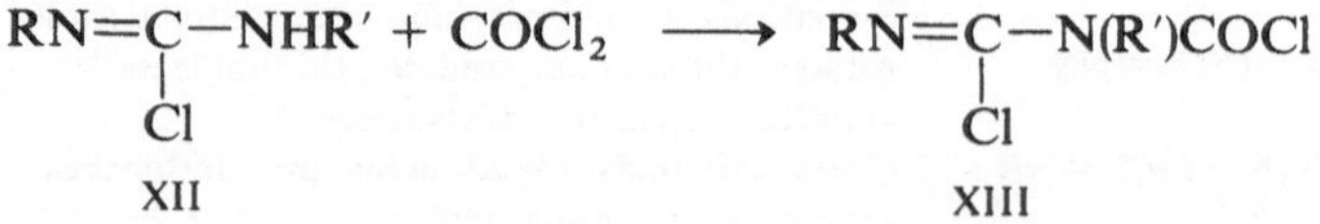

$$\underset{\underset{Cl}{|}}{RN}=C-NHR' + COCl_2 \longrightarrow \underset{\underset{Cl}{|}}{RN}=C-N(R')COCl$$

$$XII \qquad\qquad\qquad\qquad\qquad XIII$$

The electrophilic substitution reactions involving imidoyl halides are centered around the displacement on carbon–hydrogen bonds. For example,

the well-known Gattermann, Houben–Hoesch, and Vilsmeier–Haack aldehyde and ketone syntheses involve imidoyl halides as reactive intermediates.
Since these reactions are the subject of numerous review articles, no attempt
has been made to be comprehensive; recent literature related to these
reactions is cited in Chapter 3.

Another interesting application of iminium chlorides in organic
synthesis involves their use as catalysts in the conversion of carboxylic
acids to carboxylic acid chlorides ([1]), carbamates to isocyanates ([22]), amino-
acids to isocyanato acid chlorides ([23]) and arylamine hydrochlorides to
isocyanates ([24]). In these reactions chlorodimethylformiminium chloride is
being used, because it is readily available. However, other imidoyl halides
could react in a like manner.

Perhaps more important than the substitution reactions are the elimination reactions of imidoyl halides. Three types of reactions are encountered,
depending upon the substituents attached to the imidoyl moiety. For
example, if R′ = H, elimination of hydrogen chloride occurs quite readily
and the corresponding nitrile is obtained.

$$R-\underset{\underset{Cl}{|}}{C}=NH \xrightarrow{\Delta} RCN + HCl$$

In the absence of α-hydrogens in the substituent R, with R′ = alkyl,
elimination of alkyl halide is often observed.

$$R-\underset{\underset{Cl}{|}}{C}=NR' \xrightarrow{\Delta} RCN + R'Cl$$

If an α-halogen is available hydrogen chloride can be eliminated,
thereby generating a cumulative double bond system or a 1,3-dipole.

$$RCH_2-\underset{\underset{Cl}{|}}{C}=NR' \xrightarrow{\Delta} RCH=C=NR' + HCl$$

The classical example of this type of reaction is the v. Braun method of
degradation, by which a secondary amine can be transformed into a primary
amine, and finally into ammonia, via the imidoyl chloride intermediates.
This reaction has been used for the structure elucidation of alkaloids.
Although degradative studies are not conducted to any extent today due to
the convenience of spectral methods, the v. Braun elimination is still used
in organic synthesis to prepare nitriles as well as alkyl halides.

Far more interesting are the elimination sequences which lead to the
generation of reactive heterocumulenes (ketenimines, carbodiimides, isocyanates, isothiocyanates) or 1,3-dipoles (nitrile oxides, nitrile imides, nitrile

ylides). In Table III the various products obtained from imidoyl halides *via* elimination procedures are shown. The elimination can be conducted by using a base as the hydrogen chloride scavenger, or by thermolysis, preferentially in an inert solvent. Both methods are quite useful, and Quilico and his co-workers ([11]) and recently Huisgen and his students ([5]) have used hydroxamoyl halides and hydrazidoyl halides extensively for the generation of nitrile oxides, nitrile imides, and nitrile ylides, respectively. The 1,3-dipolar addition of fulminic acid, for example, has settled the question regarding the structure of this species ([6,12]).

The elimination of hydrogen chloride from chloroformamidine hydrochlorides is another example illustrating the utility of imidoyl halides. Thermolysis of diarylchloroformamidine hydrochlorides (XIV), obtained from N,N'-disubstituted thioureas and phosgene, is the best method of synthesis of unsymmetrical carbodiimides (XV) because the generated

TABLE III
Elimination Reactions of Imidoyl Halides

Starting material	Products
$R-C(Cl)=NR$	Nitriles, ketenimines, ynamines
$RO-C(Cl)=NR$	Isocyanates
$RS-C(Cl)=NR$	Isothiocyanates
$RNH-C(Cl)=NR$	Carbodiimides
$R-C(Cl)=NOH$	Nitrile oxides
$R-C(Cl)=NNHR$	Nitrile imides, nitrile ylides

hydrogen chloride can be removed with nitrogen ([15]). We have shown that N-sulfonylcarbodiimides also can be obtained by this method ([21]). N-Sulfonylcarbodiimides are the precursors of the corresponding sulfonylureas, which are widely used oral antidiabetics.

$$RNH-CCl=\overset{\oplus}{N}HR']Cl^{\ominus} \longrightarrow RN=C=NR' + 2HCl\uparrow$$
$$\text{XIV} \qquad\qquad\qquad\qquad \text{XV}$$

Another elimination reaction sequence, utilizing chlorodimethylformiminium chloride as catalyst, involves the synthesis of isocyanates by phosgenating carbamates XVI ([22]). Thus, phosgene is used to generate the iminium chloride XVII, which reacts with XVI to form the chloroimidate XVIII. The isolated isocyanate is formed by thermolysis of XVIII. Since N,N-dimethylformamide is regenerated, only a catalytic amount is necessary.

$$RNHCOOR + ClCH{=}\overset{\oplus}{N}(CH_3)_2]Cl^{\ominus} \longrightarrow RN{=}C(Cl)OR + HCl$$

XVI   XVII   XVIII

$$+ OHCN(CH_3)_2$$

$$RN{=}C(Cl)OR' \xrightarrow{\Delta} RNCO + R'Cl$$

A variety of other reactions of imidoyl halides are described in the literature, for example addition reactions to the C=N double bond, the Michaelis–Arbuzov reaction, Grignard reactions, and reduction to the corresponding azomethine. The latter is another useful method for the synthesis of aldehydes.

In conclusion it can be stated that imidoyl halides are an unusual, versatile class of compounds, which are of considerable importance in organic synthesis.

IV. PHYSICOCHEMICAL DATA

The regular imidoyl halides are colorless liquids or solids, which are often sensitive to atmospheric moisture. This fact combined with the reactivity and thermal instability of imidoyl halides has served virtually to prevent widespread use in organic synthesis. However, improvements in the handling of hygroscopic chemicals, such as the use of a dry box, as well as the identification of molecules using spectral methods, have made it quite easy to synthesize and characterize sensitive compounds. It is often not necessary to isolate imidoyl halides because *in situ* generation and subsequent reaction can afford the desired derivative in high yield.

Depending upon the basicity of the nitrogen atom, imidoyl halides can form the corresponding hydrohalides, or iminium salts, which are also often quite hygroscopic.

$$RC(Cl){=}NR' + HCl \longrightarrow RC(Cl){=}\overset{\oplus}{N}HR']Cl^{\ominus}$$

In fact, in the synthesis of imidoyl halides derived from N-alkylsubstituted amides, the corresponding iminium salts are obtained as the sole products. The hydrogen halide can be readily removed using an organic base, such as triethylamine, as the hydrogen halide scavenger.

If the imidoyl halides are generated from N,N'-disubstituted carboxylic acid amides, or from tetrasubstituted ureas, only iminium salts are formed.

$$RCONR'_2 + COCl_2 \longrightarrow RC(Cl){=}\overset{\oplus}{N}R'_2]Cl^{\ominus} + CO_2$$

R = Alkyl or $R''_2N{-}$

Sometimes this class of compounds is referred to as immonium chlorides.

The characteristic spectral feature of the imidoyl halides and iminium salts is the stretching vibration of the C=N double bond, which occurs at about 1640 cm^{-1}. The C=N absorption of the majority of the compounds falls within the range of 1680–1600 cm^{-1} and the intensity of the absorption is relatively strong, comparable to that of a C=O bond absorption in amides and ureas. Only in the case of carbonimidoyl difluorides, chlorofluorides, and fluoroformamidines is a change to longer wavelength observed (i.e., 1808–1815, 1742, and 1725 cm^{-1}, respectively). In contrast, a change to shorter wavelength is observed in bromo thioimidates, which absorb at 1538 cm^{-1} and in arylsulfonylchlorothioimidates, which absorb at 1527–1563 cm^{-1} (the corresponding chloro compounds show their stretching vibration at 1650 cm^{-1}). Likewise, arylsulfonyl substituents attached to the nitrogen in chloroformamidines cause shifting to 1605–1546 cm^{-1}. Thus, the general trend of absorption with regard to wavelength is —C(F)=N— > —C(Cl)=N— > —C(Br)=N—.

The NMR spectra of imidoyl halides are not too well investigated. We have utilized NMR spectroscopy to elucidate the structure of arylsulfonylchloroformamidines ([21]), but no data with regard to the chemical shift of alkyl groups attached to the carbon or nitrogen atom of the —C(X)=N— system is available. Of course, differentiation of the imidoyl halide structure in α-hydrogen containing compounds from the corresponding tautomeric enamine structures is easily accomplished by NMR spectroscopy, provided that the system contains olefinic protons.

A recent study related to the structure of chlorodimethylformiminium chloride and chlorotetramethylformamidinium chloride showed that both compounds are completely in the polar form as evidenced by spin-spin coupling between the olefinic proton in the former, which appears at lower field, and the N-methyl protons, which are at higher field. Likewise, the carbon-13-proton coupling constant indicated a positive charge on the nitrogen ([9]).

Martin and Martin ([8]) have also investigated the NMR spectra of intermediate complexes formed in the Vilsmeier–Haack reaction. These authors show that the NMR spectrum of chlorodimethylformiminium chloride is similar to that of the DMF–POCl$_3$ complex. The chemical shift of the methyl groups attached to the nitrogen in $(CH_3)_2\overset{\oplus}{N}{=}C(X)H$ is approximately 4 ppm, whereas the olefinic proton appears at 10.5 ppm. Further work is needed to establish data for a general characterization of imidoyl halide structures, but NMR spectroscopy can be quite useful in the aliphatic series.

V. REFERENCES

1. Bosshard, H. H., Mory, R., Schmid, M., and Zollinger, H., *Helv. Chim. Acta* 42, 1653 (1959).
2. Brower, K. R., Way, J. W., Samuels, P., and Amstutz, E. D., *J. Org. Chem.* 19, 1830 (1954).
3. Cerniani, A., and Passerini, R., *Ann. Chim. (Rome)* 44, 3 (1954); *Chem. Abstr.* 49, 4626 (1955).
4. Hudson, R. F., and Wardill, J. E., *J. Chem. Soc.* 1729 (1950).
5. Huisgen, R., *Angew. Chem. Intern. Ed.* 2, 565 (1963).
6. Huisgen, R., and Christl, M., *Angew. Chem.* 79, 471 (1967).
7. Illuminati, G., Marino, G., and Sleiter, G., *J. Am. Chem. Soc.* 89, 3510 (1967).
8. Martin, G., and Martin, M., *Bull. Soc. Chim. France* 637 (1963); *Chem. Abstr.* 60, 391 (1964).
9. Namanworth, E., Dissertation, Yale University, 1965.
10. Ottenheym, J. H., and Garritsen, J. W., German Pat. 1,157,210 (1963). *Chem. Abstr.* 60, 6756 (1964).
11. Quilico, A., and Fusco, R., *Gazz. Chim. Ital.* 67, 589 (1937).
12. Quilico, A., and Stagno d'Alcontres, G., *Gazz. chim. Ital.* 79, 654, 703 (1949).
13. Ricci, A., Todesco, P. E., and Vivarelli, P., *Gazz. Chim. Ital.* 95, 101 (1965); *Chem. Abstr.* 63, 1676 (1965).
14. Ricci, A., Todesco, P. E., and Vivarelli, P., *Gazz. Chim. Ital.* 95, 490 (1965); *Chem. Abstr.* 63, 8171 (1965).
15. Sayigh, A. A. R., and Ulrich, H., United States Pat. 3,301,895 (1967).
16. Souchay, P., Armand, J., and Valentini, F., *Compt. Rend. Ser. C*, 262, 985 (1966).
17. Todesco, P. E., and Vivarelli, P., *Gazz. Chim. Ital.* 92, 1221 (1962); *Chem. Abstr.* 59, 396 (1963).
18. Todesco, P. E., and Vivarelli, P., *Gazz. Chim. Ital.* 94, 372 (1964); *Chem. Abstr.* 61, 6902 (1964).
19. Ugi, I., Beck, F., and Fetzer, U., *Ber.* 95, 126 (1962).
20. Ulrich, H., and Sayigh, A. A. R., *Angew. Chem. Intern. Ed.* 5, 704 (1966).
21. Ulrich, H., Tucker, B., and Sayigh, A. A. R., *Tetrahedron* 22, 1565 (1966).
22. Ulrich, H., Tucker, B., and Sayigh, A. A. R., *Angew. Chem. Intern. Ed.* 6, 636 (1967).
23. Ulrich, H., Tucker, B., and Sayigh, A. A. R., *J. Org. Chem.* 32, 4052 (1967).
24. Ulrich, H., unpublished results.
25. Young, T. E., and Amstutz, E. D., *J. Am. Chem. Soc.* 73, 4773 (1951).

Chapter 2
CARBONIMIDOYL DIHALIDES

I. INTRODUCTION

In 1874 Sell and Zierold ([137]) obtained phenylcarbonimidoyl dichloride I in the reaction of phenylisothiocyanate II with chlorine. Several years later, Nef ([108,109]) prepared carbonimidoyl dichlorides by addition of chlorine to isocyanides, such as phenyl isocyanide III. This relationship between I and isocyanides (isonitriles) led to the frequently used term isocyanide dichlorides for the compounds under discussion.

$$C_6H_5NCS \xrightarrow{Cl_2} C_6H_5N{=}CCl_2 \xleftarrow{Cl_2} C_6H_5NC$$
$$\text{II} \qquad\qquad\quad \text{I} \qquad\qquad\quad \text{III}$$

The carbonimidoyl dihalides are derivatives of the hypothetical carbonimid acid $HN{=}C(OH)_2$, and *Chemical Abstracts* uses the term imidocarbonyl dihalides for compounds I. Recently, the more appropriate name "carbonimidoyl dihalide" was advanced ([107]), and it will be used in this monograph.

The salts of the parent compound $Cl_2C{=}NH$ (IV) were synthesized by Allenstein and Schmidt ([1]) in 1964, by adding hydrogen chloride to the antimony pentachloride complexes of cyanogen halides V. The compounds obtained (IV) are stable at room temperature, with the exception of the iodine derivative (IV, X=I)([1]).

$$X{-}C{\equiv}N \rightarrow SbCl_5 + 2HCl \longrightarrow X{-}\underset{\underset{\text{IV}}{Cl}}{\overset{\oplus}{C}}{=}\overset{\oplus}{N}H_2]Cl^{\ominus}$$
$$\text{V}$$

In addition to the two classical methods of synthesis of carbonimidoyl dichlorides, several new methods were developed more recently. Thus, chlorination of imidoyl chlorides, generated from monosubstituted formamides and thionyl chloride, provides a useful new method to produce a great variety of carbonimidoyl dichlorides ([14,86,95a]). Likewise,

13

high-temperature chlorination of carbamoyl chlorides and N-methyl amines affords carbonimidoyl dichlorides, often in excellent yield ([49]). The reaction of isocyanates and phosphorus pentachloride is another method of synthesis of aliphatic carbonimidoyl dichlorides ([85,148]).

The carbonimidoyl dichloride structure VI was established recently for tetrameric cyanogen chloride ([69,72,131,133]) a compound which can also be synthesized by chlorinating 2-dimethylamino-4,6-dichloro-1,3,5-triazine VII ([49]).

$$4\ \text{ClCN} \longrightarrow$$

N=CCl$_2$ triazine ring with Cl, N substituents — **VI**

$$\longleftarrow$$

N(CH$_3$)$_2$ triazine ring with Cl, N substituents — **VII**

Tetrameric cyanogen chloride has been used to synthesize a great variety of 1,3,5-triazine derivatives ([69-83,131,133]).

The synthesis of carbonimidoyl dihalides is not limited to compounds having the $N=CX_2$ group attached to aliphatic and aromatic hydrocarbyl and heterocyclic moieties. Acyl ([4,114,125]), aroyl ([4,55,56,114,125]), sulfonyl ([5,7,32,37,111,115]), and phosphoryl carbonimidoyl dichlorides([2,3,19,64]) have been synthesized, and a variety of other combinations can certainly be visualized. A recent survey of syntheses of carbonimidoyl dichlorides has been published by Kühle and his co-workers ([95a]).

The availability of several synthetic procedures has prompted extensive investigation of the reactive carbonimidoyl dihalides, especially by Holtschmidt ([49,54a]) and Kühle ([86,95a]) at Farbenfabriken Bayer, A.G., and by Kodama et al. ([69-83]) and Mukaiyama and his colleagues ([35,104-106]) in Japan.

The reactions of carbonimidoyl dihalides can be divided into nucleophilic substitution reactions, whereby the halo groups are replaced stepwise by other nucleophiles, and addition reactions. The substitution reactions are often accompanied by elimination, depending upon the stability of the formed imidoyl derivatives. For example, the monosubstitution products can eliminate hydrogen halide, alkyl halide, or sulfenyl chloride, i.e.,

$$\underset{\overset{|}{\text{Cl}}}{\text{RN}=\text{C}-\text{XR}'} \longrightarrow \text{RN}=\text{C}=\text{X} + \text{R}'\text{Cl}$$

X = NR, O, S   R' = H, Alkyl

Similarly, the disubstitution products sometimes undergo elimination to yield the corresponding heterocumulene.

The addition to the C=N bond is limited to reactive species, such as hydrogen halides, carbonyl fluoride and organometallics, and seems to proceed well only with carbonimidoyl difluorides. The dehalogenation of carbonimidoyl dihalides by triphenylphosphine provides a useful method to generate isocyanides, which are sometimes difficult to obtain by Hofmann's classical method.

All in all a great variety of compounds can be synthesized from carbonimidoyl dihalides, and their use as agricultural chemicals has been indicated in the patent literature ([17,27,29,48,65,88,94,99,144]).

II. SYNTHESIS

A. Addition of Halogen to Isocyanides

The addition of halogen to isocyanides to produce carbonimidoyl dichlorides was demonstrated by Nef ([108,109]) and by Guillemard ([39]). The reaction proceeds in chloroform at room temperature; however, no yields are reported. In the aliphatic series it is sometimes advantageous to conduct the reaction in diethyl ether, using sulfuryl chloride as the chlorinating agent. Thus, ethylcarbonimidoyl dichloride VIII is obtained in the reaction of ethyl isocyanide with sulfuryl chloride ([108]). However, addition of chlorine to long-chain aliphatic isocyanides does not produce the corresponding carbonimidoyl dichlorides ([60]).

$$EtNC + SO_2Cl_2 \longrightarrow EtN{=}CCl_2 + SO_2$$
$$VIII$$

Similarly, addition of bromine and iodine to phenyl isocyanide yields the corresponding crude carbonimidoyl dihalides ([145a]); however, Nef ([108]) could not purify the obtained materials. Smith and Kalenda ([141]) obtained carbonimidoyl dichlorides on addition of chlorine to dialkylaminoalkyl isocyanides.

N-(3-Dimethylaminopropyl)carbonimidoyl dichloride ([141]). To a solution of 1.44 g of 3-dimethylaminopropyl isocyanide in 11 ml of chloroform chlorine is added with ice-cooling until a yellow color is observed. The precipitated 1.45 g (62%) of N-(3-dimethylaminopropyl)carbonimidoyl dichloride, m.p. 130.5–132°C, is collected and dried *in vacuo*. The compound is very hygroscopic.

B. Halogenation of Isothiocyanates

The chlorination of isothiocyanates is the classical synthesis of carbonimidoyl dichlorides. Sell and Zierold ([137]) in 1874 obtained phenylcarbonimidoyl dichloride (IX) upon chlorination of phenyl isothiocyanate in chloroform at 0°C, and Nef ([108]) observed that under the conditions employed by Sell and Zierold appreciable chlorination of the ring occurred. However, if the reaction is conducted in carbon tetrachloride, no chlorination of the ring is observed ([14,107]).

$$C_6H_5NCS + 2Cl_2 \longrightarrow C_6H_5N{=}CCl_2 + SCl_2$$
$$\text{IX}$$

Phenylcarbonimidoyl dichloride ([107]). To a solution of 67.5 g (0.5 mole) of phenyl isothiocyanate in 200 ml of carbon tetrachloride chlorine is added with stirring and ice-cooling below 3°C until the chlorine is no longer absorbed (approximately three to four hours). The solvent and sulfur dichloride, formed in the reaction, are removed by distillation, and vacuum distillation of the residue yields 74–78 g (85–90%) of phenylcarbonimidoyl dichloride, b.p. 103–106°C (31 mm).

In the aliphatic series, the chlorination of isothiocyanates proceeds exceedingly well, provided that the crude reaction products are treated with aqueous sodium sulfite and sodium carbonate in order to remove impurities which can give rise to decomposition during vacuum distillation ([136]).

In addition to isothiocyanates, compounds having the configuration $RN{=}C(SX)_2$ (X = alkyl or K) react rapidly with chlorine to produce carbonimidoyl dichlorides. Thus, Neidlein and his co-workers ([110,111,115]) and Gompper and Kunz([37]) obtained arylsulfonylcarbonimidoyl dichlorides X in the reaction of S,S-dialkyl-N-sulfonyldithiocarbonimides XI with chlorine.

$$RSO_2N{=}C(SR')_2 + 2\,Cl_2 \longrightarrow RSO_2N{=}CCl_2 + 2\,R'SCl$$
$$\text{XI} \qquad\qquad\qquad\qquad \text{X}$$

Methanesulfonylcarbonimidoyl dichloride ([37]). To a solution of 19 g of S,S-dimethyl N-methanesulfonylcarbonimide in 180 ml of glacial acetic acid chlorine is added below 15°C with cooling for a period of one hr. Evaporation of the solvent and vacuum distillation of the residue yields 14 g (83%) of methanesulfonylcarbonimidoyl dichloride, m.p. 73–76°C (m.p. 79–80°C is listed in Reference 95; the b.p. can not be established because the compound sublimes under vacuum).

Arenesulfonylcarbonimidoyl dichlorides are obtained by the same procedure in 92–98% yield ([37]).

The reaction of XI with one equivalent of bromine or sulfuryl chloride yields the arenesulfonylhalothioformimidates XII (X = Br or Cl) ([110]).

$$RSO_2N{=}C(SR')_2 + X_2 \longrightarrow RSO_2N{=}\underset{\underset{\text{XII}}{X}}{C}{-}SR' + R'SX$$
$$\text{XI} \qquad\qquad\qquad\qquad$$

Similarly, carbonimidoyl dichlorides X are obtained upon chlorination of dithiocarbamate salts XIII ([5,32,93,98]) or benzoyldithiocarbamates ([21]).

$$RSO_2N{=}C(SK)_2 + 4\,Cl_2 \longrightarrow RSO_2N{=}CCl_2 + 2\,SCl_2 + 2\,KCl$$
$$\text{XIII} \qquad\qquad\qquad\qquad \text{X}$$

The chlorination of isothiocyanates and dithiocarbamates is a general method and good yields of carbonimidoyl dichlorides are obtained. However, on attempted conversion of o-nitrophenyl isothiocyanate a reaction mixture was obtained, which decomposed violently upon heating ([24]).

The chlorination of benzyl isothiocyanate affords benzylcarbonimidoyl dichloride XIV and α,α-dichlorobenzylcarbonimidoyl dichloride XV, depending upon the amount of chlorine being used ([17,54a]).

$$C_6H_5CH_2NCS + 2\,Cl_2 \longrightarrow C_6H_5CH_2N{=}CCl_2 + SCl_2$$
$$\text{XIV}$$

$$C_6H_5CH_2NCS + 4\,Cl_2 \longrightarrow C_6H_5CCl_2N{=}CCl_2 + SCl_2 + 2\,HCl$$
$$\text{XV}$$

Similarly, acyl isothiocyanates add chlorine to produce the corresponding acylcarbonimidoyl dichlorides (XVI) ([4,114,125]).

$$RCONCS + 2\,Cl_2 \longrightarrow RCON{=}CCl_2 + SCl_2$$
$$\text{XVI}$$

Dichloroacetylcarbonimidoyl dichloride ([95a]). To an amount of 724 g dichloroacetyl isothiocyanate, containing 2 ml of titanium tetrachloride, 800 g of chlorine is added below 20°C. Rectification affords 771 g dichloroacetylcarbonimidoyl dichloride, b.p. 80–82°C/15 mm.

Isothiocyanato groups attached to phosphorus are also readily converted to the corresponding carbonimidoyl dichlorides. For example, chlorination of the phosphorus isothiocyanates XVII yields the carbonimidoyl chlorides XVIII ([2,3,19,64]).

$$(RO)_2P{-}NCS + 2\,Cl_2 \longrightarrow (RO)_2P{-}N{=}CCl_2 + SCl_2$$
$$\quad\;\downarrow \qquad\qquad\qquad\qquad\quad\;\downarrow$$
$$\quad\;O \qquad\qquad\qquad\qquad\qquad O$$
$$\text{XVII} \qquad\qquad\qquad\qquad\quad \text{XVIII}$$

The mechanism of the chlorination of isothiocyanates involves initial addition of one mole of chlorine to the $C{=}S$ bond to produce the 1-(chlorothio)formimidoyl chlorides XIX ([26,56a,95a]). Recently Ottmann and Hooks ([120,121]) isolated a variety of alkyl and aryl 1-(chlorothio)formimidoyl chlorides and demonstrated that addition of a second mole of chlorine converts the intermediate XIX to the carbonimidoyl dichloride.

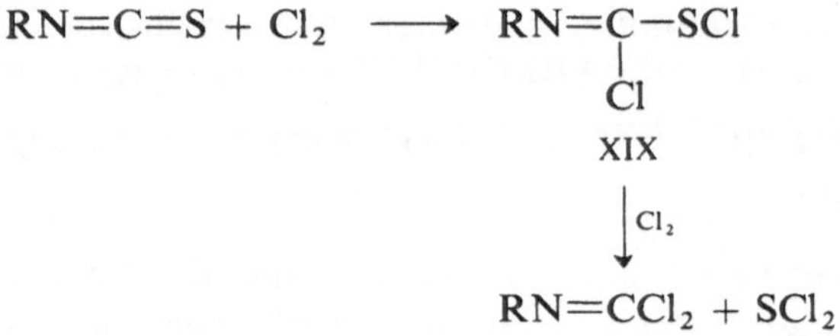

$$RN{=}C{=}S + Cl_2 \longrightarrow RN{=}\underset{\underset{XIX}{\overset{|}{Cl}}}{\overset{|}{C}}{-}SCl$$

$$\downarrow Cl_2$$

$$RN{=}CCl_2 + SCl_2$$

The 1-(chlorothio)formimidoyl chlorides XIX can be added to olefins, and 1:1 adducts are obtained in good yield ([95a,122]) (see Chapter 5).

The intermediates XIX were also isolated by Ivanova et al. ([56]) in the reaction of aroyl isothiocyanates with chlorine.

The addition of halogen to the C=S rather than the C=N bonds, as for example postulated by Gumpert ([41]) for the bromination of phenyl isocyanate, is in line with the favored participation of the C=S bond in addition reactions of isothiocyanates ([151]). One is tempted to speculate that the second step involves α-elimination of sulfur dichloride, followed by addition of chlorine to the generated isocyanide. However, further work is needed to establish the mechanism of this reaction.

Earlier, Helmers ([47]) and Dyson and Harrington ([24–26]) isolated the by-product (XX), which most likely is formed by addition of XIX to un-reacted isothiocyanate.

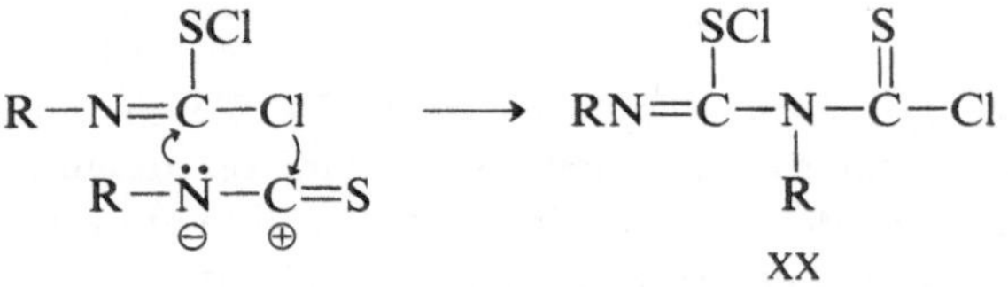

However, Kühle and his co-workers ([95a]) propose the cyclic structures (A) or (B) for the 1:1 adducts, obtained from isothiocyanates and 1-(chlorothio)-formimidoyl chlorides.

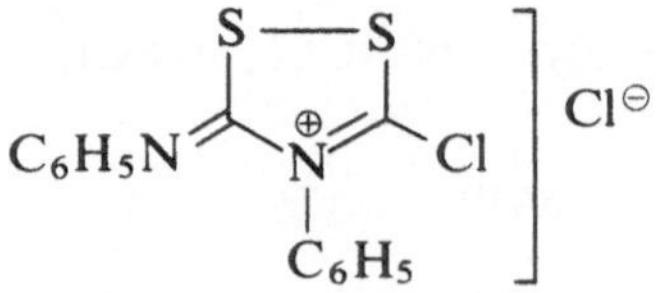

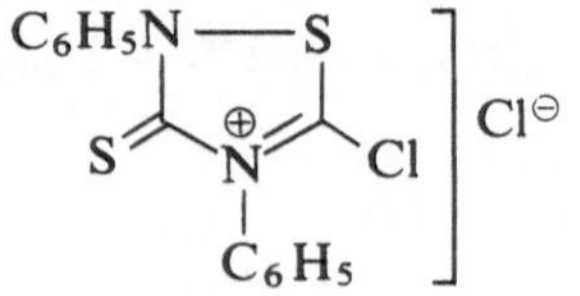

A general method of synthesis of carbonimidoyl difluorides from alkyl and aryl isothiocyanates was recently described by Sheppard ([139]). Thus, upon reaction of isothiocyanates with mercuric fluoride, the corresponding carbonimidoyl difluorides were obtained in good yield. The reaction may occur *via* the four-centered transition state XXI, followed by rearrangement and elimination of mercuric sulfide ([139]).

$$RNCS + HgF_2 \longrightarrow \left[\begin{array}{c} F \text{------} Hg\text{--}F \\ | \quad\quad | \\ R\text{--}N{=}C \text{------} S \end{array} \right] \longrightarrow$$

XXI

$$\begin{array}{c} F\text{------}Hg \\ | \quad\quad | \\ RN{=}C\text{------}S \\ | \\ F \end{array} \longrightarrow RN{=}CF_2 + HgS$$

Both the formation of XX as well as the reaction of mercuric fluoride with isothiocyanates are examples of the well-known insertion reaction of heterocumulenes ([151]).

C. Halogenation of Isocyanates and Carbamoyl Chlorides

The reaction of isocyanates with phosphorus pentachloride, and the reaction of secondary carbamoyl chlorides with chlorine, yield carbonimidoyl dichlorides. Gumpert ([41]) in 1885 investigated the reaction of phenyl isocyanate and phosphorus pentachloride, and on the basis of reduction of the intractable reaction mixture to N-methylaniline XXII the initial formation of phenylcarbonimidoyl dichloride was suggested.

$$C_6H_5N{=}C{=}O + PCl_5 \longrightarrow [C_6H_5N{=}CCl_2] \longrightarrow C_6H_5NCH_3$$

XXII

A recent patent ([85]) describes the formation of carbonimidoyl dichlorides in the reaction of several aliphatic and aromatic isocyanates with phosphorus pentachloride. This reaction can be conducted using phosphorus oxychloride as solvent, or without a solvent; however, only from aliphatic isocyanates ([85,148]) and diisocyanates ([123]) are good yields of carbonimidoyl dichlorides obtained.

Methylcarbonimidoyl dichloride ([85]). To a suspension of 70 g of phosphorus pentachloride and 70 ml of phosphoryl chloride, 18 g (0.31 mole) of methyl isocyanate is added dropwise. After stirring at room temperature for some time the reaction mixture is fractionally distilled to yield 38 g of a distillate, b.p. 60–85°C. Redistillation yields 29.5 g (83.5%) of methylcarbonimidoyl dichloride, b.p. 79–80°C.

We have demonstrated that the reason for the low yields obtained in the aromatic series is an alternate reaction sequence leading to trichlorophosphazenes XXIII and carbonyl chloride ([147,148,151]).

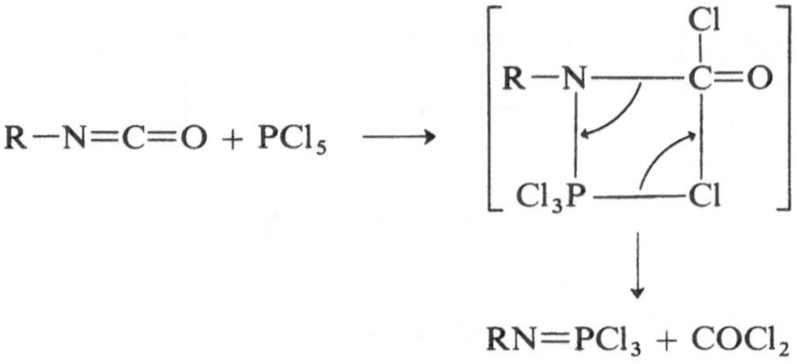

$$R-N=C=O + PCl_5 \longrightarrow \begin{bmatrix} R-N\underset{}{\overset{Cl}{\rule{0pt}{0pt}}}\ \ C=O \\ \\ Cl_3P\ \rule{0pt}{0pt}\ Cl \end{bmatrix}$$

$$\downarrow$$

$$RN=PCl_3 + COCl_2$$
XXIII

The generated trichlorophosphazene (XXIII) adds to the starting isocyanate to produce the corresponding carbodiimide (XXIV) and phosphoryl chloride ([147–150]). Thus, yields of up to 50% of diarylcarbodiimide are obtained in the reaction of aryl isocyanates and phosphorus pentachloride, whereas no carbodiimides are formed in the reaction of alkyl isocyanates and phosphorus pentachloride ([148]).

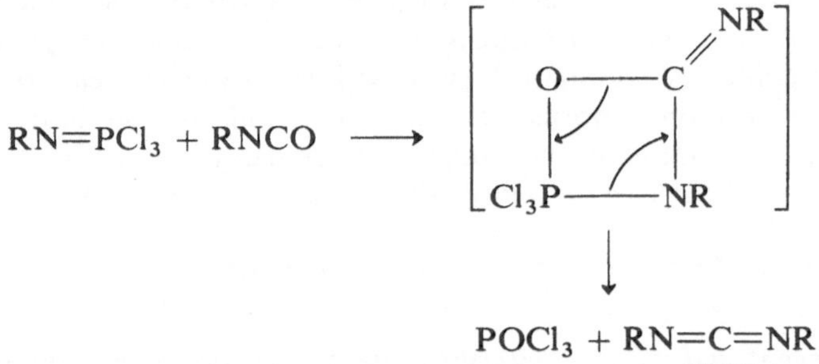

$$RN=PCl_3 + RNCO \longrightarrow \begin{bmatrix} O\ \rule{0pt}{0pt}\ C\overset{NR}{\diagup} \\ \\ Cl_3P\ \rule{0pt}{0pt}\ NR \end{bmatrix}$$

$$\downarrow$$

$$POCl_3 + RN=C=NR$$
XXIV

In the reaction of benzoyl isocyanate with phosphorus pentachloride reaction occurs exclusively on the C=O group attached to the isocyanate moiety, and the chloroimidoyl-N-carbonyl chloride XXV is obtained in good yield ([112]).

$$C_6H_5CONCO + PCl_5 \longrightarrow C_6H_5-\underset{Cl}{\overset{}{C}}=N-COCl$$
XXV

Compound XXV can also be obtained in the chlorination of benzyl isocyanate ([49]). The reaction of aliphatic haloacyl isocyanates with phosphorus pentachloride proceeds in a similar manner; however, infrared evidence indicates an equilibrium mixture of imidoyl chloride (1634 and 1770 cm^{-1}) and α-halo isocyanate (2260–2270 cm^{-1}). Thus, from chloroacetyl isocyanate and phosphorus pentachloride the trichloro compound XXVI is obtained in 85% yield ([112]).

$$ClCH_2CONCO + PCl_5 \longrightarrow$$

$$\underset{\underset{Cl}{|}}{ClCH_2-C}=N-COCl \longleftrightarrow ClCH_2CCl_2NCO$$

XXVI

The α,α-dichlorocarbonimidoyl dichlorides, which can not be obtained by the above reaction, are readily formed from acylcarbonimidoyl dichlorides and phosphorus pentachloride. Thus, in the reaction of N-benzoylcarbonimidoyl dichloride with phosphorus pentachloride the α,α-dichloro derivative XXVII is obtained in good yield ([112]).

$$C_6H_5CON=CCl_2 + PCl_5 \longrightarrow C_6H_5CCl_2N=CCl_2$$
XXVII

In contrast, phosphoryl isocyanates are attacked by phosphorus pentachloride on the isocyanato group, with concomitant cleavage of the alkoxy groups by the generated hydrogen chloride. Thus, dichlorophosphorylcarbonimidoyl dichloride (XXVIII) is obtained in the reaction of the phosphoryl isocyanate and phosphorus pentachloride ([64]).

$$\underset{\underset{O}{\downarrow}}{(RO)_2P}-NCO + PCl_5 \longrightarrow \underset{\underset{O}{\downarrow}}{Cl_2P}-N=CCl_2$$

XXVIII

While alkyl and aryl isocyanates are only converted to carbonimidoyl dichlorides by phosphorus pentachloride, arylsulfonyl isocyanates can be directly chlorinated to produce arylsulfonylcarbonimidoyl dichlorides. For example, chlorination of benzenesulfonyl isocyanate at 60°C yields benzenesulfonylcarbonimidoyl dichloride (XXIX) in unspecified yield ([32]).

$$C_6H_5SO_2N=C=O + Cl_2 \longrightarrow C_6H_5SO_2N=CCl_2 + COCl_2$$
XXIX

In addition to isocyanates secondary carbamoyl chlorides are readily converted to carbonimidoyl dichlorides by high-temperature chlorination, provided one of the alkyl substituents is a methyl group ([31,49–52,54]).

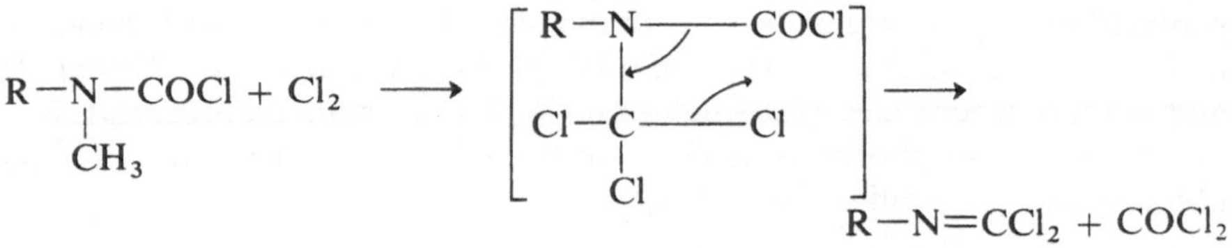

$$R-N-COCl + Cl_2 \longrightarrow \left[\begin{array}{c} R-N-\!\!\!\!-COCl \\ | \\ Cl-C-\!\!\!\!-Cl \\ | \\ Cl \end{array} \right] \longrightarrow$$

$$R-N=CCl_2 + COCl_2$$

with the left reactant showing CH_3 below the N.

Thus, chloromethylcarbonimidoyl dichloride (**XXX**) can be obtained by the chlorination of either N,N-bis(chloromethyl)carbamoyl chloride (**XXXI**) or N,N-dimethylcarbamoyl chloride (**XXXII**) at 190–200°C. Upon further chlorination dichloromethyl- (**XXXIII**) and trichloromethylcarbonimidoyl dichloride (**XXXIV**) are obtained ([31,49–54,99]).

$$(CH_3)_2N-COCl \longrightarrow ClCH_2N=CCl_2 \longleftarrow (CH_2Cl)_2N-COCl$$

          XXXII   XXX   XXXI

$$\downarrow$$

$$CHCl_2N=CCl_2 + CCl_3N=CCl_2$$

                    XXXIII   XXXIV

Similarly, N-methylacetanilid (**XXXV**) yields pentachlorophenylcarbonimidoyl dichloride (**XXXVI**) upon high-temperature chlorination ([52]).

$$C_6H_5N(CH_3)COCH_3 \xrightarrow{Cl_2} C_6Cl_5N=CCl_2$$

          XXXV   XXXVI

N,N-dialkylimidoyl derivatives are also converted by high-temperature chlorination into carbonimidoyl dichlorides ([49]). For example, 4-dimethylamino-2,6-dichloro-1,3,5-triazine (**XXXVII**) reacts with chlorine *via* two pathways to yield the traizinylcarbonimidoyl dichloride **XXXVIII** and chloroform, or cyanuric chloride (**XXXIX**) and trichloromethylcarbonimidoyl dichloride ([49]).

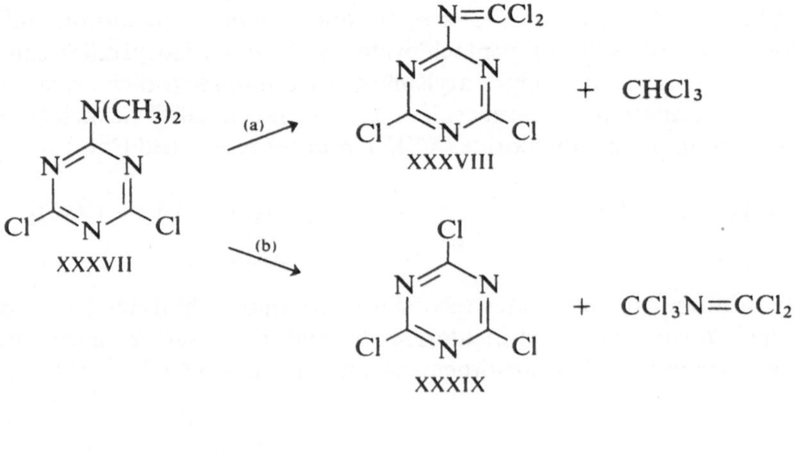

However, the reaction proceeds preferentially *via* route (a), as evidenced by the isolation of 72% of the tetrameric cyanogen chloride XXXVIII. The triazine derivative XXXVIII can also be obtained by chlorination of the corresponding isothiocyanate ([76]). A somewhat lower yield of the triazinyl-bis-carbonimidoyl chloride was obtained in the chlorination of 2,6-bis(di-methylamino)-4-chloro-1,3,5-triazine ([49]). If cyclic carbamoyl chlorides, or carbamoyl chlorides having alkyl substituents other than methyl are chlorinated, the expected imidoyl chlorides are obtained ([49,144]). For example, N-ethyl-N-phenylcarbamoyl chloride (XL) affords pentachloro-phenyltrichloroacetimidoyl chloride (XLI) ([144]). The latter compound shows strong acaricidal activity ([144]).

$$C_6H_5N(COCl)CH_2CH_3 \xrightarrow{\text{Cl}_2} C_6Cl_5N{=}\underset{\underset{\text{Cl}}{|}}{C}{-}CCl_3$$

$$\text{XL} \qquad\qquad\qquad\qquad \text{XLI}$$

Perfluorocarbamoyl fluorides likewise eliminate carbonyl fluoride to afford the corresponding carbonimidoyl difluorides ([45,46,155]). For example, heating of bis(trifluoromethyl)carbamoyl fluoride (XLII) at 490–575°C produced an almost quantitative yield of trifluorcarbonimidoyl difluoride (XLIII) and carbonyl fluoride (XLIV) ([46,155]).

$$(CF_3)_2NCOF \longrightarrow CF_3N{=}CF_2 + COF_2$$

$$\text{XLII} \qquad\qquad \text{XLIII} \qquad \text{XLIV}$$

D. Halogenation of Imidoyl Chlorides

The chlorination of formanilide with chlorine and sulfur dichloride, in the presence of phosphorus trichloride, phosphorus oxychloride, or thionyl chloride, was investigated in 1920 by Bly, Perkins, and Lewis ([14]). In addition to 2,4-dichloroformanilide, a mixture of phenyl-, 4-chloro-phenyl-, and 2,4-dichlorophenylcarbonimidoyl dichlorides was obtained. Recently, Kühle ([84,86,95a]) synthesized a great number of aromatic carbon-imidoyl dichlorides by adding the corresponding arylformamides to one equivalent of sulfuryl chloride, dissolved in excess thionyl chloride at 15–20°C. The thionyl chloride converts the arylformamides XLV to the corresponding imidoyl chloride intermediates XLVI, which are chlorinated by the sulfuryl chloride to afford the carbonimidoyl dichlorides.

$$RNHCHO + SOCl_2 \longrightarrow R\overset{\oplus}{N}H{=}CHCl\big]Cl^{\ominus} + SO_2$$

$$\text{XLV} \qquad\qquad\qquad \text{XLVI}$$

$$R\overset{\oplus}{N}H{=}CHCl\big]Cl^{\ominus} + SO_2Cl_2 \longrightarrow RN{=}CCl_2 + 2\,HCl + SO_2$$

The yields are good, especially when the aromatic ring is deactivated by the presence of substituents, such as chloro, nitro, and carboxyl groups. However, 2,5-dichloro-4-nitrophenylformamide (XLVII) reacts with elimination of cyanogen chloride, as evidenced by the isolation of substantial amounts of 1,2,4,5-tetrachlorobenzene (XLVIII) ([86]).

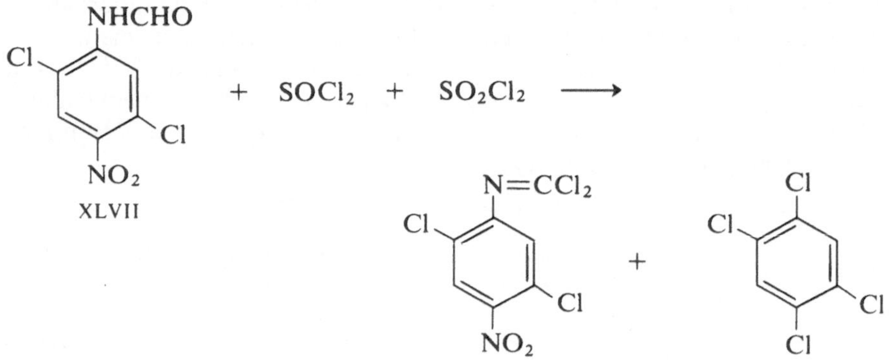

4-Chlorophenylcarbonimidoyl dichloride ([86]). To a solution of 67.5 g (0.5 mole) of sulfuryl chloride in 150 ml of thionyl chloride, 77.5 g (0.5 mole) of N-(4-chlorophenyl)formamide is added portionwise at 15–20°C. After stirring for several hours at room temperature the reaction mixture is heated slowly to 80°C, and the thionyl chloride is removed by distillation. Vacuum distillation of the residue yields 89 g (86%) of 4-chlorophenyl-carbonimidoyl dichloride, b.p. 110–113°/10 mm.

Aliphatic formamides, such as cyclohexylformamide, react likewise. However, the reaction has to be conducted stepwise, i.e., the initial conversion to XLVI has to be complete, otherwise chlorination of the formamide occurs faster, giving rise to the formation of isocyanates XLIX ([86]).

$$RNHCHO + SO_2Cl_2 \longrightarrow [RNHCOCl] \longrightarrow RNCO + HCl$$
$$\text{XLIX}$$

The reaction of phenylthioformamide (L) with bromine yields 4-bromo-phenylcarbonimidoyl dibromide (LI), most likely by a similar mechanism ([8]).

$$C_6H_5NH-\underset{\underset{\text{L}}{\overset{\|}{S}}}{C}-H + Br_2 \longrightarrow 4\text{-}BrC_6H_4N{=}CBr_2$$
$$\text{LI}$$

In the absence of the chlorinating agent reaction of the imidoyl chloride XLVI (R = aryl) with unreacted arylformamide occurs with formation of formamidines ([43,57,58,86,134,152]) (see Chapter 3).

E. Halogenation of Amines

Tertiary amines, such as N,N-dimethylaniline, can be chlorinated under controlled conditions to yield 2,4,6-trichlorophenylcarbonimidoyl dichloride (LII), which can be further chlorinated in the presence of $FeCl_3$ to yield pentachlorophenylcarbonimidoyl dichloride (LIII). The yields of LII and LIII are 82–87% and 96–98%, respectively ([49]). It is absolutely necessary to conduct the initial chlorination below 60°C, otherwise considerable amounts of intractable tars are formed due to side reactions ([49]).

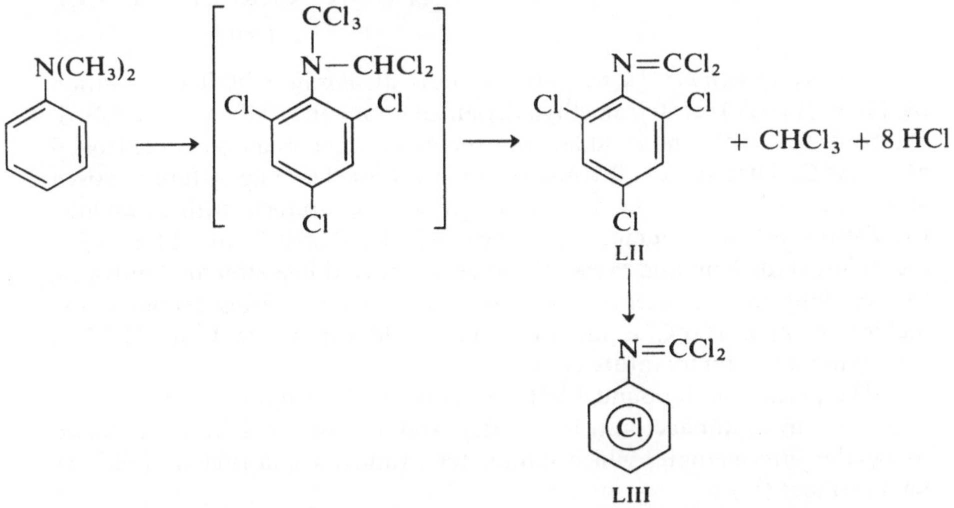

Upon reaction of LII with two equivalents of dimethylamine the corresponding chloroformamidine (LIV) can be obtained, which upon high-temperature chlorination yields the chloroimidoylcarbonimidoyl dichloride derivative LV ([49]).

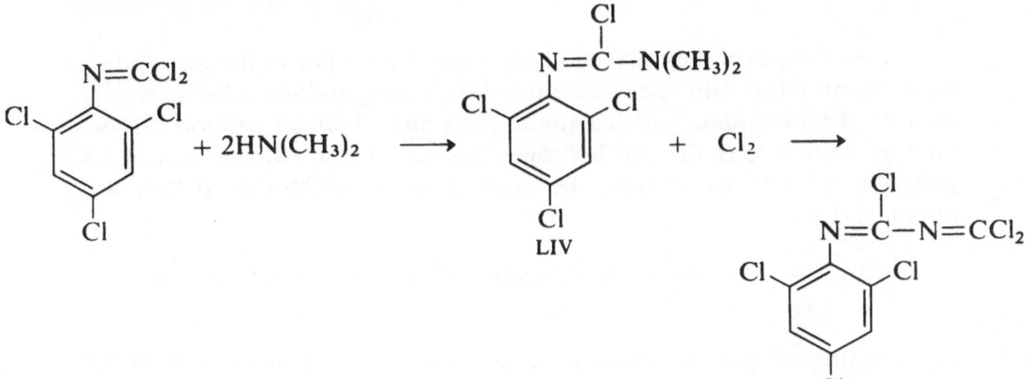

The chlorination of aliphatic tertiary amines is less uniform, and hitherto only good results have been obtained in the chlorination of N,N,N′,N′-tetramethylethylenediamine (LVI), in which all the carbon atoms are in α-position to the nitrogen. Thus, chlorination of LVI in trichlorobenzene produces the bis-carbonimidoyl dichloride LVII in good yield ([49,53]).

$$(CH_3)_2NCH_2CH_2N(CH_3)_2 + Cl_2 \longrightarrow$$
$$\text{LVI}$$
$$Cl_2C{=}N{-}CCl_2CCl_2{-}N{=}CCl_2$$
$$\text{LVII}$$

Tetrachloroethane 1,2-bis-carbonimidoyl dichloride ([49]). To a solution of 116 g (1 mole) of tetramethylethylenediamine in 1000 ml of 1,2,4-trichlorobenzene chlorine is added and the exothermic reaction is controlled at 70–80°C. After the exothermic reaction ceases the temperature is raised at the rate of 10–15°C/hr and chlorination is continued with ultraviolet irradiation. When the temperature has reached 190–210°C, the chlorination is continued for 5 hr, and excess chlorine is removed in a stream of nitrogen. On cooling and concentration tetrachloroethane 1,2-biscarbonimidoyl dichloride, m.p. 166°C is obtained. The yield can range from 30–71%, depending upon temperature control.

The perhalo compound LVII was also obtained upon chlorination of N,N-dimethylaminoacetonitrile (LVIII), and the nitrile LIX is supposed to be the intermediate, which eliminates cyanogen chloride to yield the final product ([49]).

$$(CH_3)_2NCH_2CN + Cl_2 \longrightarrow [Cl_2C{=}NCCl_2CN] \longrightarrow$$
$$\text{LVIII} \qquad\qquad\qquad \text{LIX}$$
$$Cl_2C{=}N{-}CCl_2CCl_2{-}N{=}CCl_2 + ClCN$$
$$\text{LVII}$$

One example of the conversion of a secondary amine to the corresponding carbonimidoyl dihalide was reported by Young and his colleagues ([158]), who heated bis(trifluoromethyl)amine (LX) and obtained trifluorocarbonimidoyl difluoride (LXI) and hydrogen fluoride. Upon chlorination of LX with phosphorus trichloride, the corresponding dichloride (LXII) was obtained ([158]).

$$CF_3N{=}CCl_2 \xleftarrow{\ PCl_3\ } (CF_3)_2NH \xrightarrow{\ \Delta\ } CF_3N{=}CF_2 + HF$$
$$\text{LXII} \qquad\qquad \text{LX} \qquad\qquad \text{LXI}$$

Heating of perfluorodimethylamine ([9]) or the mercury salt of LX with sulfur ([156]) also yielded LXI.

Trifluoromethylcarbonimidoyl difluoride can also be obtained from N-halo-bis(trifluoromethyl)amine and iron pentacarbonyl ([37a]).

$$(CF_3)_2NCl + Fe(CO)_5 \longrightarrow CF_3N{=}CF_2$$

Refluxing of perfluorodimethylcyclohexylamine with $AlCl_3$ in carbon tetrachloride gives a mixture of products containing the carbonimidoyl dichloride LXIII ([143]).

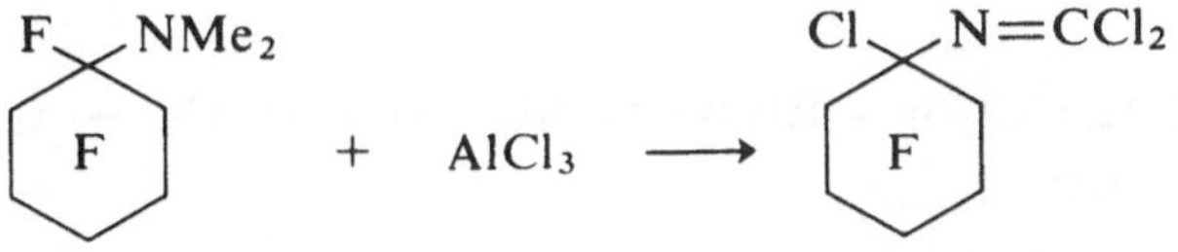

LXIII

Upon fluorination of LXIII with hydrogen fluoride in the presence of SbF_5 the corresponding perfluorinated carbonimidoyl difluoride was obtained ([143]).

F. Miscellaneous Methods

Perhaps the most direct formation of a carbonimidoyl dihalide involves the tetramerization of cyanogen chloride. Although cyanuric chloride, the trimer, is the main product of the oligomerization of cyanogen chloride, yields up to 39% of the tetramer LXIV can be obtained under controlled conditions ([60,69,72,131,133]).

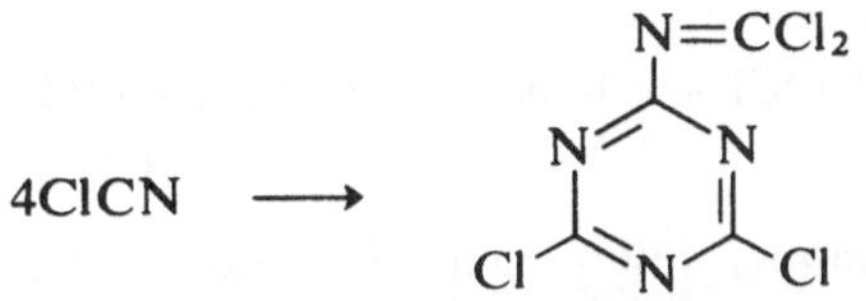

LXIV

2,4-Dichloro-1,3,5-triazine-6-carbonimidoyl dichloride (tetrameric cyanogen chloride) ([131]). To 450 g moist chloroform containing 6 g of N,N-dimethylformamide, hydrogen chloride is added at 0–5°C until the mixture is saturated. Then dropwise and with ice-cooling 223 g cyanogen chloride is added, and the cooling bath is removed. The temperature slowly rises to 50–55°C, and after refluxing for 16 hr the chloroform is removed by distillation. Addition of 300 g pet. ether and cooling to 0–5°C precipitates cyanuric chloride (trimeric cyanogen chloride), which is removed by filtration. Evaporation of the pet. ether and vacuum distillation yields 87 g of crude 2,4-dichloro-1,3,5-triazine-6-carbonimidoyl dichloride, b.p. 117–142°C/13 mm. On standing for three days co-distilled cyanuric chloride separates, and after removal by filtration, tetrameric cyanogen chloride is obtained in 97–98.5% purity.

An interesting new method of synthesis of carbonimidoyl dichlorides involves the addition of dichlorocarbene to aliphatic carbodiimides. Thus, reaction of phenyl(bromodichloromethyl)mercury (LXV) (the dichlorocarbene precursor) with diisopropylcarbodiimide produces isopropyl-carbonimidoyl dichloride (LXVI) in 63% yield ([138]).

$$C_6H_5HgCCl_2Br + RN{=}C{=}NR \longrightarrow \left[\begin{array}{c} RN{-}C\overset{\displaystyle NR}{\diagup} \\ \diagdown C \diagup \\ Cl \quad\quad Cl \end{array} \right]$$

LXV

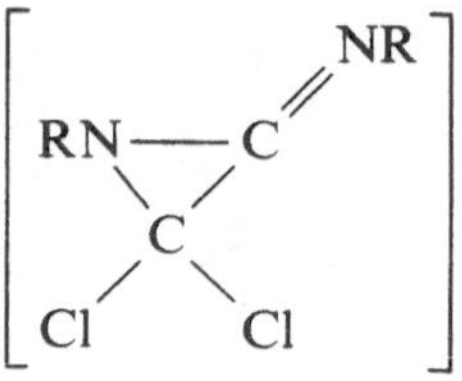

$$R = iso\text{-}C_3H_7 \qquad\qquad C_6H_5HgBr + RN{=}CCl_2$$

LXVI

The addition of cyanogen chloride to carboxylic acid chlorides is another useful method of synthesis for acyl and aroylcarbonimidoyl dichlorides. For example, trichloroacetyl chloride reacts readily with cyanogen chloride at 50°C in an autoclave to yield the corresponding carbonimidoyl dichloride (LXVII) ([125]). However, on standing, gradual dissociation into the starting materials was observed ([125]).

$$CCl_3COCl + ClCN \longrightarrow CCl_3CON{=}CCl_2$$

LXVII

In a similar manner sulfur dichloride adds cyanogen chloride to produce chlorosulfenylcarbonimidoyl dichloride (LXVIII) ([10]), which is also obtained from dithiocyanogen and chlorine ([10,63]).

$$SCl_2 + ClCN \longrightarrow ClS{-}N{=}CCl_2$$

LXVIII

Other sulfur halides, such as SF_5Cl, react likewise to yield the corresponding inorganic carbonimidoyl dichlorides ([146]). For example, reaction of SF_5Cl with cyanogen chloride in the presence of ultraviolet irradiation gives rise to the formation of the carbonimidoyl dichloride LXIX ([146]).

$$SF_5Cl + ClCN \longrightarrow SF_5N{=}CCl_2$$

LXIX

The thermal reaction of trichloronitrosomethane (LXX) yields trichloromethylcarbonimidoyl dichloride among other reaction products ([130]).

$$3\ CCl_3NO \longrightarrow CCl_3N{=}CCl_2 + CCl_3NO_2 + NOCl$$
LXX

Cyclic N-methylamides, such as N-methylpyrrolidone (LXXI), are readily chlorinated to give the carbonimidoyl dichloride LXXII ([49]).

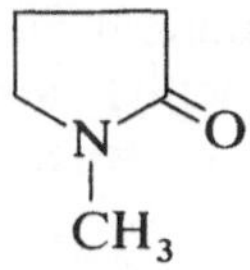

$$+\ Cl_2 \longrightarrow ClCO(CH_2)_3N{=}CCl_2$$
LXXII

LXXI

Likewise, chlorination of N-methylisatoic anhydride (LXXIII) affords the carbonimidoyl dichloride LXXIV in 42% yield. The position of the chlorine group in the ring has not been established ([49]).

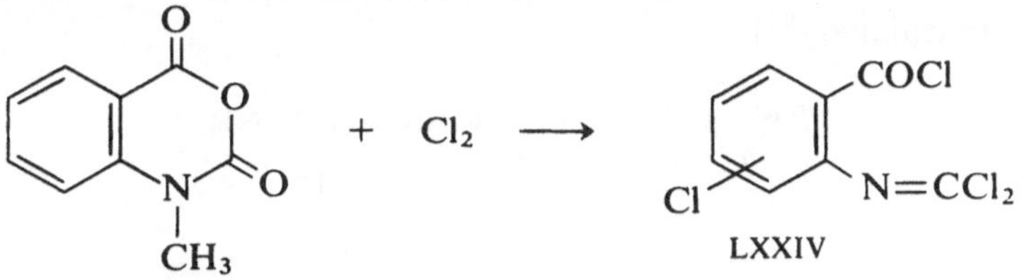

$$+\ Cl_2 \longrightarrow$$

LXXIII

LXXIV

A variety of fragmentation reactions lead to perfluorocarbonimidoyl difluorides. For example, addition of trifluoronitrosomethane to tetrafluoroethylene yields the oxazetidine derivative LXXV, which upon thermolysis ([13,97]) or photolysis ([11]) gives trifluoromethylcarbonimidoyl difluoride (LXXVI) and carbonyl fluoride.

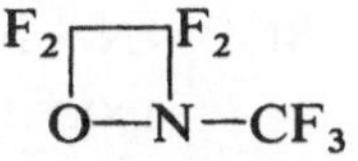

$$\longrightarrow CF_3N{=}CF_2 + COF_2$$
LXXVI

LXXV

Trifluoromethylcarbonimidoyl difluoride (LXXVI) was also obtained upon heating NF_3 and trifluoroacetonitrile over caesium fluoride at 520°C ([23]), and via pyrolysis of perfluoroazoalkanes ([159]).

The fragmentation of perfluoroazides can also lead to carbonimidoyl difluorides. Thus, heating of the azide LXXVII gives rise to the formation of LXXVIII ([68]).

$$CF_3CHFCF_2N_3 \overset{\Delta}{\longrightarrow} CF_3CHFN{=}CF_2 + N_2$$
LXXVII LXXVIII

Also pyrolysis of perfluorotriethylamine yields pentafluorethyl-carbonimidoyl difluoride (LXXIX) ([125a]).

$$(C_2F_5)_3N \xrightarrow{\Delta} CF_3CF_2N{=}CF_2$$

LXXIX

Another useful synthesis of aromatic carbonimidoyl difluorides involves the reaction of the corresponding carbonimidoyl dichlorides with hydrogen fluoride. The intermediate trifluoromethyl amine (LXXX) eliminates hydrogen fluoride upon heating in the presence of potassium fluoride ([66,67,128]).

$$RN{=}CCl_2 + 3\,HF \longrightarrow RNHCF_3 \longrightarrow RN{=}CF_2 + HF$$

LXXX

Instead of the monomeric carbonimidoyl difluorides often its dimers LXXXI are isolated ([67]).

$$2\,RN{=}CF_2 \longrightarrow RN(CF_3)\underset{\underset{F}{|}}{C}{=}NR$$

LXXXI

Thiele ([145]) synthesized isocyanogen tetrabromide (LXXXII) by allowing hydrazotetrazole (LXXXIII) to react with bromine, and Scott and Cronin ([136a]) obtained carbonimidoyl dibromides in the reaction of substituted tetrazoles with bromine.

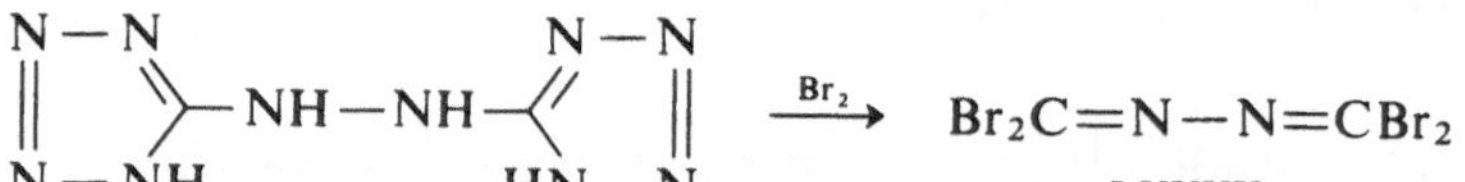

$$\xrightarrow{Br_2} Br_2C{=}N{-}N{=}CBr_2$$

LXXXIII   LXXXII

The halo groups attached to carbon in chloromethylcarbonimidoyl dichlorides are reactive enough to undergo the Friedel–Crafts reaction. For example, benzylcarbonimidoyl dichloride is obtained from chloromethylcarbonimidoyl dichloride and benzene in the presence of aluminum chloride ([95a]).

The addition of cyanogen chloride to N-chloro-bis(trifluoromethyl)-amine in the presence of light to produce the carbonimidoyl dichloride LXXXIV has been reported by Dobbie and Emeleus ([20a]).

$$(CF_3)_2NCl + ClCN \longrightarrow (CF_3)_2N-N=CCl_2$$

LXXXIV

The simultaneous addition of cyanogen chloride and chlorine to cyclohexene yields the carbonimidoyl dichloride LXXXV ([95a]).

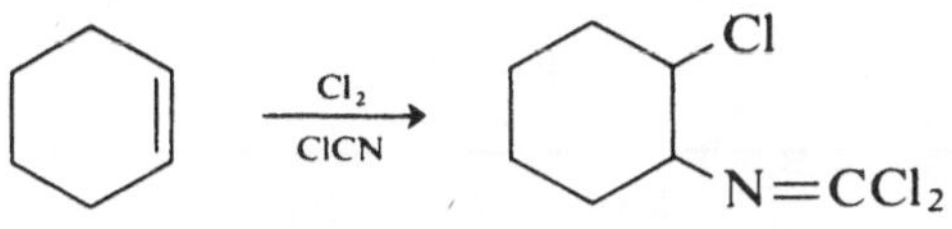

LXXXV

In Tables I–VII the hitherto-reported carbonimidoyl dihalides are listed.

TABLE I
Carbonimidoyl Difluorides*

$RN=CF_2$

R	Method of preparation	B.p., °C/mm (M.p., °C)	Yield, %	Reference
CF_3	E, F	−33	96	9, 11, 13, 23, 45, 46, 68, 155–158
CF_3CHF	F	24		68
CF_3CF_2	F	−6		13, 46, 125a
$ClCH_2CH_2$	F	81–82/10†		128
C_2H_5	B	17–18		95a
C_3F_7	F	25.6		13
$(CF_3)_2CH$	F	36		68
C_6F_{11}	F	(93–94)		143
C_6H_5	F	128–129/2.5‡ 49/12		128 95a
$4\text{-}CH_3C_6H_4$	F	74–78/50		95a
$4\text{-}ClC_6H_4$	F	82–83/15		95a
$2,5\text{-}Cl_2C_6H_3$	F	92–95/15		95a
$2,4,6\text{-}Cl_3C_6H_2$	F	130–137/23		95a
C_6Cl_5	F	(54–58)		95a
$F_2C=N$	F	3		102
$BrFC=N$	F	68		102

* A mixed carbonimidoyl dihalide, $CF_2ClN=CFCl$, b.p. 55°C, was obtained by method F ([97]).
† Along with dimer or trimer, b.p. 122–123/14 mm; m.p. 47–48°C.
‡ Dimer.

TABLE II
Aliphatic Carbonimidoyl Dichlorides

$$RN{=}CCl_2$$

R	Method of preparation	B.p., °C/mm (M.p., °C)	Yield, %	Reference
ClS	B, F	50–51/18	80, 27	10, 61
CF_3	E	46		158
CH_2Cl	C	36–37/12		50, 51
CCl_3	C	59–60/12	87	32, 50, 51, 100
CH_3	B	77.5–78	85.7	93, 127
	C	79–80	83.5	85
CH_3OCH_2	B	30–32/12		95a
$C_2H_5OCH_2CH_2$	B	51–53/14	77	95a
$C_3H_7OCH_2$	B	57–59/10	72	136
$CCl_3CH_2OCH_2OCH_2$	B	70–72/0.2	57	95a
$C_6H_{11}OCH_2$	B	70–75/10	32	95a
$C_6H_5OCH_2$	B	87–89/0.2	41	95a
$2,4\text{-}Cl_2C_6H_3OCH_2$	B	134–138/0.5		95a
CCl_3CCl_2	C	120–121/13		52
$C(CN)Cl_2CCl_2$	C	122–124/15		52
$C(Cl_2C{=}N)Cl_2CCl_2$	C	(166)		100
$C(COCl)Cl_2(CCl_2)_2$	E	188–190/15 (98–101)		49
CHF_2CH_2	B	63/100		154, 15
CH_2ClCH_2	B	62/12		15
$Et_2NCH_2CH_2$	A	(120–122)	64	141
C_2H_5	A	100–101		85
	C	102		109
$n\text{-}C_3H_7$	B	125–127		93
$Me_2NCH_2CH_2CH_2$	A	(130.5–132)	62	141
$n\text{-}C_4H_9$	B	40–45/13		93
	C	45–45.5/14	50	85
$t\text{-}C_4H_9$	A	129–131	52	95a
$Cl_2C{=}N(CH_2)_4$	C	74–76/0.03	56	123, 124

Table II—*continued*

R	Method of preparation	B.p., °C/mm (M.p., °C)	Yield, %	Reference
$Cl_2C=N(CH_2)_6$	C	118–120/0.75	60	123, 124
$n\text{-}C_4H_9OCH_2CH_2CH_2$	B	99–102/10	85	136
$t\text{-}C_8H_{17}$	B	78–79/10	88	136
$C_{10}H_{17}{}^*$	B	73–75/0.01	91	136
$C_{12}H_{25}$	C	163–166/15	26.5	85
	B	107/0.002	95	136
$C_{14}H_{29}$	B	140–141/0.01	95	136
$C_{16}H_{33}$	B	164–166/0.02	95	136
$C_{18}H_{37}$	B	159–161/0.003	91	136
C_6H_{11}	C	82–83.5/12	46, 50	85, 148
	B	76–77/10	90	35, 136
$2\text{-}ClC_6H_{10}$	F	85/0.8	50	95a
$4\text{-}OCNC_6H_{10}$	C	95/0.45	21	123, 124
$4\text{-}Cl_2C=NC_6H_{10}$	C	108/0.05 (38–39)	47	123, 124
$C_6H_5Cl_5$	F	(138–143)	84	132
C_6ClF_{10}	F	†		143
$C_6H_5CH_2$	B	117–118/17	79	17
$C_6H_5CCl_2$	F	(58–59)	72	112
	B	150–160/15	64	17, 54a
$2\text{-}ClC_6H_4CCl_2$	B	99–109/0.2	78	54a
$3\text{-}ClC_6H_4CCl_2$	B	154–156/3.5	70	54a
$4\text{-}ClC_6H_4CCl_2$	B	109–113/0.05	76	54a
$2,6\text{-}Cl_2C_6H_3CCl_2$	B	(54)	87	54a
$3,4\text{-}Cl_2C_6H_3CCl_2$	B	132–136/0.08	95	17, 54a
$3\text{-}CF_3C_6H_4CCl_2$	B	84–86/0.08	77	54a

* Derived from isocamphanyl isothiocyanate.
† Mixture with $C_6F_{10} = NCCl_3$

TABLE III
Aromatic Carbonimidoyl Dichlorides

$$RN{=}CCl_2$$

R	Method of preparation	B.p., °C/mm (M.p., °C)	Yield, %	Reference
C_6H_5	A	204–205	—	108
	B	83/10	82	121
		103–106/31	85–90	107, 132
	D	94–99/14	51	86
$2\text{-}CH_3C_6H_4$	B	125–130/15	—	24
	A	214–25	—	108
$3\text{-}CH_3C_6H_4$	B	130/10	—	24
$4\text{-}CH_3C_6H_4$	B	121–124/20	—	24
	A	225–226	—	108
$4\text{-}FC_6H_4$	D	82–87/11	71.5	86
$2\text{-}ClC_6H_4$	D	104–106/10	86.7	84
$4\text{-}ClC_6H_4$	A	135–141/31	61	107
	C	117–120/14	—	85
	D	110–113/10	86	84
$4\text{-}BrC_6H_4$	B	122–124/15	—	18, 24
	D	131–137/12	72.5	86
$4\text{-}CH_3OC_6H_4$	B	155–160/15	—	24
$3\text{-}O_2NC_6H_4$	B	165–170/15 (68)	—	24
	D	120–125/0.5 (69)	83	86
$4\text{-}O_2NC_6H_4$	B	(80)	—	24
	D	127/0.35 (78–82)	88	86
$4\text{-}ClOCC_6H_4$	D	152–158/11	49	84
$4\text{-}CH_3OOCC_6H_4$	D	157–159/12 (96–98)	88.5	86
$4\text{-}C_6H_5COC_6H_4$	D	170–180/0.2	73.5	84
$4\text{-}C_6H_5N{=}NC_6H_4$	D	(82–84)	31	84
$3\text{-}Cl_2C{=}NC_6H_4$	C	160–161/16	—	85
$4\text{-}Cl_2C{=}NC_6H_4$	D	172–180/14 (96–98)	51.7	86
$4\text{-}Cl_2C{=}NC_6Cl_4$	E	(154–156)	85	49
$2\text{-}C_6H_5SO_2C_6H_4$	D	210/0.3	39	84
$4\text{-}C_6H_5C_6H_4$	D	172–177/12	72	86
$2,6\text{-}(CH_3)_2C_6H_3$	A	72–76/0.4	—	95a
$2,6\text{-}(i\text{-}C_3H_7)_2C_6H_3$	D	104–120/0.4	80.5	86
$2,4\text{-}CH_3(Cl)C_6H_3$	D	121–123/10	96.6	84

Table III—*continued*

R	Method of preparation	B.p., °C/mm (M.p., °C)	Yield, %	Reference
2,6-CH$_3$(Cl)C$_6$H$_3$	D	110–112/11	92.5	86
2,3-Cl$_2$C$_6$H$_3$	D	142–149/14	73.5	84
2,4-Cl$_2$C$_6$H$_3$	D	126–132/10	85.3	84
2,5-Cl$_2$C$_6$H$_3$	D	128–135/15	78.4	84
3,4-Cl$_2$C$_6$H$_3$	D	141–148/13	80	84
3,5-Cl$_2$C$_6$H$_3$	D	147–155/13	53	86
2,4-Br$_2$C$_6$H$_3$	D	161–162/12	85.5	86
2,5-Cl(CF$_3$)C$_6$H$_3$	D	101–102/12	75	86
2,4-CF$_3$(Cl)C$_6$H$_3$	D	104–110/10	78.3	84
2,4-Cl(NO$_2$)C$_6$H$_3$	D	131–133/0.1	22	84
4,3-Cl(NO$_2$)C$_6$H$_3$	D	118/0.11	79.5	86
2,4-CH$_3$O(Cl)C$_6$H$_3$	D	161–166/11 (81)	37	86
2,5-CH$_3$O(NO$_2$)C$_6$H$_3$	D	140–141/0.17	82.5	86
2,5-CH$_3$(NO$_2$)C$_6$H$_3$	D	119–120/0.15 (59)	94	86
4,3-CH$_3$(NO$_2$)C$_6$H$_3$	D	166–168/13	82.5	86
2,4-Cl(4-ClC$_6$H$_4$O)C$_6$H$_3$	D	172–180/0.25 (46–48)	81.7	86
4,2,6-CH$_3$(C$_2$H$_5$)$_2$C$_6$H$_2$	D	151–161/10	55.6	84
5,2,4-CH$_3$(Cl)$_2$C$_6$H$_2$	D	143–146/10	84	86
4,2,5-O$_2$N(Cl)$_2$C$_6$H$_2$	D	136–138/0.15 (83)	20	86
4,2,5-C$_2$Cl$_5$(Cl)$_2$C$_6$H$_2$	E	190/0.6	60.6	49
2,4,5-Cl$_3$C$_6$H$_2$	D	162–165/16	81	84
2,4,6-Cl$_3$C$_6$H$_2$	E	88–90/0.3	87	49
2,3,4-CH$_3$(Cl)(Cl$_2$C=N)C$_6$H$_2$	B	50–51/0.05	70	121, 124
C$_6$Cl$_5$	E, C	134–136/0.2 (73–75)	98	49, 52
α-C$_{10}$H$_7$	D	183–190/14	20	84
β-C$_{10}$H$_7$	B	153–156/2	—	61
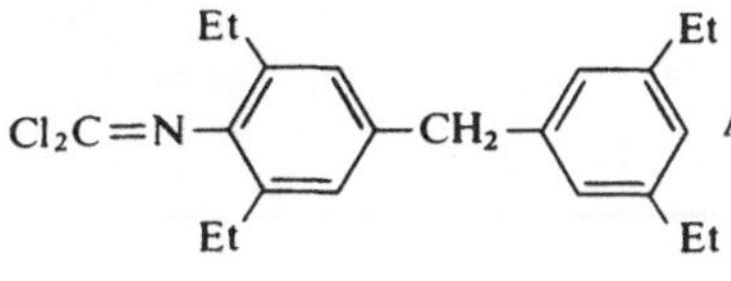	A	205–210/0.2	—	95a

TABLE IV
Acyl and Aroylcarbonimidoyl Dichlorides

$$RCON{=}CCl_2$$

R	Method of preparation	B.p., °C/mm (M.p., °C)	Yield, %	Reference
EtOOC	B	57–59/10	—	114
CH_2Cl	B	76.5/15	54	4
	F	80/20	60–70	125
$CHCl_2$	B	36–38/0.09	—	4
	F	82/14	70–80	125
CCl_3	B	37/0.07	76	4
	F	85/18	70–80	125
CH_3	B	28/9	—	16
$2,4\text{-}Cl_2C_6H_3OCH_2$	B	140–141/0.09	87	95a
$2,4,5\text{-}Cl_3C_6H_2OCH_2$	B	—*	94	95a
$Cl(CH_2)_2$	B	39/0.03	86.7	4
$ClCO(CH_2)_2$	F	101–103	28	125
$CH_3CH_2CCl_2$	B	93–95/14	82.2	95a
$CH_3(CH_2)_6$	B	80–81/0.07	39	4
$CH_3(CH_2)_{10}$	B	125–130/0.06	54	4
$CH_3(CH_2)_{12}$	B	130–140/0.04	—	4
$CH_3(CH_2)_{14}$	B	(34–36)	—	4
C_6H_5	B	124–126/12	81–89	16, 55, 56
		62–64/0.02	95	4, 114
	F	114/20	20–30	125
$4\text{-}CH_3C_6H_4$	B	85–88/0.02	—	114
$2\text{-}ClC_6H_4$	B	160–162/20	81–89	55, 56
$4\text{-}ClC_6H_4$	B	153–154/17	81–89	56
		90–93/0.02		114
$3,4\text{-}Cl_2C_6H_3$	B	113/0.09	85	95a
$2,5\text{-}Cl_2C_6H_3$	B	118–121/0.07	94	95a
$2,6\text{-}Cl_2C_6H_3$	B	120–123/0.1	37.2	95a
$4\text{-}O_2NC_6H_4$	B	(62–66)	90	95a
$3\text{-}O_2N\text{-}4\text{-}ClC_6H_3$	B	(58–61)	91.5	95a
$4\text{-}Cl_2C{=}NCOC_6H_4$	B	(83)		4
C_6H_5O	B	127–130/11	85	95a
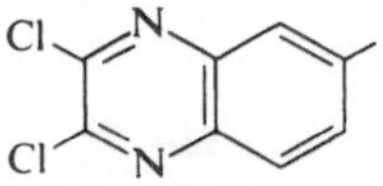	B	(91–94)	77	95a

* Not distillable.

TABLE V
Arenesulfonylcarbonimidoyl Dichlorides

$$RSO_2N=CCl_2$$

R	Method of preparation	B.p., °C/mm (M.p., °C)	Yield, %	Reference
$(CH_3)_2N$	B	(64)	—	7, 32
CH_3	B	(79–80)	82	111, 115
		(76–78)	73	5, 32
		(73–76)	83	37
$C_6H_5CH_2$	B	(78)	—	32
		(79–80)	80	115
		(83–87)	—	95a
C_6H_5	B	110–112/0.01	63	111, 115
		122–125/0.1	—	32
		175–178/22	92	37
$2\text{-}CH_3C_6H_4$	B	Oil	—	7, 32
$4\text{-}CH_3C_6H_4$	B	116–118/0.01	73	111, 115
		(83–85)		
		100/0.005	97	37
$4\text{-}ClC_6H_4$	B	(75–77)	52	5
		115–116/0.1	98	37
$3,4\text{-}Cl_2C_6H_3$	B	(78)	—	7, 32

TABLE VI
Phosphorylcarbonimidoyl Dichlorides

$$R_2P\underset{\underset{O}{\downarrow}}{-}N=CCl_2$$

R	Method of preparation	B.p., °C/mm (M.p., °C)	Yield, %	Reference
Cl	F	83–86/25	—	19
EtO	B	81–82/0.1	83	64
		81.5–82/1.0	—*	2, 3
PrO	B	94–96/1.0	—*	2
i-PrO	B	62–63/0.05	81	64
		83–84/1.5	—*	2
BuO	B	111–113/1.0	—*	2
		104–108/0.08	—*	3
i-BuO	B	100–102/1.0	—*	2
C_6H_5O	B	157/0.05	70	64

* Reported yields range from 61–89 %.

TABLE VII

Carbonimidoyl Dibromides

$RN=CBr_2$

R	Method of preparation	B.p., °C/mm (M.p., °C)	Yield, %	Reference
C_6H_5	A	76.5–77.5/0.15	76	95a
	D	128–129/11	46	84
$4\text{-}BrC_6H_4$	F	(115)	—	8
$C_6H_5CH=N$	F	(15–17)	19	136a
$4\text{-}ClC_6H_4CH=N$	F	(92–93)	80	136a
$4\text{-}BrC_6H_4CH=N$	F	(98–99)	75	136a
$4\text{-}CH_3OC_6H_4CH=N$	F	(60–61)	85	136a
$4\text{-}O_2NC_6H_4CH=N$	F	(145–147)	78	136a
$Br_2C=N$	F	(42)	—	145

III. PHYSICOCHEMICAL PROPERTIES

Carbonimidoyl dihalides are colorless or light yellow (nitro derivatives) liquids, or low-melting solids which can be distilled *in vacuo* without decomposition. They are generally easily soluble in inert organic solvents, such as chloroform, carbon tetrachloride, benzene, halogenated benzene, etc. Their characteristic spectral feature is the $C=N$ double bond, which is easily recognized in the infrared region at approximately $1600–1650 \text{ cm}^{-1}$ and the asymmetrical C–Cl stretching at approximately 880 cm^{-1} for the dichlorides. The replacement of chlorine by fluorine affects the $C=N$

TABLE VIII

Asymmetrical Vibration Stretching of the
$C=N$ Group in Carbonimidoyl Dihalides

Class of compounds	$v_{C=N} (\text{cm}^{-1})$	Reference
$R_FN=CF_2$*	1808–1815	13, 155
$R_FN=CClF$	1742	13
$R_FN=CCl_2$	1669	13
$RN=CCl_2$	1600–1650	5, 107, 112
$RSO_2N=CCl_2$	1600–1605	115
$RCON=CCl_2$	1640	111
$D_2\overset{\oplus}{N}=CCl_2$	1594	1

* R_F = perfluoroalkyl.

stretching vibration markedly. The $CF_3N{=}CClF$ absorbs at $1742\ cm^{-1}$ [13], and carbonimidoyl difluorides absorb at $1783{-}1815\ cm^{-1}$ [13,155].

The hydrochloride of the deuterated parent compound,

$$Cl_2C{=}ND\overset{\oplus}{\underset{2}{}}]Cl^{\ominus}$$

absorbs at $1594\ cm^{-1}$ [1]. Alkyl and arylcarbonimidoyl dichlorides absorb at approximately $1600{-}1650\ cm^{-1}$, while trifluoroalkylcarbonimidoyl dichlorides absorb at $1669\ cm^{-1}$. Aroyl and arenesulfonyl groups attached to the $C{=}N$ bond cause only a slight shift, as evidenced by their absorption at $1640\ cm^{-1}$ [111] and $1600{-}1605\ cm^{-1}$ [111,115].

The $C{=}N$ stretching vibrations for several groups of carbonimidoyl dihalides are listed in Table VIII.

IV. CHEMICAL BEHAVIOR

A. Reaction with Oxygen–Hydrogen Bonds

The reaction of carbonimidoyl dichlorides with excess water results in complete hydrolysis with formation of the corresponding amine and carbon dioxide [19]. This reaction affords isocyanate as an intermediate, as evidenced by the isolation of 1,3-diphenylurea from the reaction mixture, obtained by treating phenylcarbonimidoyl dichloride with water [24,108]. In the reaction of tetrameric cyanogen chloride with water at 20°C 2-amino-4,6-dichloro-1,3,5-triazine (LXXXVI) is obtained, whereas at 100°C total hydrolysis to cyanuric acid (LXXXVII) occurs [131,140].

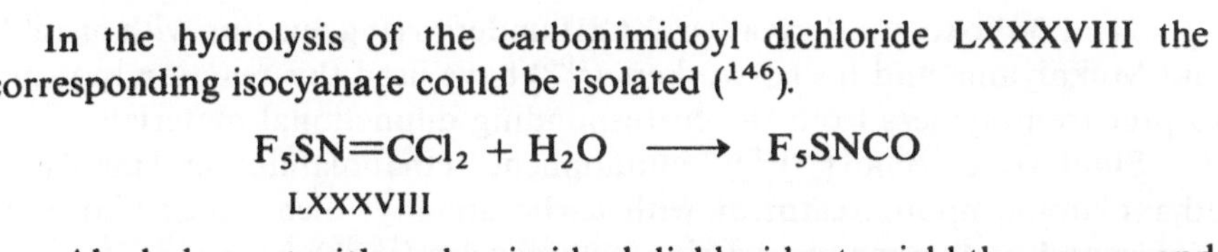

In the hydrolysis of the carbonimidoyl dichloride LXXXVIII the corresponding isocyanate could be isolated [146].

$$F_5SN{=}CCl_2 + H_2O \longrightarrow F_5SNCO$$

LXXXVIII

Alcohols react with carbonimidoyl dichlorides to yield the corresponding carbamates (LXXXIX) [25,109,137]; however, with alkoxides stepwise replacement of the chloro groups is observed. Thus, the intermediate imidochloroformate XC can be isolated in good yield if only one equivalent of alkoxide or phenoxide is being used [61,87,96]. Further reaction with a second equivalent of alkoxide ([70,82,87,96,140,142,153]) or phenoxide ([19,25,44, 64]) yields the isocyanate O,O-acetals XCI.

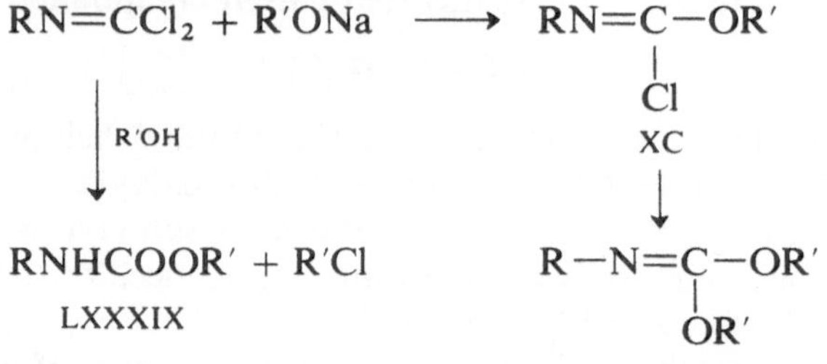

$$RN{=}CCl_2 + R'ONa \longrightarrow \begin{array}{c} RN{=}C{-}OR' \\ | \\ Cl \\ XC \end{array}$$

$$\downarrow \text{R'OH}$$

$$\begin{array}{c} RNHCOOR' + R'Cl \\ LXXXIX \end{array} \qquad \begin{array}{c} R{-}N{=}C{-}OR' \\ | \\ OR' \\ XCI \end{array}$$

The carbamates LXXXIX are also obtained upon treatment of XC with a variety of acids, such as hydrogen chloride, carboxylic acids, and phosphoric acid ([104]). Reaction of XC with amines yields the O,N-acetals XCII. For example, the monosubstituted products, derived from phenols (XC, R' = aryl) upon reaction with alkyl amines afford the corresponding O,N-acetals, which are useful herbicides ([61,88]).

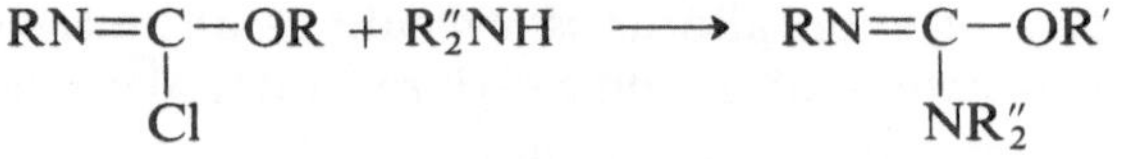

$$\begin{array}{c} RN{=}C{-}OR \\ | \\ Cl \end{array} + R''_2NH \longrightarrow \begin{array}{c} RN{=}C{-}OR' \\ | \\ NR''_2 \\ XCII \end{array}$$

From ethylene glycol and carbonimidoyl dichlorides the cyclic acetal XCIII is obtained ([16,35,40]), which upon heating in the presence of lithium chloride ([40]) or aluminum chloride ([106]) isomerizes to the oxazolidinone derivative XCIV.

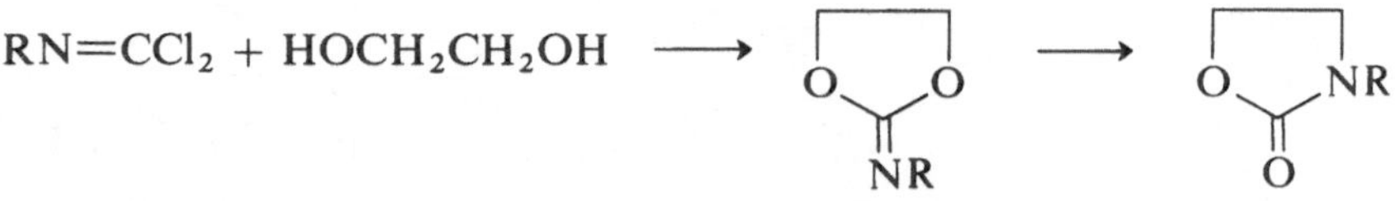

$$RN{=}CCl_2 + HOCH_2CH_2OH \longrightarrow$$

XCIII

XCIV

The 1,3-dioxolane derivatives XCIII undergo ring opening with phenols, and Mukaiyama and his co-workers ([106]) have used this reaction sequence to prepare polymers from the corresponding difunctional materials.

Similarly, catechol ([16,38]), 2-aminophenol, ethanolamine, and mercapto-ethanolamine upon treatment with carbonimidoyl dichlorides afford the corresponding five-membered ring heterocycles ([16,86]).

The reaction of phenylcarbonimidoyl dichloride with acetic acid was investigated by Sell and Zierold ([137]), and by Nef([109]), who obtained acetani-lide from the reaction mixture. Dyson and Harrington ([24]) isolated substan-tial amounts of 1,3-diphenylurea and thus concluded that phenyl isocyanate is not an intermediate in this reaction. However, both products can also be obtained from phenyl isocyanate and acetic acid. Furthermore, the reaction of certain carbonimidoyl dichlorides with formic acid produces isocyanate ([19,64]), and therefore the following reaction mechanism is indicated:

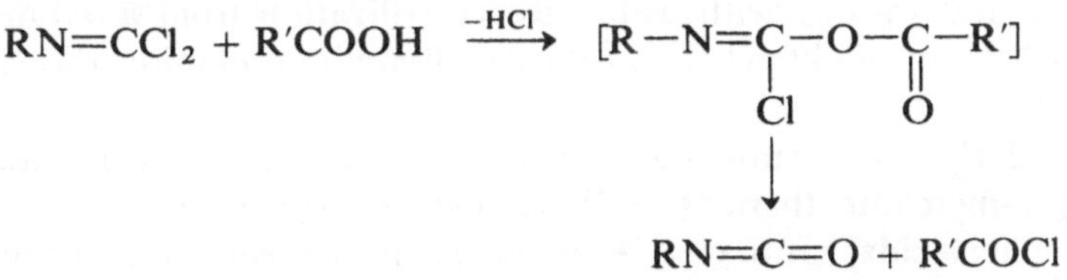

$$RN=CCl_2 + R'COOH \xrightarrow{-HCl} [R-N=\underset{\underset{Cl}{|}}{C}-O-\underset{\underset{O}{\|}}{C}-R']$$

$$\downarrow$$

$$RN=C=O + R'COCl$$

However, from arylcarbonimidoyl dichlorides and mercaptoacetic acid (XCV) the five-membered heterocycle XCVI is obtained ([86]).

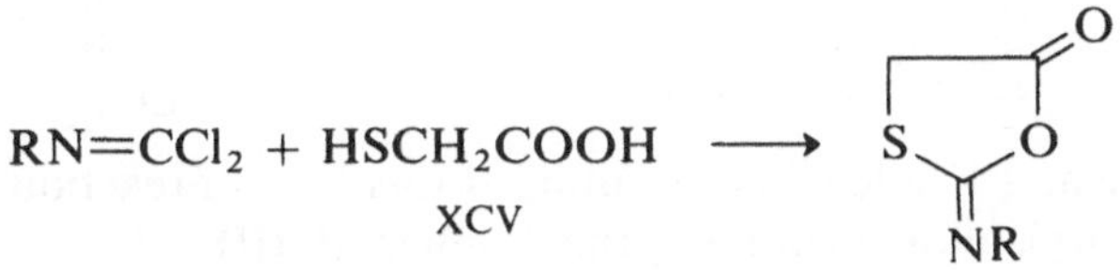

$$RN=CCl_2 + HSCH_2COOH \longrightarrow$$

XCV

XCVI

B. Reaction with Sulfur–Hydrogen Bonds

The reaction of carbonimidoyl dichlorides with hydrogen sulfide gives rise to the formation of isothiocyanates in certain instances ([77,146]). For example, the sulfenyl isothiocyanate XCVIII is obtained upon treatment of XCVII with hydrogen sulfide ([146]).

$$F_5SN=CCl_2 + H_2S \longrightarrow F_5SN=C=S$$

XCVII   XCVIII

In the reaction of carbonimidoyl dichlorides with thiolate ion both the monosubstitution product XCIX and the S,S-acetal C can be obtained, depending upon the amount of the thiolate ion being used ([2,6,19,30,74,83,86, 114,115,140]).

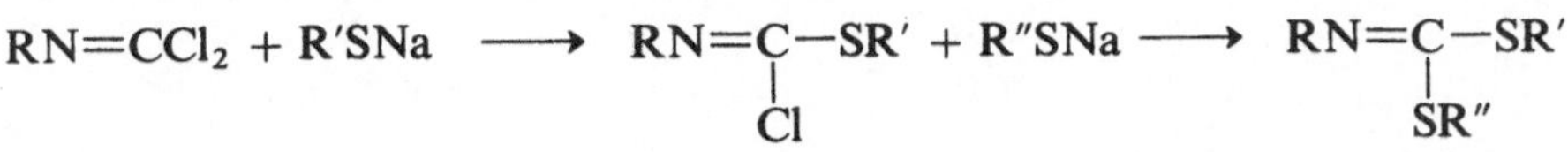

$$RN=CCl_2 + R'SNa \longrightarrow RN=\underset{\underset{Cl}{|}}{C}-SR' + R''SNa \longrightarrow RN=\underset{\underset{SR''}{|}}{C}-SR'$$

XCIX   C

The intermediate XCIX can also be obtained from arylformamide and the corresponding sulfenyl chloride ([86,114,140]).

The thiolate ion can be generated by using an equivalent amount of a tertiary amine, the base serving also as hydrogen chloride scavenger ([2,30]). From the intermediate XCIX and primary or secondary amines the corresponding S,N-acetals are obtained ([2]).

Reaction of p-toluenesulfonylcarbonimidoyl dichloride with thiophenol [37]. To a solution of 0.92 g of sodium in 60 ml of *n*-butanol is added 4.4 g thiophenol and a solution of 5 g of *p*-toluenesulfonylcarbonimidoyl dichloride in 20 ml of acetone. The precipitate, which is formed after some time, is collected and washed with water. Recrystallization from *n*-butanol yields 6.5 g (82%) of diphenyl N-*p*-toluenesulfonyliminodithiocarbonate, m.p. 159–160°C.

Similarly, five-membered heterocycles are obtained from thioglycols ([16,101]), 2-mercaptoethanol ([16,86]) and mercaptoacetic acid ([86]).

The herbicidal xanthates CI are formed in the reaction of the potassium salt CII and carbonimidoyl dichlorides ([27,94]).

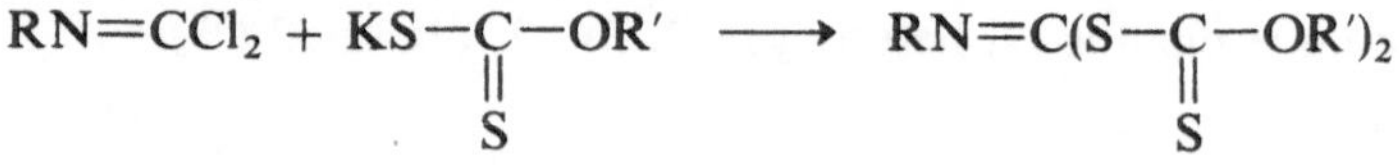

$$RN{=}CCl_2 + KS{-}\underset{\underset{S}{\|}}{C}{-}OR' \longrightarrow RN{=}C(S{-}\underset{\underset{S}{\|}}{C}{-}OR')_2$$

$$CII \qquad\qquad\qquad\qquad CI$$

Dickore and Kühle ([20]) have utilized this type of reaction to prove the structure of arylsulfonyl isothiocyanate dimers (CIII).

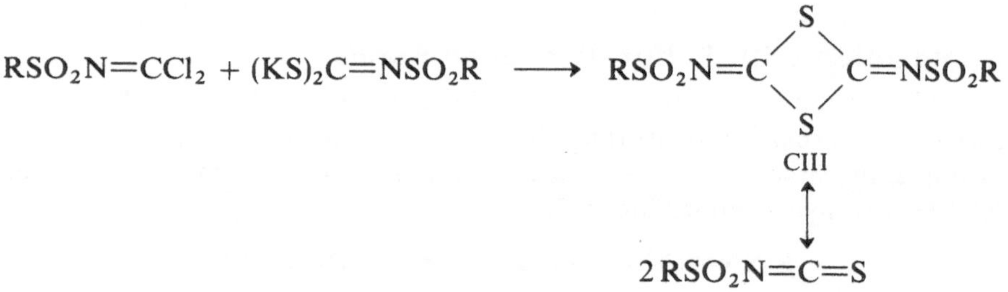

$$RSO_2N{=}CCl_2 + (KS)_2C{=}NSO_2R \longrightarrow RSO_2N{=}C\underset{S}{\overset{S}{\diagup\diagdown}}C{=}NSO_2R$$

$$CIII$$

$$\updownarrow$$

$$2\,RSO_2N{=}C{=}S$$

The potassium salts of mercaptothiophosphates CIV can be treated with carbonimidoyl dichlorides to produce the corresponding imides CV which have insecticidal, nematocidal, and fungicidal activity ([99]).

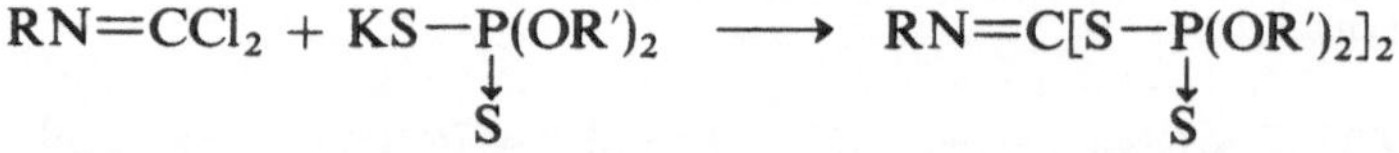

$$RN{=}CCl_2 + KS{-}\underset{\underset{S}{\downarrow}}{P}(OR')_2 \longrightarrow RN{=}C[S{-}\underset{\underset{S}{\downarrow}}{P}(OR')_2]_2$$

$$CIV \qquad\qquad\qquad\qquad CV$$

The reaction of carbonimidoyl dichlorides with sodium sulfide ([86,91]), ammonium sulfide ([90]), and phosphorus pentasulfide ([89]) yields the corresponding isothiocyanates.

$$RN{=}CCl_2 + Na_2S \longrightarrow RN{=}C{=}S + 2\,NaCl$$

This reaction has been used to convert tetrameric cyanogen chloride to the corresponding triazinyl isothiocyanate ([77]), which upon reaction with alcohols ([80]) and amines ([79]) affords the expected triazine thiocarbamates and thioureas.

The formiminium chloride CVI, obtained from formamide and thionyl chloride, undergoes rapid reaction with sulfenyl chlorides to yield thiocyanates (CVII) ([86]).

$$H_2NCHO + SOCl_2 \longrightarrow H_2\overset{\oplus}{N}=CHCl]Cl^{\ominus} + RSCl \longrightarrow RSCN + 3\ HCl$$
$$CVI\phantom{=CHCl]Cl^{\ominus} + RSCl \longrightarrow }CVII$$

The reaction of isocyanogen tetrabromide with benzenesulfinate produces a tetrasulfone, which is readily oxidized to bis(phenylsulfonyl)diazomethane ([21a]).

C. Reaction with Nitrogen–Hydrogen Bonds

The reaction of carbonimidoyl dihalides with primary and secondary amines proceeds stepwise to yield chloroformamidines CVIII and guanidines CIX, respectively ([19,24,28,36,42,56,64,73,92,95a,108,110,114–117,124,133,137,140]).

$$RN=CCl_2 + R_2'NH \longrightarrow \underset{\underset{CVIII}{Cl}}{RN=C-NR_2'} + R_2'NH \longrightarrow \underset{\underset{CIX}{NR_2'}}{RN=C-NR_2'}$$

5-Chlorotolylene-2,4-bis(1,3-diethyleneguanidine) ([124]). To a solution of 7.5 g of ethyleneimine and 17.3 g of triethylamine in 75 ml of toluene a solution of 12.5 g of 5-chlorotolylene-2,4-bis(carbonimidoyl dichloride) is added dropwise and with stirring, controlling the reaction at 15°C. After stirring for 2 hours at room temperature triethylamine hydrochloride is removed by filtration and, after evaporation of the solvent and crystallization from hexane, 5-chlorotolylene-2,4-bis(1,3-diethyleneguanidine), m.p. 84–85°C is obtained in 32% yield.

A similar procedure can be used for almost any primary or secondary amine and the yields are normally high. In the case of the sensitive aziridine derivatives a loss in yield is encountered due to polymerization. Of course, the amines can be used in excess as both reagents and hydrogen chloride scavengers (for example, see Reference [115]).

Tertiary amines react with elimination of alkyl halide, thereby forming the corresponding chloroformamidine (CVIII) ([86]).

In contrast, reaction of primary amine hydrochlorides with arylcarbonimidoyl dichlorides and arenesulfonylcarbonimidoyl dichlorides produces the corresponding carbodiimides (CX) ([5,30,113,115,117]).

$$RN=CCl_2 + R'NH_2 \cdot HCl \longrightarrow RN=C=NR' + 3\ HCl$$
$$CX$$

Benzenesulfonylcyclohexylcarbodiimide ([115]). A mixture of cyclohexyl-amine hydrochloride (6.8 g; 0.5 mole) and benzenesulfonylcarbonimidoyl dichloride (11.9 g; 0.5 mole) in 50 ml chlorobenzene is heated under reflux until the evolution of hydrogen chloride ceases. Evaporation of the solvent and vacuum distillation of the residue yields 6.1 g (46%) of benzene-sulfonylcyclohexylcarbodiimide, b.p. 175–180°C/0.01 mm.

Derkach and Liptuga ([19]) demonstrated the stepwise mechanism leading to carbodiimides by treating phosphorylcarbonimidoyl dichloride (CXI) with a deficient amount of arylamines, such as aniline and *p*-bromo-aniline. The elimination of hydrogen chloride from the intermediate CXII occurs in diethylether at room temperature, using triethylamine as the hydrogen chloride scavenger.

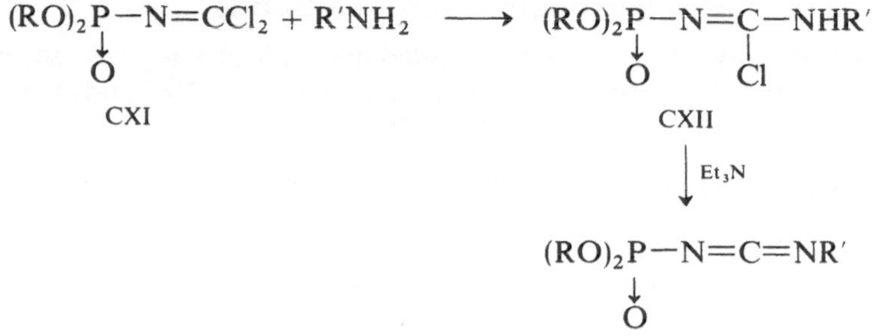

$$(RO)_2P-N=CCl_2 + R'NH_2 \longrightarrow (RO)_2P-N=C-NHR'$$

CXI

CXII

$$\downarrow Et_3N$$

$$(RO)_2P-N=C=NR'$$

CXIII

Thus, the carbodiimides CXIII, having a phosphorus atom attached to the cumulative double bond system, were obtained for the first time ([19]).

The compound *o*-phenylenediamine, upon reaction with arylcarbon-imidoyl dichlorides, yields the corresponding benzimidazoles (CXIV) ([86,107]).

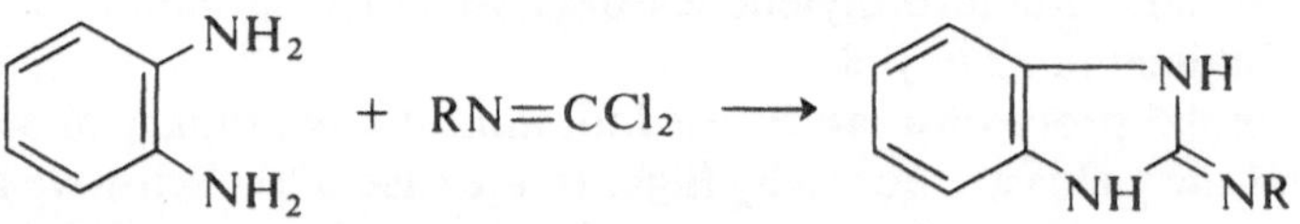

CXIV

The reaction of α-halocarbonimidoyl dichlorides with amidines yields 1,3,5-triazine derivatives CXV ([135]).

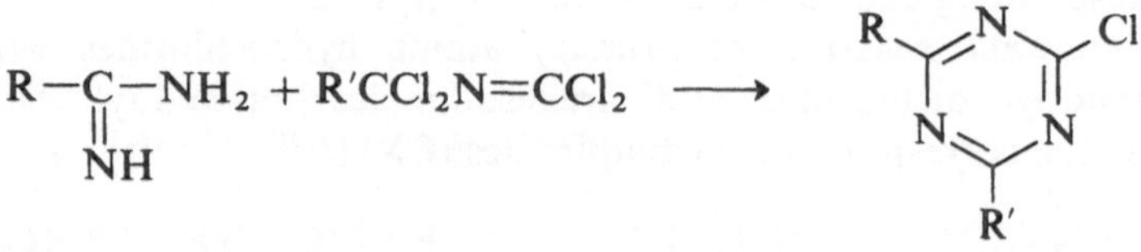

$$R-C-NH_2 + R'CCl_2N=CCl_2 \longrightarrow$$

NH

CXV

Aromatic acid hydrazides, semicarbazides, and thiosemicarbazides react with carbonimidoyl dichlorides with elimination of hydrogen chloride and formation of the corresponding heterocycles ([103,117]). Thus, from benzhydrazide and phenylcarbonimidoyl dichloride, a 91% yield of 5-phenyl-2-phenylamino-1,3,4-oxadiazole (CXVI) is obtained ([103]).

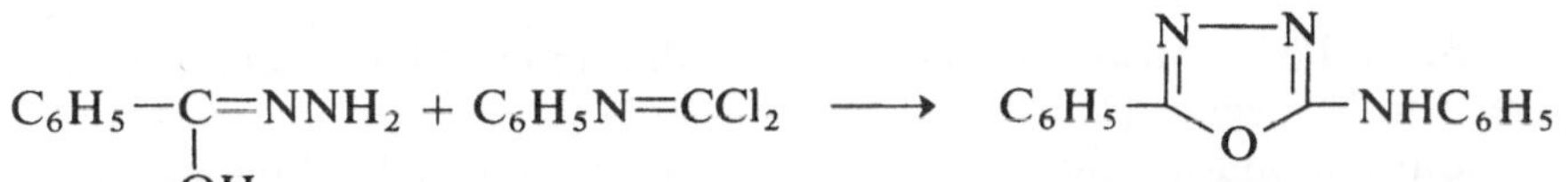

$$C_6H_5-\underset{\underset{OH}{|}}{C}{=}NNH_2 + C_6H_5N{=}CCl_2 \longrightarrow$$

CXVI

However, from benzaldoxime and phenylcarbonimidoyl dichlorides a mixture of phenyl isocyanate, benzonitrile, and aniline hydrochloride is obtained.

$$C_6H_5CH{=}NOH + C_6H_5N{=}CCl_2 \longrightarrow C_6H_5NCO + C_6H_5CN$$
$$+ C_6H_5NH_2{\cdot}HCl$$

The reaction of phenylcarbonimidoyl dichloride with sodium azide yields 5-chlorophenyltetrazole (CXVII) and the corresponding azide (CXVIII), depending upon the stoichiometry ([62,126]).

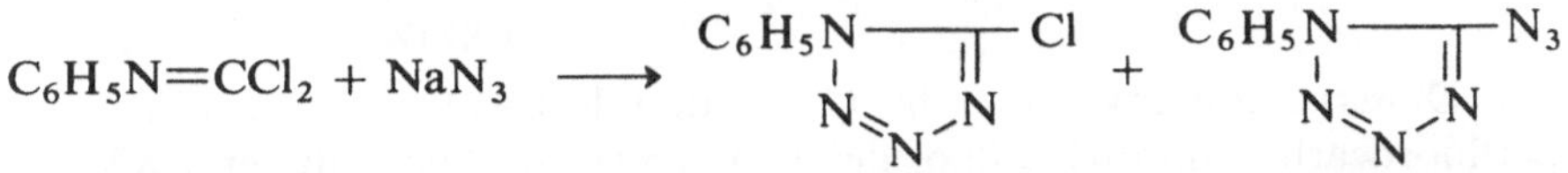

$$C_6H_5N{=}CCl_2 + NaN_3 \longrightarrow$$

CXVII   CXVIII

Thus, reaction of carbonimidoyl dihalides with a wide variety of nitrogen-containing substrates can be achieved, and the products obtained range from linear chloroformamidines, guanidines, and carbodiimides to five- and six-membered heterocycles.

D. Addition to the C=N Bond

The reaction of carbonimidoyl dichlorides with hydrogen fluoride results in replacement of the chloro groups by fluorine and addition of hydrogen fluoride to the C=N group with formation of trifluoromethylamines CXIX ([66,128]). Upon heating of CXIX in the presence of potassium fluoride, elimination of hydrogen fluoride occurs and the corresponding carbonimidoyl difluorides are obtained ([66,128]). This method provides a smooth transformation of carbonimidoyl dichlorides into carbonimidoyl difluorides. The reaction of CXIX with aqueous potassium fluoride affords the corresponding carbamoyl fluorides ([67,129]).

$$RN{=}CCl_2 + 3\,HF \longrightarrow RNHCF_3 + KF \longrightarrow RN{=}CF_2 + HF$$
$$\text{CXIX}$$

For example, addition of hydrogen fluoride to trifluoromethylcarbonimidoyl difluoride yields bis-trifluoromethylamine (CXX) ([128]), and addition of hydrogen chloride yields the 1:1 adduct CXXI ([156]).

$$CF_3NHCF_3 \xleftarrow{HF} CF_3N{=}CF_2 \xrightarrow{HCl} CF_3NHCF_2Cl$$

 CXX   CXXI

Several of the trifluoromethylamines thus obtained are wood preservatives and herbicides ([65]).

Carbonimidoyl difluorides, such as trifluoromethylcarbonimidoyl difluoride, undergo reaction with silver difluoride to yield perfluoroamines CXXII ([156]), while with trifluoromethylsulfurpentafluoride the tertiary amines CXXIII are formed ([22]).

$$(CF_3)_2NF \xleftarrow{AgF_2} CF_3N{=}CF_2 \xrightarrow{CF_3SF_5} (CF_3)_3N$$

 CXXII   CXXIII

The addition of carbonyl fluoride to trifluoromethylcarbonimidoyl difluoride to yield the carbamoyl fluoride CXXIV was observed by Fawcett and his colleagues ([34]).

$$CF_3N{=}CF_2 + COF_2 \longrightarrow (CF_3)_2NCOF$$

 CXXIV

Dimerization involving the C=N double bond occurs readily with perfluorocarbonimidoyl difluorides ([45]). Thus, the linear dimer CXXV is obtained from trifluoromethylcarbonimidoyl difluoride, and its structure was proven by Young and his co-workers ([157]). Pyrolysis of CXXV at 500°C regenerates the monomeric carbonimidoyl difluoride ([97]).

$$2\,CF_3N{=}CF_2 \rightleftharpoons (CF_3)_2N{-}CF{=}NCF_3$$

 CXXV

Insertion of the C=N bond into the mercury–fluorine bond in mercuric fluoride was observed by Young et al. ([156]), who obtained the mercury derivative CXXVI in the reaction of trifluoromethylcarbonimidoyl difluoride with mercuric difluoride.

$$2\,CF_3N{=}CF_2 + HgF_2 \longrightarrow (CF_3)_2N{-}Hg{-}N(CF_3)_2$$

 CXXVI

In the reaction of the mercury derivative CXXVI with $ClNO_2$, a mixture of the nitramine CXXVII and the carbonimidoyl dihalide CXXVIII is obtained ([97]).

$$(CF_3)_2N{-}Hg{-}N(CF_3)_2 + ClNO_2 \longrightarrow (CF_3)_2NNO_2 + CF_2ClN{=}CFCl$$

 CXXVI   CXXVII   CXXVIII

The vapor phase oxidation of trifluoromethylcarbonimidoyl difluoride affords N-nitroso-bis(trifluoromethyl)amine and carbonyl fluoride ([158]).

E. Miscellaneous Reactions

Phenylcarbonimidoyl dichloride readily adds chlorine to the phenyl ring to produce 1,2,3,4,5,6-hexachlorocyclohexylcarbonimidoyl dichloride (CXXIX) in high yield ([132]).

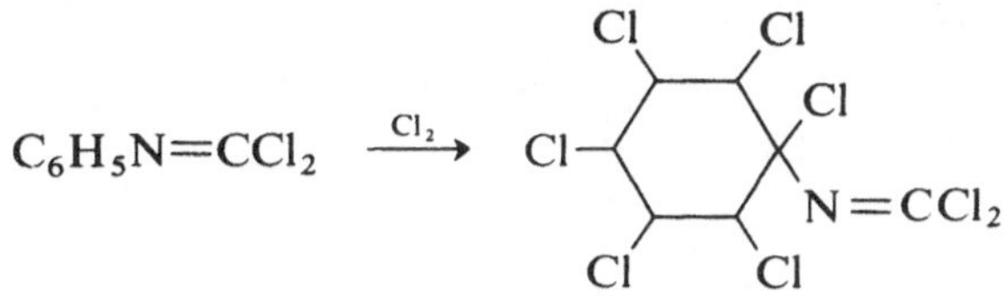

$$C_6H_5N{=}CCl_2 \xrightarrow{Cl_2}$$

CXXIX

The chlorination of acylcarbonimidoyl dichlorides with phosphorus pentachloride yields α,α-dichlorocarbonimidoyl dichlorides CXXX ([112]), which were also obtained by chlorination of the corresponding acylisothiocyanates ([114]).

$$RCON{=}CCl_2 + PCl_5 \longrightarrow RCCl_2N{=}CCl_2$$

CXXX

The reaction of carbonimidoyl dichlorides with acid anhydrides produces isocyanates ([86]), and since isothiocyanates are readily converted into carbonimidoyl dichlorides, a convenient two-step transformation of isothiocyanates into isocyanates is realized.

$$RNCS + Cl_2 \longrightarrow RN{=}CCl_2 + (R'CO)_2O \longrightarrow$$
$$RNCO + 2\,R'COCl$$

The reverse of this reaction is feasible in the aliphatic series, i.e. carbonimidoyl dichlorides can be obtained from isocyanates and phosphorus pentachloride, and subsequent reaction of the carbonimidoyl dichloride with a variety of sulfur compounds produces isothiocyanates.

Likewise, trifluoromethyl isocyanate (CXXXI) was obtained from trifluoromethylcarbonimidoyl difluoride by means of water ([12]) or hydrated fluorides ([157]).

$$CF_3N{=}CF_2 + CuF_2{\cdot}H_2O \longrightarrow CF_3NCO$$

CXXXI

The conversion of carbonimidoyl dichlorides to isocyanates using dimethylsulfoxide was suggested by Johnson and his co-workers ([59]).

Perhaps the best method for conversion of carbonimidoyl dichlorides into isocyanates consists of their reaction with one equivalent of sodium methoxide, followed by thermolysis. Thus, Kodama and Sekiba ([75]) obtained the triazinyl isocyanate CXXXII from tetrameric cyanogen chloride *via* the intermediate CXXXIII. Since N-heterocyclic isocyanates cannot be synthesized by direct phosgenation, this method may have general applicability.

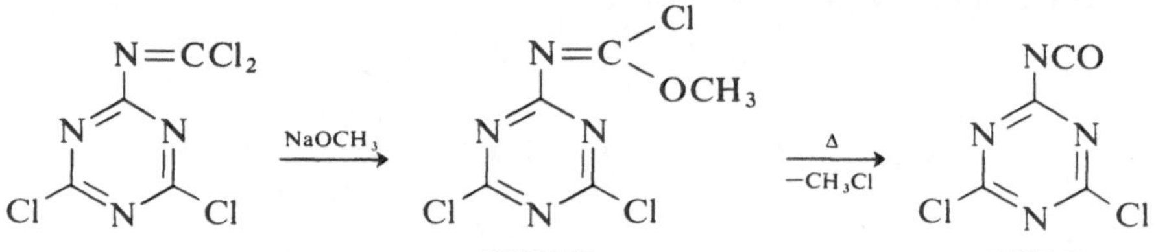

CXXXIII   CXXXII

The triazinyl isocyanate CXXXII has been converted into ureas ([71,78]) and carbamates ([71,81]) by the usual procedures.

The Friedel–Crafts reaction of carbonimidoyl dichlorides with benzene, in the presence of aluminum chloride, yields the imidoyl chloride CXXXIV ([25]).

$$RN{=}CCl_2 + C_6H_6 \xrightarrow{AlCl_3} RN{=}\underset{\underset{Cl}{|}}{C}{-}C_6H_5$$

CXXXIV

The pyrolysis of 2,4-dichloro-6-trichloromethylphenylcarbonimidoyl dichloride (CXXXV), at 450–500°C yields pentachloroindolenine (CXXXVI) ([33]).

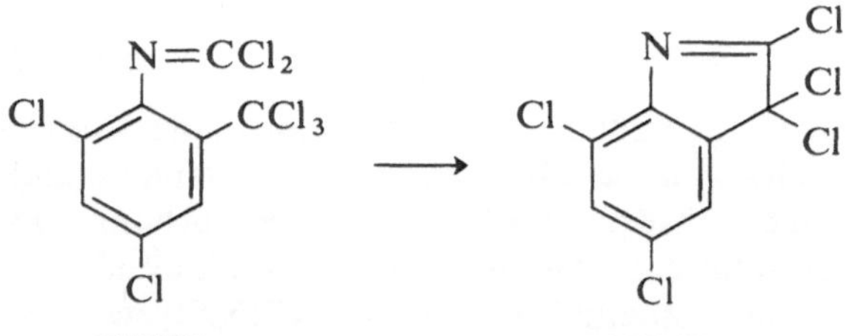

CXXXV   CXXXVI

In the reaction of carbonimidoyl dichlorides with CuCN the corresponding substitution product (CXXXVII) is obtained ([95]).

$$RN{=}CCl_2 + CuCN \longrightarrow RN{=}\underset{\underset{Cl}{|}}{C}{-}CN$$

CXXXVII

The reaction of carbonimidoyl dichlorides with epoxides produces 2-oxazolidinones after hydrolysis. For example, 4-chlorophenylcarbonimidoyl dichloride reacts with ethylene oxide in the presence of zinc chloride as the catalyst to form the 2-oxazolidinone derivative CXXXVIII in low yield ([119]).

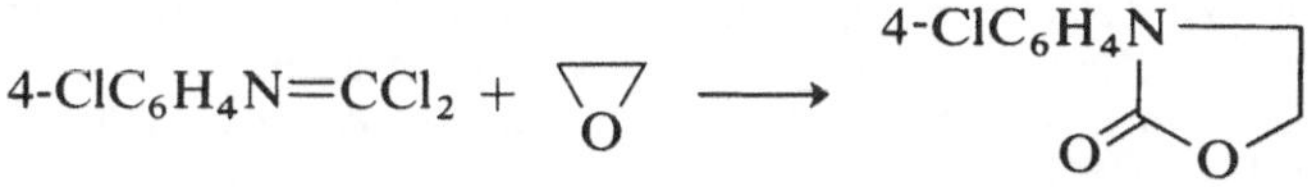

$$4\text{-ClC}_6\text{H}_4\text{N}{=}\text{CCl}_2 + \text{[epoxide]} \longrightarrow$$

CXXXVIII

Abstraction of the halo groups from carbonimidoyl dihalides (reversal of their syntheses) to yield isocyanides can be achieved by the use of triphenyl and tributylphosphine ([86,101]).

$$\text{RN}{=}\text{CCl}_2 + (\text{C}_6\text{H}_5)_3\text{P} \longrightarrow \text{RNC} + (\text{C}_6\text{H}_5)_3\text{PCl}_2$$

In contrast, phosphites undergo the Michaelis–Arbuzov reaction with carbonimidoyl dichlorides to produce the corresponding phosphonates (CXXXIX) ([86,99,101,115]).

$$\text{RN}{=}\text{CCl}_2 + \text{P(OR}')_3 \longrightarrow \text{RN}{=}\text{C[P(OR}')_2]_2$$
$$\downarrow$$
$$\text{O}$$

CXXXIX

Reaction of 4-chlorophenylcarbonimidoyl dichloride with triethyl phosphite ([99]). To 21 g (0.01 mole) 4-chlorophenylcarbonimidoyl dichloride 33.2 g (0.02 mole) triethyl phosphite is added dropwise and with stirring. The reaction proceeds, as indicated by the evolution of ethyl chloride, and the exothermic reaction is controlled by the rate of addition (maximum temperature approximately 40°C). Heating to 50–60°C until the generation of ethyl chloride ceases, and rectification under vacuum, yields 33.6 g (82%) of the corresponding bis-phosphonate, b.p. 190–195°C/0.05 mm.

A special synthesis of carbonimidoyl dichlorides involves the addition of the sulfenyl chloride CXL to an olefin. For example, CXL adds to cyclohexene to produce the carbonimidoyl dichloride CXLI in 80% yield ([10]).

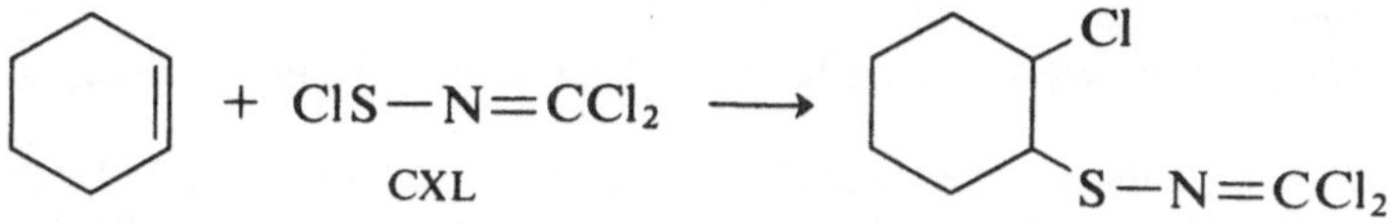

$$\text{[cyclohexene]} + \text{ClS}{-}\text{N}{=}\text{CCl}_2 \longrightarrow$$

CXL

CXLI

V. REFERENCES

1. Allenstein, E., and Schmidt, A., *Ber.* **97**, 1286 (1964),
2. Alimov, P., and Lerkova, L. N., *Izv. Akad. Nauk SSR, Ser. Khim.* 932 (1964); *Chem. Abstr.* **61**. 5502 (1964).
3. Anders, B., Kühle, E., and Malz, H., German Pat. 1.173,469 (1964); *Chem. Abstr.* **61**, 11892 (1964).
4. Anders, B., and Kühle, E., German Pat. 1,178,422 (1964); *Chem. Abstr.* **61**, 14539 (1964).
5. Anders, B., and Kühle, E., *Angew. Chem. Intern. Ed.* **4**, 430 (1965).
6. Anders, B., Federmann, M., and Kühle, E., German Pat. 1,195,750 (1965); *Chem. Abstr.* **63**, 11371 (1965).
7. Anders, B., Kühle, E., and Dickore, K., United States Pat. 3,299,134 (1967).
8. Antia, M. B., and Pandit, M. I., *Vikram, J. Vikram Univ.* **5**, 106 (1961); *Chem. Abstr.* **59**, 3798 (1963).
9. Attaway, J. A., Groth, R. H., and Bigelow, L. A., *J. Am. Chem. Soc.* **81**, 3599 (1959).
10. Bacon, R. G. R., Irwin, R. S., Pollock, J. M., and Pullin, A. D. E., *J. Chem. Soc.* 764 (1958).
11. Banks, R. E., Haszeldine, R. N., and Sutcliffe, H., *J. Chem. Soc.* 4066 (1964).
12. Barr, D. A., and Haszeldine, R. N., *J. Chem. Soc.* 3428 (1956).
13. Barr, D. A., Haszeldine, R. N., and Willis, C. J., *J. Chem. Soc.* 1351 (1961).
14. Bly, R. S., Perkins, G. A., and Lewis, W. L., *J. Am. Chem. Soc.* **44**, 2896 (1922).
15. Brintzinger, H., Pfannstiel, K., and Koddebusch, H., *Ber.* **82**, 389 (1949).
16. Burkhardt, J., Feinauer, R., Gulbins, E., and Hamann, K., *Ber.* **99**, 1912 (1966).
17. Degener, E., Holtschmidt, H., and Schmelzer, H. G., German Pat. 1.163,803 (1964); *Chem. Abstr.* **60**, 14432 (1964).
18. Dennstedt, M., *Ber.* **13**, 228 (1880).
19. Derkach, G. I., and Liptuga, N. I., *Zh. Obshch. Khim.* **36**, 461 (1966); *Chem. Abstr.* **65**, 634 (1966).
20. Dickore, K., and Kühle, E., *Angew. Chem. Intern. Ed.* **4**, 430 (1965).
20a. Dobbie, R. C., and Emeleus, H. J., *J. Chem. Soc.* A, 933 (1966).
21. Douglass, I. B., and Johnson, T. B., *J. Am. Chem. Soc.* **60**, 1486 (1938).
21a. Diekmann, J., *J. Org. Chem.* **28**, 2933 (1963).
22. Dresdner, R. D., *J. Am. Chem. Soc.* **79**, 69 (1957).
23. Dresdner, R. D., Tlumac, F. N., and Young, Y. A., *J. Am. Chem. Soc.* **82**, 5831 (1960).
24. Dyson, G. M., and Harrington, T., *J. Chem. Soc.* 191 (1940).
25. Dyson, G. M., and Harrington, T., *J. Chem. Soc.* 150 (1942).
26. Dyson, G. M., and Harrington, T., *J. Chem. Soc.* 374 (1942).
27. Eue, L., Hack, H., Kühle, E., and Holtschmidt, H., Belgian Pat. 659,017 (1965); *Chem. Abstr.* **64**, 1970 (1966).
28. Farbenfabriken Bayer, A. G., British Pat. 888,646 (1962); *Chem. Abstr.* **57**, 13696 (1962).
29. Farbenfabriken Bayer, A. G., British Pat. 981,107 (1965); *Chem. Abstr.* **62**, 11094 (1965).
30. Farbenfabriken Bayer, A. G., British Pat. 1,022,040 (1966).
31. Farbenfabriken Bayer, A.G., Netherlands Appl. 6,409,122 (1965); *Chem. Abstr.* **63**, 8327 (1965).
32. Farbenfabriken Bayer, A.G., Netherlands Appl. 6,414,309 (1965); *Chem. Abstr.* **64**, 2016 (1966).
33. Farbenfabriken Bayer, A.G., Netherlands Appl. 6,516,261 (1966); *Chem. Abstr.* **65**, 18567 (1966).
34. Fawcett, F. S., Tullock, C. W., and Coffman, D. D., *J. Am. Chem. Soc.* **84**, 4275 (1962).
35. Fujisawa, T., Tamura, Y., and Mukaiyama, T., *Bull. Chem. Soc. Japan* **37**, 793 (1964); *Chem. Abstr.* **61**, 16166 (1964).

36. Gauss, W., and Kühle, E., French Pat. 1,363,063 (1964); *Chem. Abstr.* **62**, 2763 (1965).
37. Gompper, R., and Kunz, R., *Ber.* **99**, 2900 (1966).
37a. Green, M., and Tipping, A. E., *J. Chem. Soc.* 5774 (1965).
38. Gross, H., Rusche, J., and Bornowski, H., *Ann.* **675**, 142 (1964).
39. Guillemard, H., *Ann. Chim.* [8] **14**, 224 (1904).
40. Gulbins, K., and Hamann, K., *Angew. Chem.* **73**, 434 (1961).
41. Gumpert, F., *J. prakt. Chem.* **31**, 119 (1885).
42. Gysin, H., Knuesli, E., and Rumpf, J., United States Pat. 3,053,843 (1962); *Chem. Abstr.* **58**, 6846 (1963).
43. Hagedorn, I., Etling, H., and Lichtel, E., *Ber.* **99**, 520 (1966).
44. Hantzsch, A., and Mai, L., *Ber.* **28**, 977 (1895).
45. Hauptschein, M., Braid, M., and Lawlor, F. E., *J. Org. Chem.* **23**, 323 (1958).
46. Hauptschein, M., United States Pat. 2,966,517 (1960); *Chem. Abstr.* **55**, 10480 (1961).
47. Helmers, O., *Ber.* **20**, 786 (1887).
48. Homeyer, B., Kühle, E., and Malz, H., German Pat. 1,147,797 (1963); *Chem. Abstr.* **59**, 1044 (1963).
49. Holtschmidt, H., *Angew. Chem. Intern. Ed.* **1**, 632 (1962).
50. Holtschmidt, H., German Pat. 1,141,278 (1962); *Chem. Abstr.* **59**, 456 (1963).
51. Holtschmidt, H., Belgian Pat. 617,414 (1962); *Chem. Abstr.* **58**, 12430 (1963).
52. Holtschmidt, H., and Zecher, W., Belgian Pat. 622,381 (1962); *Chem. Abstr.* **59**, 11526 (1963).
53. Holtschmidt, H., and Zecher, W., Belgian Pat. 622,382 (1962); *Chem. Abstr.* **59**, 11534 (1963).
54. Holtschmidt, H., United States Pat. 3,190,918 (1965).
54a. Holtschmidt, H., Degener, E., and Schmelzer, H. G., *Ann.* **701**, 107 (1967).
55. Ivanova, Zh. M., Derkach, G. I., and Kirsanova, N. A., *Zh. Obshch. Khim.* **34**, 3516 (1964); *Chem. Abstr.* **62**, 2725 (1965).
56. Ivanova, Zh. M., Kirsanova, N. A., and Derkach, G. I., *Zh. Organ. Khim.* **1**, 2186 (1965); *Chem. Abstr.* **64**, 11123 (1966).
56a. Jarovenko, N. N., Motornyi, S. P., and Kirenskaja, L. I., *Zh. Obshch. Khim.* **29**, 3789 (1959).
57. Jentzsch, W., *Ber.* **97**, 1361 (1964).
58. Jentzsch, W., *Ber.* **97**, 2755 (1964).
59. Johnson, H. W., Jr., and Daughhetee, P. H., Jr., *J. Org. Chem.* **29**, 246 (1964).
60. Jungermann, E., and Smith, F. W., *J. Am. Oil Chemist's Soc.* **36**, 388 (1959); *Chem. Abstr.* **54**, 289 (1960).
61. Kaji, A., and Miyazaki, K., *Nippon Kagaku Zasshi*, **87**, 727 (1966); *Chem. Abstr.* **65**, 15255 (1966).
62. Kauer, J. C., and Sheppard, W. A., *J. Org. Chem.* **32**, 3580 (1967).
63. Kaufmann, H. P., and Liepe, J., *Ber.* **57**, 923 (1924).
64. Kirsanov, A. V., Derkach, G. I., and Liptuga, N. I., *Zh. Obshch. Khim.* **34**, 2812 (1964); *Chem. Abstr.* **61**, 14514 (1964).
65. Klauke, E., and Kühle, E., Belgian Pat. 632,365 (1963); *Chem. Abstr.* **61**, 6952 (1964).
66. Klauke, E., and Kühle, E., German Pat. 1,170,414 (1964).
67. Klauke, E., *Angew. Chem. Intern. Ed.* **5**, 848 (1966).
68. Knunyants, I. L., Bykhovskaya, E. G., and Frosin, V. N., *Doklady Akad. Nauk SSSR* **132**, 357 (1960); *Chem. Abstr.* **54**, 20841 (1960).
69. Kodama, Y., *Yuki Gosei Kagaku Kyokai Shi* **21**, 525 (1963); *Chem. Abstr.* **59**, 8748 (1963).
70. Kodama, Y., and Sekiba, T., *Yuki Gosei Kagaku Kyokai Shi* **21**, 788 (1963); *Chem. Abstr.* **59**, 15290 (1963).

71. Kodama, Y., and Sekiba, T., *Yuki Gosei Kagaku Kyokai Shi* **22**, 467 (1964); *Chem. Abstr.* **61**, 7015 (1964).

72. Kodama, Y., Sekiba, T., and Ho, T., *Yuki Gosei Kagaku Kyokai Shi* **22**, 567 (1964); *Chem. Abstr.* **61**, 9160 (1964).

73. Kodama, Y., Sekiba, T., and Kato, M., *Yuki Gosei Kagaku Kyokai Shi* **22**, 749 (1964); *Chem. Abstr.* **61**, 12002 (1964).

74. Kodama, Y., *Yuki Gosei Kagaku Kyokai* **23**, 57 (1965); *Chem. Abstr.* **62**, 9132 (1965).

75. Kodama, Y., and Sekiba, T., Japanese Pat. 11,389 (1965); *Chem. Abstr.* **63**, 13293 (1965).

76. Kodama, Y., and Sekiba, T., Japanese Pat. 20,394 (1965); *Chem. Abstr.* **63**, 18124 (1965).

77. Kodama, Y., and Sekiba, T., Japanese Pat. 20,395 (1965); *Chem. Abstr.* **63**, 18124 (1965).

78. Kodama, Y., Sekiba, T., and Kato, M., Japanese Pat. 28,098 (1965); *Chem. Abstr.* **64**, 11231 (1966).

79. Kodama, Y., and Shinohara, S., Japanese Pat. 28,099 (1965); *Chem. Abstr.* **64**, 11231 (1966).

80. Kodama, Y., and Shinohara, S., Japanese Pat. 28,100 (1965); *Chem. Abstr.* **64**, 11231 (1966).

81. Kodama, Y., Sekiba, T., and Kato, M., Japanese Pat. 28,101 (1965); *Chem. Abstr.* **64**, 11232 (1966).

82. Kodama, Y., and Sekiba, T., Japanese Pat. 430 (1966); *Chem. Abstr.* **64**, 11230 (1966).

83. Kodama, Y., and Sekiba, T., Japanese Pat. 431 (1966); *Chem. Abstr.* **64**, 11231 (1966).

84. Kühle, E., German Pat. 1,094,737 (1960); *Chem. Abstr.* **55**, 25860 (1961).

85. Kühle, E., and Wegler, R., German Pat. 1,126,371 (1959); *Chem. Abstr.* **57**, 11102 (1962).

86. Kühle, E., *Angew. Chem. Intern. Ed.* **1**, 647 (1962).

87. Kühle, E., German Pat. 1,126,380 (1962).

88. Kühle, E., Eue, L., and Bayer, O., German Pat. 1,137,000 (1962).

89. Kühle, E., and Sasse, K., German Pat. 1,174,772 (1964); *Chem. Abstr.* **61**, 10627 (1964).

90. Kühle, E., German Pat. 1,192,189 (1965); *Chem. Abstr.* **63**, 6918 (1965).

91. Kühle, E., United States Pat. 3,235,580 (1966).

92. Kühle, E., and Eue, L., United States Pat. 3,267,097 (1966); *Chem. Abstr.* **65**, 13729 (1966).

93. Kühle, E., and Anders, B., French Pat. 1,405,732 (1965); *Chem. Abstr.* **64**, 9603 (1966).

94. Kühle, E., Belgian Pat. 660,171 (1965); *Chem. Abstr.* **64**, 603 (1966).

95. Kühle, E., and Anders, B., German Pat. 1,224,306 (1966); *Chem. Abstr.* **65**, 18536 (1966).

95a. Kühle, E., Anders, B., and Zumach, G., *Angew. Chem.* **79**, 663 (1967).

96. Lonza, Ltd., British Pat. 1,052,567 (1966); *Chem. Abstr.* **66**, 55525 (1967).

97. Makarov, S. P., Shpanskii, V. A., Ginsburg, V. A., Shchekotikhin, A. I., Filatov, A. S., Martynova, L. L., Parlovskaya, I. V., Golovaneva, A. F., and Yakubovich, A. Y., *Dokl. Akad. Nauk SSSR* **142**, 596 (1962); *Chem. Abstr.* **57**, 4528 (1962).

98. Malinovski, M. S., *Sci. Records Gorky State Univ.* **7**, 29 (1939); *Chem. Abstr.* **35**, 442 (1941).

99. Malz, H., Kühle, E., and Bayer, O., German Pat. 1,138,389 (1962).

100. Malz. H., Holtschmidt, H., Kühle, E., and Bayer, O., Belgian Pat. 627,486 (1963); *Chem. Abstr.* **60**, 10557 (1964).

101. Malz, H., and Kühle, E., German Pat. 1,158,501 (1963); *Chem. Abstr.* **60**, 6795 (1964).

102. Mitsch, R. A., and Ogden, P. H., *J. Org. Chem.* **31**, 3833 (1966).

103. Moeckel, K., and Gehlen, H., *Z. Chem.* **4**, 388 (1964); *Chem. Abstr.* **62**, 1647 (1965).

104. Mukaiyama, T., Fujisawa, T., and Hyugaji, T., *Bull. Chem. Soc. Japan* **35**, 687 (1962).

105. Mukaiyama, T., Fujisawa, T., and Mitsunobo, O., *Bull. Chem. Soc. Japan* **35**, 1104 (1962).

106. Mukaiyama, T., Fujisawa, T., Nohira, H., and Hyugaji, T., *J. Org. Chem.* **27**, 3337 (1962).

107. Murphy, D. B., *J. Org. Chem.* **29**, 1613 (1964).
108. Nef, J. U., *Ann.* **270**, 267 (1892).
109. Nef, J. U., *Ann.* **280**, 291 (1894).
110. Neidlein, R., and Haussmann, W., *Angew. Chem. Intern. Ed.* **4**, 521 (1965).
111. Neidlein, R., and Haussmann, W., *Tetrahedron Letters* 1753 (1965).
112. Neidlein, R., and Haussmann, W., *Tetrahedron Letters* 2423 (1965)
113. Neidlein, R., and Heukelbach. E., *Tetrahedron Letters* 2665 (1965).
114. Neidlein, R., and Haussmann, W., *Ber.* **99**, 239 (1966).
115. Neidlein, R., Haussmann, W., and Heukelbach, E., *Ber.* **99**, 1252 (1966).
116. Neidlein, R., and Haussmann, W., *Tetrahedron Letters* 2217 (1966).
117. Neidlein, R., and Heukelbach, E., *Arch. Pharm.* **299**, 944 (1966).
118. Neidlein, R., and Haussmann, W., *Z. Naturforsch.* **21**, 898 (1966).
119. Oda. R.. Hamada. T.. Ito. Y.. and Okano. M.. *Bull. Inst. Chem. Res. Kyoto Univ.* **44**. 227 (1966); *Chem. Abstr.* **66**, 37806f (1967).
120. Ottmann, G., and Hooks, H., *Angew. Chem. Intern. Ed.* **4**, 432 (1965).
121. Ottmann, G., and Hooks. H., *J. Org. Chem.* **31**, 838 (1966).
122. Ottmann, G., and Hooks, H., *Angew. Chem. Intern. Ed.* **5**, 250 (1966).
123. Ottmann, G., and Hooks, H., United States Pat. 3,267,144 (1966); *Chem. Abstr.* **65**, 20029 (1966).
124. Ottmann, G., and Hooks, H., *J. Med. Chem.* **9**, 962 (1966).
125. Pawellek, D., *Angew. Chem. Intern. Ed.* **5**, 845 (1966).
125a. Pearlson, W. A., and Hals, L. J., United States Pat. 2,643,267 (1950).
126. Pel'kis, P. S., and Dunaevska, T. S., *Mem. Inst. Chem. Acad. Sci. Ukrain. SSSR* **6**, 163 (1940); *Chem. Abstr.* **34**, 5829 (1940).
127. Petrov, K. A., and Neimysheva, A. A., *Zh. Obshch. Khim.* **29**, 2165 (1959); *Chem. Abstr.* **54**, 10911 (1960).
128. Petrov, K. A., and Neimysheva, A. A., *Zh. Obshch. Khim.* **29**, 2169 (1959); *Chem. Abstr.* **54**, 10912 (1960).
129. Petrov, K. A., and Neimysheva, A. A., *Zh. Obshch. Khim.* **29**, 2695 (1959); *Chem. Abstr.* **54**, 10912 (1960).
130. Prandtl, W., and Sennewald, K., *Ber.* **62**, 1764 (1929).
131. Riethman, J., and Wegmüller, H., German Pat. 1,132,559 (1962).
132. Rosen, I., and Stallings, J. P., *J. Org. Chem.* **25**, 1484 (1960).
133. Sallmann, A., and Pfister, R., German Pat. 1,163,841 (1964); *Chem. Abstr.* **64**, 5119 (1966).
134. Sayigh, A. A. R., and Ulrich, H., *J. Chem. Soc.* 3146 (1963).
135. Schmelzer, H. G., and Degener, E., German Pat. 1,178,437 (1964); *Chem. Abstr.* **61**, 16081 (1964).
136. Schmidt, E., and Thulke, K., *Ann.* **663**, 46 (1963).
136a. Scott, F. L., and Cronin, D. A., *Chem. and Ind.* 1757 (1964).
137. Sell, E., and Zierold, G., *Ber.* **7**, 1228 (1874).
138. Seyferth, D., and Damrauer, R., *Tetrahedron Letters* 189 (1966).
139. Sheppard, W. A., *J. Am. Chem. Soc.* **87**, 4338 (1965).
140. Shipton, G. O., British Pat. 945,855; *Chem. Abstr.* **60**, 10700 (1964).
141. Smith, P. A. S., and Kalenda, N. W., *J. Org. Chem.* **23**, 1599 (1958).
142. Smith, W. R., *Am. Chem. J.* **16**, 372 (1894).
143. Sokolov, S. V., and Mazalov, S. A., and Gerasimov, S. I., *Zh. Vses. Khim. Obshchestva im. D. I. Mendeleeva* **10**, 234 (1965).
144. Tarnow, H., Holtschmidt, H., and Unterstenhoefer, G., German Pat. 1,197,078 (1965); *Chem. Abstr.* **63**, 14774 (1965).
144a. Thiele. J.. *Ber.* **26**. 2645 (1893).

145. Thiele, J., *Ann.* **303**, 5770 (1898).

145a. Tscherniak, M., *Bull. Soc. Chim. France* [2] **30**, 185 (1878).

146. Tullock, C. W., Coffman, D. D., and Mutterties, E. L., *J. Am. Chem. Soc.* **86**, 357 (1964).

147. Ulrich, H., and Sayigh, A. A. R., *Angew. Chem. Intern. Ed.* **1**, 595 (1962).

148. Ulrich, H., and Sayigh, A. A. R., *J. Chem. Soc.* 5558 (1963).

149. Ulrich, H., and Sayigh, A. A. R., *Angew. Chem. Intern. Ed.* **3**, 585 (1964).

150. Ulrich, H., and Sayigh, A. A. R., *J. Org. Chem.* **30**, 2779 (1965).

151. Ulrich, H., *Cycloaddition Reactions of Heterocumulenes*, Academic Press, New York (1967).

152. Warren, W. H., and Wilson, F. E., *Ber.* **68**, 957 (1935).

153. Wellcome Found. Ltd., Belgian Pat. 667,875 (1966); *Chem. Abstr.* **65**, 5398 (1966).

154. Yarovenko, N. N., Motornyi, S. P., and Kirenskaya, L. I., *Zh. Obshch. Khim.* **29**, 3789 (1959); *Chem. Abstr.* **54**, 19479 (1960).

155. Young, J. A., Simmons, T. C., and Hoffmann, F. W., *J. Am. Chem. Soc.* **78**, 5637 (1956).

156. Young, J. A., Tsoukalas, S. N., and Dresdner, R. D., *J. Am. Chem. Soc.* **80**, 3604 (1958).

157. Young, J. A., Durell, W. S., and Dresdner, R. D., *J. Am. Chem. Soc.* **81**, 1587 (1959).

158. Young, J. A., Tsoukalas, S. N., and Dresdner, R. D., *J. Am. Chem. Soc.* **82**, 396 (1960).

159. Young, J. A., and Dresdner, R. D., *J. Org. Chem.* **28**, 833 (1963).

Chapter 3

IMIDOYL HALIDES

I. INTRODUCTION

The reaction of substituted carboxylic acid amides with a variety of halogenating agents, such as phosphorus pentachloride, phosphorus penta-bromide, thionyl chloride, and carbonyl chloride, produces imidoyl halides or iminium halides, depending upon the structure of the amide. For example, a monosubstituted carboxylic acid amide can react with carbonyl chloride to produce the imidoyl chloride I, or its hydrochloride II, if R′ is aliphatic.

$$
\underset{\text{II}}{\left[R\!-\!\overset{\oplus}{\underset{\overset{|}{Cl}}{\ddot{C}}}\!-\!\ddot{N}HR' \right]Cl^{\ominus}} \quad \underset{-CO_2}{\overset{COCl_2}{\longleftarrow}} \quad RCONHR' \quad \underset{-CO_2}{\overset{COCl_2}{\longrightarrow}} \quad \underset{\text{I}}{R\!-\!\underset{\overset{|}{Cl}}{C}\!=\!NR' + HCl}
$$

The naming of both reaction products has caused problems over the years. While the early literature refers to both classes of compounds as imido and amido chlorides, respectively, *Chemical Abstracts* has named compounds II alkylene ammonium chlorides (when R = alkyl), while compounds I have been named imidoyl chlorides. In order to differentiate the reactive compounds II from ammonium chlorides, which are chemically non-related, I prefer the name "iminium chlorides." Compounds I and II are intraconvertible, and their chemical properties are quite interrelated.

The reaction of carboxylic acid amides with phosphorus pentachloride, to yield compounds I and II, was investigated extensively by Wallach and his students in 1876–1882, and later by v. Braun, who utilized this reaction to degrade secondary amines to ammonia. The highly reactive imidoyl chlorides have been used, especially by v. Braun, as synthetic tools to produce a variety of compounds which could not easily be synthesized by other procedures. However, imidoyl chlorides never made the headlines in synthetic organic chemistry, most likely because of the inherent difficulty involved in handling these highly reactive compounds.

Often the synthesized imidoyl chlorides were characterized only by hydrolysis to the starting carboxylic acid amides, which proceeds very readily at room temperature. It is now clear that imidoyl chlorides are intermediates in a wide variety of standard synthetic procedures, such as the Gattermann aldehyde synthesis, the Houben–Hoesch ketone synthesis, and the Beckmann rearrangement, to name only a few. Their chemistry is also related to the chemistry of enamines, their tautomers, provided α-hydrogens are adjacent to the C=N double bond. Furthermore, the general acid chloride reactions can be conducted, using imidoyl chlorides, thereby forming a variety of interesting classes of compounds. For example, dechlorination and dehydrochlorination of imidoyl chlorides affords ketenimines.

Iminium chlorides have become valuable synthetic tools, as evidenced by the work of Zollinger and his colleagues in Switzerland, Arnold and his co-workers in Czechoslovakia, and Eilingsfeld *et al.* in Germany. The intermediacy of iminium chlorides in the Vilsmeier–Haack aldehyde synthesis is well established.

Recently, iminium chlorides have been used as catalysts in the transformation of carboxylic and sulfonic acids to the corresponding chlorides. Similarly, the chlorination of hydroxy compounds, using carbonyl chloride, is catalyzed by iminium chlorides.

The phosgenation of isatoic anhydride and aminocarboxylic and aminosulfonic acids, to afford isocyanatoarylcarbonyl chlorides and isocyanatoarlsylfonyl chlorides, is also catalyzed by iminium chlorides.

The substituents in I and II are generally alkyl and aryl groups. However, recently imidoyl chlorides having N-carbethoxy and N-cyano groups have been synthesized. Dichlorination of aliphatic isocyanates in the α-position yields N-chlorocarbonyl-substituted imidoyl chlorides, and C-cyano-substituted imidoyl chlorides were found to be intermediates in the 1,2-rearrangement of N-nitrosoamides, using phosphorus pentachloride.

The listed imidoyl and iminium halides are compounds which have been characterized at least by boiling point or melting point. In addition to the listed compounds, many more have been prepared as reactive intermediates.

II. SYNTHESIS

A. Halogenation of Thioamides

The reaction of thioamides with suitable chlorinating agents, such as phosphorus halides, thionyl chloride, carbonyl chloride, etc., is perhaps the most general method of synthesis of imidoyl chlorides. However, this method has been used only in very few instances because thioamides are less readily available than amides.

For example, Dimroth ([106]) treated thioamides, derived from iso-thiocyanates and malonates, with phosphorus pentachloride, and he obtained the corresponding imidoyl chlorides (III), which can be dehydro-chlorinated to produce the ketenimines IV.

$$C_6H_5NH-\underset{\underset{S}{\|}}{C}-CH(COOEt)_2 + PCl_5$$

$$\longrightarrow \quad C_6H_5NH-\underset{\underset{Cl}{|}}{C}=C(COOEt)_2$$

$$\updownarrow$$

$$\underset{IV}{C_6H_5N=C=C(COOEt)_2} \quad \xleftarrow{-HCl} \quad \underset{III}{C_6H_5N=\underset{\underset{Cl}{|}}{C}-CH(COOEt)_2}$$

Halogenation of N-alkyl thiobenzamide with elementary chlorine similarly yields the corresponding imidoyl chlorides. For example, from N-methylthiobenzamide (V), methylbenzimidoyl chloride (VI) was obtained ([152]).

$$\underset{\underset{\underset{V}{S}}{\|}}{C_6H_5-\underset{\|}{C}-NHCH_3} \quad \xrightarrow{Cl_2} \quad \underset{VI}{C_6H_5\underset{\underset{Cl}{|}}{C}=NCH_3} + S + HCl$$

B. Halogenation of Amides

The classical synthesis of imidoyl chlorides involves the reaction of N-substituted carboxylic acid amides with phosphorus pentahalides. Wallach and his students ([288–292]) in 1876–1882 investigated this reaction extensively, and even today their method is the standard synthetic procedure for imidoyl chlorides. Since the compounds obtained are often sensitive to moisture and to heat, their isolation and purification is rather difficult, and in many instances the obtained imidoyl halides were hydrolyzed to the corresponding amides for characterization. This method of identification is of course useful only if further transformations have been achieved, because otherwise the starting amides are regenerated. J. v. Braun and his students investigated the reaction of carboxylic acid chlorides with phosphorus pentahalides extensively, and utilized imidoyl halides in degradative studies directed toward the elucidation of structures of complex N-heterocycles, such as alkaloids. H. v. Pechmann ([227]) observed that the imidoyl chloride, obtained in the reaction of N-methylbenzamide and phosphorus pentachloride, was quite labile, and that benzonitrile and methyl chloride was obtained upon attempted distillation. J. v. Braun, who had used cyanogen bromide for the degradation of tertiary amines to secondary amines, realized that secondary amines can be converted to primary amines *via* the corresponding imidoyl chlorides. Thus, reaction of a secondary amine with benzoyl chloride yielded the corresponding amide, which was converted to the imidoyl chloride upon heating with phosphorus pentachloride. Thermal degradation resulted in an alkyl halide and an N-monosubstituted amide, which can be hydrolyzed to the primary amine ([67]). This step can be repeated, thereby converting a primary amine into an alkyl halide and ammonia ([67]).

In contrast, unsubstituted amides react with phosphorus pentachloride in a different manner, yielding the corresponding trichlorophosphazenes (VII). The correct structural assignment of the obtained products was performed by Kirsanov in 1955 ([182]). The trichlorophosphazenes eliminate phosphorus oxychloride on heating to afford the corresponding nitriles. However, imidoyl chlorides VIII are obtained in the reaction of amides IX with phosphorus pentachloride ([104,183–185]).

$$RCONH_2 + PCl_5 \xrightarrow{-2\,HCl} \underset{VII}{RCON{=}PCl_3} \xrightarrow{\Delta} RCN + POCl_3$$

$$\underset{IX}{RCONHPCl_2} + PCl_5 \longrightarrow \underset{VIII}{R{-}\overset{|}{\underset{Cl}{C}}{=}N{-}\overset{|}{\underset{O}{PCl_2}}} + POCl_3 + HCl$$

In the reaction of N-substituted carboxylic acid amides with phosphorus pentachloride generally the corresponding free imidoyl chlorides (I) or their hydrochlorides (II) (iminium chlorides) are formed. If the iminium chlorides are obtained initially, dehydrochlorination can be achieved by simply heating or using a stronger base, such as triethylamine, as the hydrogen chloride scavenger.

$$RCONHR' + PCl_5 \longrightarrow \underset{II}{R{-}\overset{|}{\underset{Cl}{C}}{=}\overset{\oplus}{N}HR']Cl^{\ominus}} + POCl_3$$

$$\downarrow {-HCl}$$

$$\underset{I}{R{-}\overset{|}{\underset{Cl}{C}}{=}NR'}$$

In the reaction of aliphatic carboxylic acid amides having hydrogens adjacent to the C=N bond, self-condensation of the generated imidoyl chloride has been observed ([56]). This self-condensation is apparently due to the isomerization of the imidoyl chloride to the corresponding vinyl amine, which is attacked by the imidoyl chloride to form the amidine derivative X ([51,55,56]).

$$\underset{\overset{|}{Cl}}{CH_3-C}{=}NR \longleftrightarrow \underset{\overset{|}{Cl}}{CH_2{=}C-NHR} \xrightarrow{-HCl} \underset{\overset{||}{NR}}{CH_3-C}-N(R)-\underset{\overset{|}{Cl}}{C}{=}CH_2$$

X

If an ortho-substituted benzene ring is attached to the nitrogen, the self-condensation is minimized due to steric hindrance and the yield of imidoyl chlorides increases in this order: $CH_3 < Cl < Br$ ([58]). Likewise, lowering of the basicity of the nitrogen prevents condensation. For example, N-sulfonimidoyl chlorides of aliphatic carboxylic acids are obtained in excellent yield ([65]). The carboxylic acid amides of α,β-unsaturated acids also do not undergo tautomerization, because the formation of allenes is not favored under mild conditions ([67]).

A different reaction occurs in the case of chloroacetanilides, because the intermediate imidoyl chlorides undergo condensation and ring closure to produce the quinoline derivative XI ([60,61]). These side reactions are discussed in more detail in Section IVC (see also [41]).

$$2\,\underset{\overset{|}{Cl}}{ClCH_2C}{=}NC_6H_5 \longrightarrow$$

XI

The isomerization of the imidoyl chlorides, having adjacent hydrogens, is also evidenced by their rapid reaction with bromine. The bromine adducts XII can eliminate hydrogen chloride to form the imidoyl bromide XIII, or eliminate hydrogen bromide to yield the imidoyl chloride XIV ([56]).

$$\underset{\overset{|}{Cl}}{R_2CH-C}{=}NR' \longleftrightarrow \underset{\overset{|}{Cl}}{R_2C{=}C-NHR'} + Br_2 \longrightarrow \underset{\overset{|}{Cl}}{R_2CBr\overset{\overset{\displaystyle Br}{|}}{C}-NHR'}$$

XII

$\swarrow^{-HBr}$ $\downarrow^{-HCl}$

$$\underset{\overset{|}{Cl}}{R_2CBrC}{=}NR' \qquad\qquad \underset{\overset{|}{Br}}{R_2CBrC}{=}NR'$$

XIV   XIII

If an excess of phosphorus pentahalides is used on aliphatic carboxylic acid amides, having α-hydrogens, further halogenation occurs. For example, the reaction of propionic acid ethylamide (XV) with 3 equivalents of phosphorus pentachloride yields the imidoyl chloride XVI, which can also be obtained from α,α-dichloropropionic acid ethylamide (XVII) and one equivalent of phosphorus pentachloride ([56,59]).

$$CH_3CH_2CONHEt \xrightarrow{3\,PCl_5} CH_3CCl_2-\underset{\underset{Cl}{|}}{C}=NEt$$

$$\underset{XV}{}$$

$$\underset{XVI}{} \xleftarrow{PCl_5} CH_3CCl_2CONHEt$$

$$\underset{XVII}{}$$

In the reaction of acetamides with 4 equivalents of phosphorus pentachloride the corresponding trichloroacetyl imidoyl chlorides (XVIII) are formed ([56]).

$$CH_3CONHR + 4\,PCl_5 \longrightarrow CCl_3-\underset{\underset{Cl}{|}}{C}=NR + POCl_3 + 3\,PCl_3 + 3\,HCl$$

$$\underset{XVIII}{}$$

Ethyltrichloroacetimidoyl chloride ([152]). To 156 g (1 mole) of N-ethyldichloroacetamide in 200 ml of carbon tetrachloride portionwise 417 g (2 moles) of phosphorous pentachloride is added and the reaction mixture is heated under reflux until the evolution of hydrogen chloride ceases. Evaporation of the solvent and vacuum distillation yields 190 g (91 %) of ethyltrichloroacetimidoyl chloride, b.p. 69–73°C/15 mm.

Further high-temperature chlorination of imidoyl chlorides to produce perchlorinated derivatives has been reported by Holtschmidt and his coworkers. For example, chlorination of ethyltrichloroacrylimidoyl chloride (XIX) affords the perchloro derivative XX in almost quantitative yield ([152]).

$$CCl_2=CCl-\underset{\underset{Cl}{|}}{C}=NC_2H_5 \xrightarrow{Cl_2} CCl_2=CClCCl_2N=\underset{\underset{Cl}{|}}{C}-CCl_3$$

$$\underset{XIX}{} \qquad\qquad \underset{XX}{}$$

A variety of benzyl and ethylhaloacetimidoyl chlorides have been chlorinated by this procedure.

Pentachloroethyltrichloroacetimidoyl chloride ([152]). To 190 g (0.91 mole) of ethyltrichloroacetimidoyl chloride with ultraviolet irradiation chlorine is added for four hours at 140–200°C and for an additional eight hours at 220°C. Fractional distillation yields 333 g (96 %) of pentachloroethyltrichloroacetimidoyl chloride, b.p. 162–165°C/14 mm; m.p. 69–71°C.

If amides of acetylene carboxylic acids are treated with phosphorus pentachloride, addition of hydrogen chloride to the triple bond occurs. ([68]).

$$C_6H_5C{\equiv}C-CONHR + PCl_5 \longrightarrow C_6H_5-\underset{\underset{Cl}{|}}{C}{=}CH-\underset{\underset{Cl}{|}}{C}{=}NR$$

In contrast, no addition of hydrogen chloride occurs when carboxylic acid amides, which contain double bonds, are treated with phosphorus pentachloride ([63,65,69]).

Cyclic lactams are converted in a like manner to the α,α-dichloro-imidoyl chlorides using 3 equivalents of phosphorus pentachloride. For example, the reaction of piperidinone (XXI) with phosphorus pentachloride affords the piperideine derivative XXII in good yield ([59]).

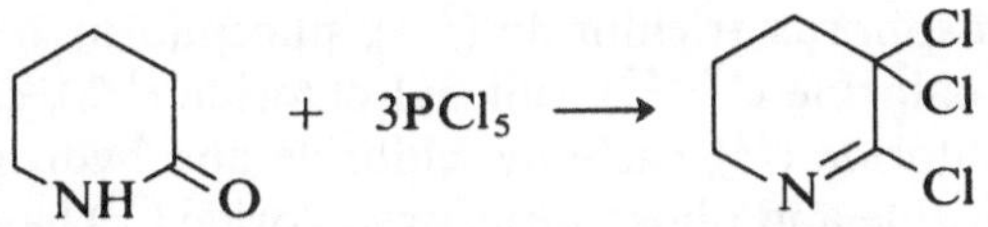

In the reaction of N-disubstituted carboxylic acid amides with phosphorus pentachloride the corresponding iminium chlorides are formed, and again partially isomerization occurs in amides with α-hydrogens. For example, the iminium chloride XXIII and the vinylamine XXIV are formed in the reaction of disubstituted acetic acid diethylamides with phosphorus pentachloride ([57,262]).

$$R_2CHCONEt_2 + PCl_5 \longrightarrow R_2CH-\underset{\underset{Cl}{|}}{C}{=}\overset{\oplus}{N}Et_2]Cl^{\ominus} + R_2C{=}\underset{\underset{Cl}{|}}{C}-NEt_2$$

XXIII — XXIV

The separation of both reaction products is quite easy because of the insolubility of the polar species XXIII in organic solvents ([57]).

The reaction of N-alkyl-N'-arylmethyloxamides XXV with phosphorus pentachloride to yield 1-alkyl-2-aryl-5-chloroimidazoles (XXVI) proceeds *via* imidoyl chloride intermediates ([124]).

$$RCH_2NHCOCONHR' + PCl_5 \longrightarrow \left[RCH_2N{=}\underset{\underset{Cl}{|}}{C}-\underset{\underset{Cl}{|}}{C}{=}NR' \right] \longrightarrow$$

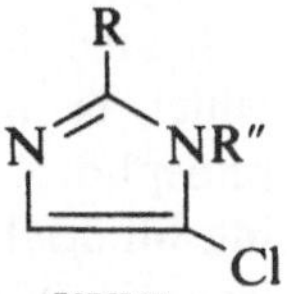

XXV

Instead of phosphorus pentachloride, other halogenating agents, such as phosphorus pentabromide ([50]), thionyl chloride ([42,64,275,281]), and carbonyl chloride ([107,109,110,140,223,239,264,281]) have been used to synthesize imidoyl halides or iminium halides, respectively. The reaction proceeds *via* an initial complex formation, followed by elimination. Chlorinating agents, such as thionyl chloride and carbonyl chloride, have the advantage of the formation of gaseous by-products, and they are therefore the preferred reagents. The initial amide complexes of N-monosubstituted amides with phosphorus oxychloride ([231,232]), *p*-toluenesulfonyl chloride ([135,144]), and thionyl chloride ([193]) are described in the literature. The complexes of N,N-dimethylformamide are especially well investigated. For example, complexes of N,N-dimethylformamide with carbonyl chloride ([9,12,16,42,73,75]), carboxylic acid chlorides ([120]), carboxylic acid bromides ([138,139]), arenesulfonyl chlorides ([4]), phosphorus trichloride ([257]), phosphorus oxychloride ([71,86, 253,256]), thionyl chloride ([42,256]), sulfenyl chloride ([190]), carbonyl chloride and aluminum chloride ([42]), carbonyl chloride and hydrogen chloride ([75]), and carbonyl chloride and phosphorus oxychloride ([75]) have been observed. The structures of the initial complexes are not too well elucidated, because they are quite labile and reaction occurs usually at room temperature. The affect of the various complexes on yield and reactivity in the Vilsmeier–Haack aldehyde synthesis is described in Section IVD.

Imidoyl chlorides of aliphatic carboxylic acids ([281]). To the solution of 0.5 mole of a carboxylic acid alkylamide in 50 ml of dry benzene, with ice-cooling and stirring, 5.45 g (0.55 mole) of phosgene is added (it is advantageous to collect the phosgene in benzene and adjust the amount of the amide accordingly, then add the benzene solution of phosgene dropwise, which assures a better control of the reaction). After standing for one hour at room temperature the reaction mixture is heated on a steam bath until a clear solution results. After purging with nitrogen the solvent is removed by evaporation and the imidoyl chloride is purified by vacuum distillation. The reported yields of distilled products are 66–68 %.

An excess of phosgene has to be avoided if α-hydrogens are present, because the tautomeric enamines react with phosgene to produce chlorovinyl N-carbonyl chlorides (see Chapter 8).

Imidoyl chlorides of aromatic carboxylic acids ([137]). An equimolar mixture of the N-monosubstituted amide and phosphorus pentachloride is heated for 15–180 min at 60–140°C. After completion of the reaction, which is indicated by the cessation of hydrogen chloride evolution, the phosphoryl chloride is removed under reduced pressure, and the remaining imidoyl chloride is distilled under vacuum. The reported yield ranges from 41–96 %, but generally yields of 80–90 % are obtained.

The mechanism of the formation of imidoyl halides or iminium halides may involve attack on oxygen or on nitrogen or on both. For example, in the reaction of N,N-dimethylformamide with carbonyl chloride the initial complexes XXVII and XXVIII can be formed, which generate the iminium chloride XXIX by loss of carbon dioxide.

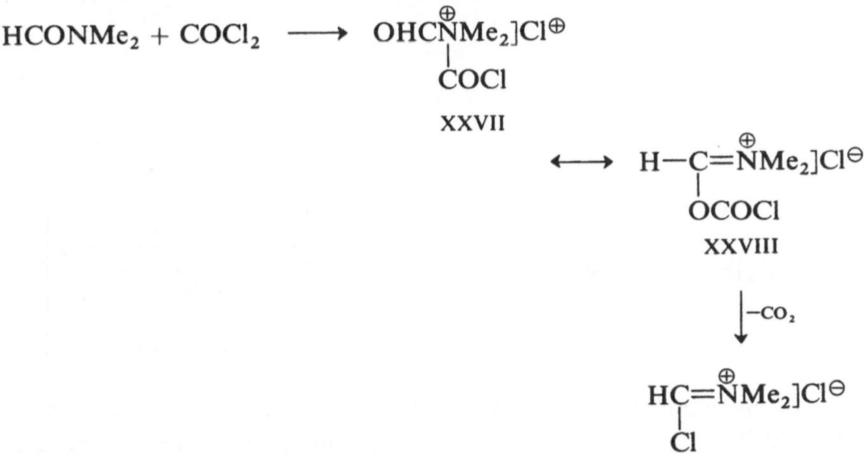

$$HCONMe_2 + COCl_2 \longrightarrow OHC\overset{\oplus}{N}Me_2]Cl^\oplus$$
$$\underset{\text{XXVII}}{\overset{|}{COCl}}$$

$$\longleftrightarrow H-C=\overset{\oplus}{N}Me_2]Cl^\ominus$$
$$\underset{\text{XXVIII}}{\overset{|}{OCOCl}}$$

$$\downarrow -CO_2$$

$$HC=\overset{\oplus}{N}Me_2]Cl^\ominus$$
$$\underset{\text{XXIX}}{\overset{|}{Cl}}$$

Chloroiminium chlorides (general procedure) ([109]). To 1 mole of the N,N-disubstituted carboxylic amide in 500–1000 ml of toluene is added approximately 1.5–2.0 moles of phosgene at 10–25°C. After stirring for several hours the precipitated chloroiminium chlorides are collected by filtration. The obtained yields range from 85–95%.

Chlorodimethylformiminium chloride ([289]). To a solution of N,N-dimethylformamide in chloroform (approximately 10% by weight), a solution of an equimolar amount of phosgene, dissolved in chloroform, is added dropwise at room temperature. It is advantageous to remove the ethanol stabilizer from the chloroform, which is easily accomplished by distillation of the solvent from an appropriate amount of p-toluenesulfonyl isocyanate. The chloroform solution of chlorodimethyliminium chloride thus obtained can be used for further reactions after trace amounts of phosgene are removed with nitrogen.

Evaporation of the solvent affords the pure chlorodimethylformiminium chloride, m.p. 140–145°C, in virtually quantitative yield.

We have observed that in the reaction of N-monosubstituted ureas with phosphorus pentachloride attack on nitrogen as well as on oxygen occurs, depending upon the substituents attached to nitrogen ([282]).

A special case involves reaction of benzoyl ureas **XXX** with phosphorus pentachloride, which most likely occurs *via* a concerted cyclic elimination process ([283]).

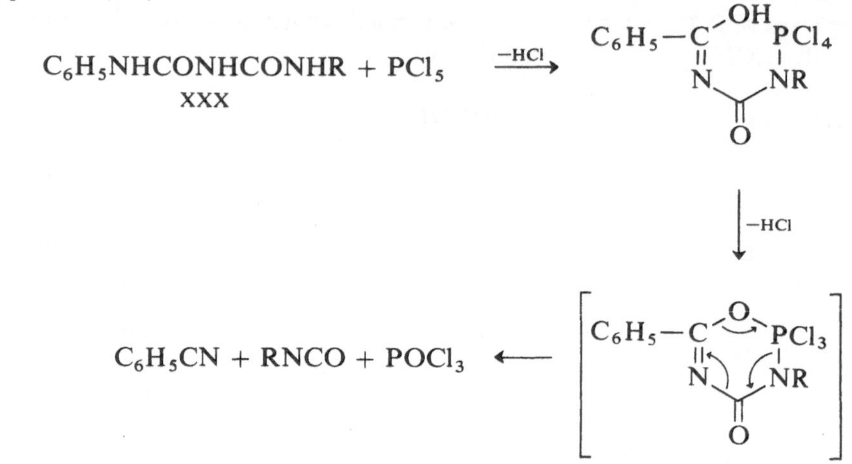

$$C_6H_5NHCONHCONHR + PCl_5 \quad \xrightarrow{-HCl} \quad$$

XXX

$$C_6H_5CN + RNCO + POCl_3 \quad \longleftarrow$$

The chlorodimethylformiminium chloride (**XXXI**) was first synthesized by Arnold ([12]), who also prepared the corresponding fluoride ([27]), bromide ([23]) and iodide ([25]). All compounds have the polar iminium halide structure with the exception of the fluoride **XXXII**, which has the depicted covalent structure.

XXXI   XXXII

Fluorodimethylformiminium fluoride ([27]). To chlorodimethylformiminium chloride excess gaseous hydrogen fluoride is added with ice-salt-cooling. After standing overnight the excess hydrogen fluoride is evaporated and vacuum distillation of the residue gives an 89% yield of fluorodimethylformiminium fluoride (1,1-difluorotrimethylamine), b.p. 80°C/30mm.

The complexes formed from N-monosubstituted formamides and halogenating agents undergo α-elimination to form isocyanides **XXXIII** in the presence of a base ([278–280]).

$$\xrightarrow{2\,Et_3N} \quad RNC + 2\,Et_3N\cdot HCl$$

XXXIII

In the absence of a base a different reaction occurs, and formamidine hydrochlorides XXXIV are isolated ([136,166,169,170,241,294]).

$$2\,HCONHR + COCl_2 \longrightarrow RNH{-}CH{=}\overset{\oplus}{N}HR]Cl^{\ominus}$$
$$XXXIV$$

This reaction proceeds according to Jentzsch ([166,169,170]) *via* N-dichloromethylformamidinium chlorides XXXV, which can be isolated under certain conditions, but our results ([241]), as well as those of Hagedorn and her students ([136]), indicate that interaction of the iminium chloride with unreacted formamide may account for the formation of XXXIV.

$$H{-}\overset{\oplus}{C}{=}NHR]Cl^{\ominus} + H{-}\underset{O}{\overset{\parallel}{C}}{-}NHR$$
$$\overset{|}{Cl}$$

$$\longrightarrow R\overset{\oplus}{NH}{=}CH{-}\underset{}{\overset{R}{\underset{|}{N}}}{-}CHO]Cl^{\ominus} + HCl$$

$$\Big\downarrow COCl_2$$

$$\underset{XXXIV}{RNH{-}CH{=}\overset{\oplus}{N}HR]Cl^{\ominus}} \xleftarrow[\text{H}_2\text{O}]{} \underset{XXXV}{\underset{\oplus}{RNH}{=}CH{-}\overset{R}{\underset{|}{N}}{-}CHCl_2]Cl^{\ominus}}$$

with the arrow labeled $-CO, HCl$

Other derivatives of amides, such as N-nitrosoamides ([214]), acylcarbamates ([219]), and N-phosphoryl amides ([78,100,183–185]) were likewise converted to the corresponding imidoyl chlorides.

A special case of the formation of an imidoyl chloride from an amide involves the reaction of trichloroacetanilide (XXXVI) with tributylphosphorus, which yields the imidoyl chloride XXXVII([261,265]).

$$\underset{XXXVI}{CCl_3CONHC_6H_5} + Bu_3P \longrightarrow \underset{\underset{XXXVII}{Cl}}{Cl_2CH{-}\overset{|}{C}{=}NC_6H_5} + Bu_3PO$$

Iminium chlorides (XXXVIII) which have no substituent on carbon, were obtained by Böhme and his students ([40]), in the reaction of aminals of formaldehyde with carbonyl chloride.

$$R_2NCH_2NR_2 + COCl_2 \longrightarrow \underset{XXXVIII}{CH_2{=}\overset{\oplus}{N}R_2]Cl^{\ominus}} + R_2NCOCl$$

Again, the polar species XXXVIII is easily separated from the carbamoyl chloride using diethyl ether as the solvent in this reaction.

C. Addition of Hydrogen Halides to Nitriles and Isonitriles

The addition of hydrogen chloride and hydrogen bromide to nitriles yields iminium halides or imidoyl halides, depending upon the stoichiometry [6,38,39,142,148,155,163–165,171,172,187,208,209,242,273,305].

$$\text{RCN} + \text{HCl} \longrightarrow \underset{\underset{\text{Cl}}{|}}{\text{R}-\text{C}}=\text{NH} \xrightarrow{\text{HCl}} \underset{\underset{\text{Cl}}{|}}{\text{R}-\text{C}}=\overset{\oplus}{\text{N}}\text{H}_2]\text{Cl}^{\ominus}$$

For the polar 2:1 adduct the term "nitrilium salt" is often used. Hantzsch ([142]) in 1931 postulated the formation of nitrilium salts, but it was not until later that iminium halides were detected and isolated ([187,208]).

Iminium halides are intermediates in the hydrolysis of nitriles ([192]) and in a variety of chemical reactions. For example, in the Ritter reaction addition of a nitrile or hydrogen cyanide to a carbonium ion occurs, leading to the intermediate formation of a nitrilium salt. Nitrilium salt intermediates have also been postulated in the Schmidt reaction and in the Beckmann rearrangement, provided the latter was performed in concentrated sulfuric acid. However, since we are predominantly concerned with imidoyl halides, these reactions are not discussed in this monograph.

The generation of imidoyl chlorides from nitriles and hydrogen halides in the presence of a suitable Lewis acid accounts for the formation of ketones in the Houben–Hoesch reaction (see Section IVD).

Schand in 1946 ([242]) demonstrated that addition of hydrogen chloride to 2-chloronitriles yields the expected 1:1 adducts. For example, 2-chloropropionitrile and hydrogen chloride combine easily to form β-chloropropionimidoyl chloride (XXXIX).

$$\text{ClCH}_2\text{CH}_2\text{CN} + \text{HCl} \longrightarrow \underset{\underset{\text{Cl}}{|}}{\text{ClCH}_2\text{CH}_2\text{C}}=\text{NH}$$

XXXIX

The 1:1 adducts slowly release hydrogen chloride on standing, and they are therefore useful for masking hydrogen chloride.

Likewise, acrylonitrile adds two equivalents of hydrogen chloride to form XXXIX ([273]).

Cyclization of α,ω-dinitriles in the presence of anhydrous hydrogen halides affords cyclic imidoyl chlorides. For example, addition of hydrogen bromide to succinonitrile causes an immediate precipitation of the imidoyl bromide XL ([44,155]).

$$\text{NC(CH}_2)_2\text{CN} + 3\ \text{HBr} \longrightarrow$$

XL

Biltz in 1892 ([38]) had obtained a similar compound in the reaction of succinonitrile with hydroiodic acid.

While substituted succinonitriles also add hydrogen bromide to produce the corresponding imidoyl halides ([305]), addition of hydrogen chloride to succinonitrile gives rise to the formation of different reaction products ([96]).

The reaction of glutaronitrile with hydrogen bromide yields the six-membered heterocycle 6-bromo-2,3,4,5-tetrahydro-2-iminopyridine dihydrobromide (XLI) ([155]).

$$NC(CH_2)_3CN + 3\,HBr \longrightarrow$$

XLI

Again, a different product is obtained when glutaronitrile is treated with hydrogen chloride ([96]).

The corresponding seven-membered ring imidoyl bromides XLII can be obtained from o-phenylene diacetonitrile (XLIII) and hydrogen bromide ([171]).

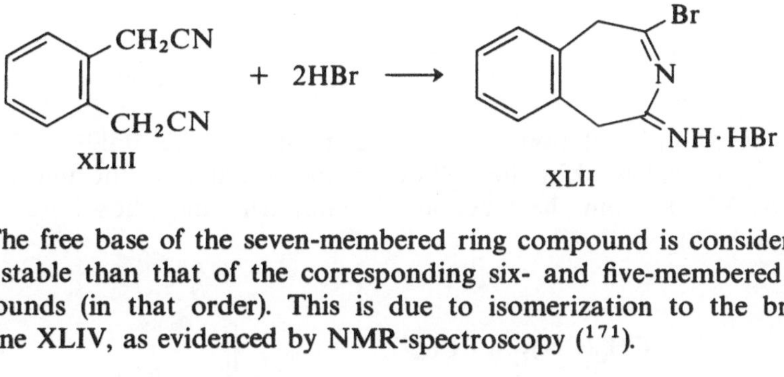

The free base of the seven-membered ring compound is considerably more stable than that of the corresponding six- and five-membered ring compounds (in that order). This is due to isomerization to the bromo enamine XLIV, as evidenced by NMR-spectroscopy ([171]).

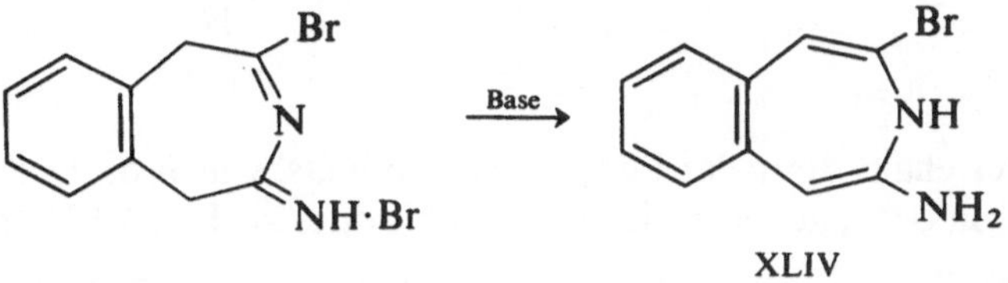

A variety of similar dinitriles react with hydrogen bromide to afford cyclic halides ([171]).

The elucidation of the structure of the hydrogen halide adducts of nitriles is rather difficult, because imidoyl chlorides and iminium chlorides which are unsubstituted on nitrogen are rather labile. They generally undergo further transformations, such as self-condensation. For example, Hinkel and Treharne ([148]) observed that the initial adduct $CH_3CN \cdot 2\,HCl$ changes on standing to a new adduct, $2\,CH_3CN \cdot HCl$. The addition of a third molecule of nitrile to the 2:1 adduct of nitriles and hydrogen halides XLV leads to the formation of 1,3,5-triazines XLVI ([128]).

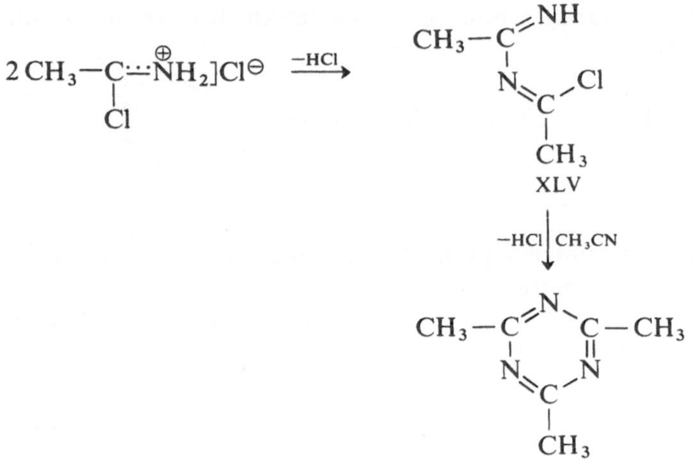

The phosphorylation of nitriles sometimes yields imidoyl chlorides ([186,251]), as evidenced by the isolation of the heterocyclic phosphorus compounds XLVII from the reaction of malonitrile and phosphorus pentachloride ([186]).

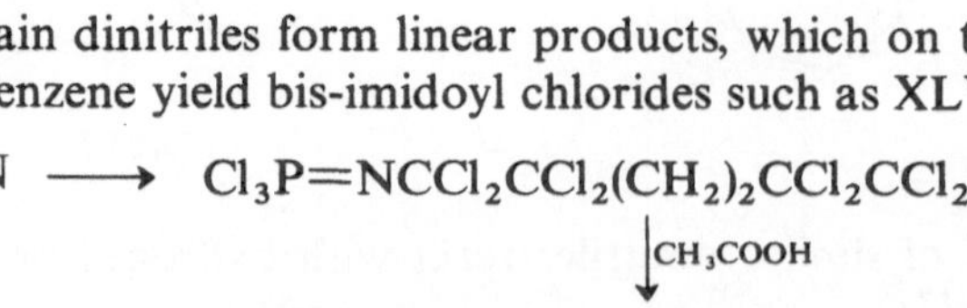

Longer chain dinitriles form linear products, which on treatment with acetic acid in benzene yield bis-imidoyl chlorides such as XLVIII ([252]).

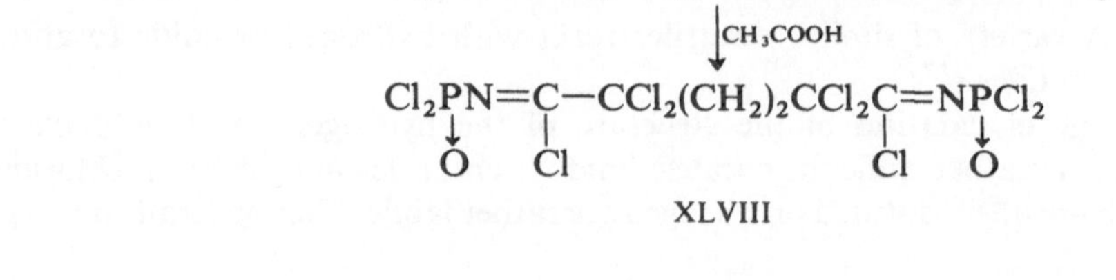

Weidinger and Kranz ([298]) treated the heterocycle XLIX (obtained from benzonitrile and sulfur trioxide) with trichloroacetonitrile in the presence of hydrogen chloride, and they obtained the triazine derivative L.

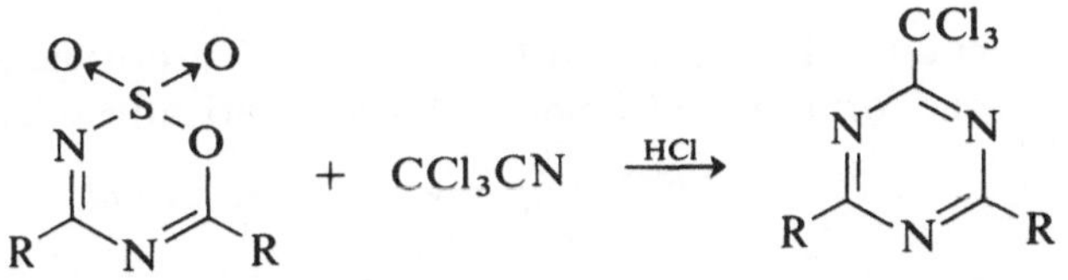

The reaction of methylene bis-thiocyanate (LI) with hydrogen bromide affords the imidoyl bromide LII in high yield ([172]).

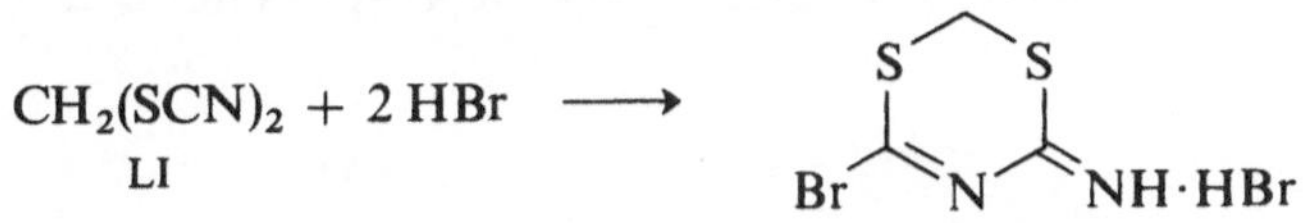

In the reaction of o-thiocyanatobenzoyl chloride with hydrogen chloride the thiazinone LIII is obtained, presumably *via* the imidoyl chloride LIV. In the absence of hydrogen chloride no reaction occurs ([254]).

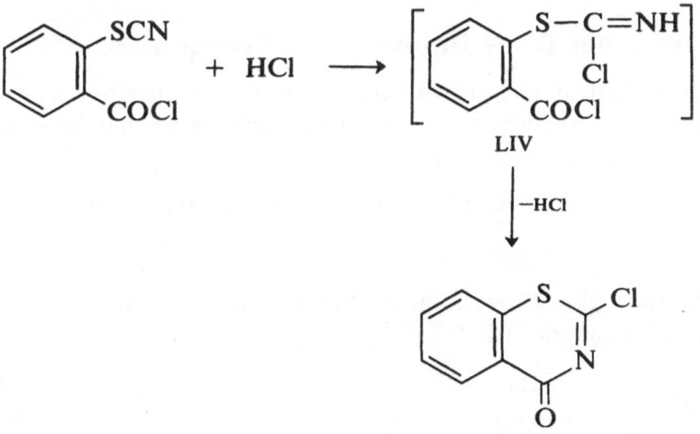

The addition of SF_5Cl to perhalo nitriles and dicyanogen yields imidoyl chlorides. For example, the bis-imidoyl chloride LV is obtained from dicyanogen and SF_5Cl ([277]).

$$NCCN + 2\,SF_5Cl \longrightarrow \underset{\underset{Cl\ \ \ Cl}{|\ \ \ \ |}}{F_5SN{=}C{-}C{=}NSF_5}$$

LV

The reaction of isocyanides (isonitriles) with carboxylic acid chlorides gives rise to the formation of imidoyl chloride derivatives LVI ([216,217]).

$$RNC + R'COCl \longrightarrow \underset{\underset{Cl}{|}}{RN{=}C{-}COR'}$$

LVI

Likewise, addition of ethyl hypochlorite to isocyanides yields the chloroimidates LVII ([218]).

$$RNC + R'OCl \longrightarrow \underset{\underset{Cl}{|}}{RN{=}C{-}OR'}$$

LVII

In the reation of isocyanides and carbonyl chloride the bis-imidoyl chlorides LVIII are obtained ([216]).

$$2\,RNC + COCl_2 \longrightarrow \underset{\underset{Cl\ \ \ \ \ \ \ \ \ Cl}{|\ \ \ \ \ \ \ \ \ \ \ |}}{RN{=}C{-}CO{-}C{=}NR}$$

LVIII

D. From other C=N Double Bond Compounds

The formation of imidoyl chlorides in the chlorination of azomethines, using *t*-butylhypochlorite as the chlorinating agent, has been reported by Paul and his colleagues ([226]).

$$RN{=}CHR' + BuOCl \longrightarrow \underset{\underset{Cl}{|}}{RN{=}CR'}$$

Addition of carboxylic acid chlorides to cyclic imines, such as LIX, yields imidoyl halides LX ([197,198]).

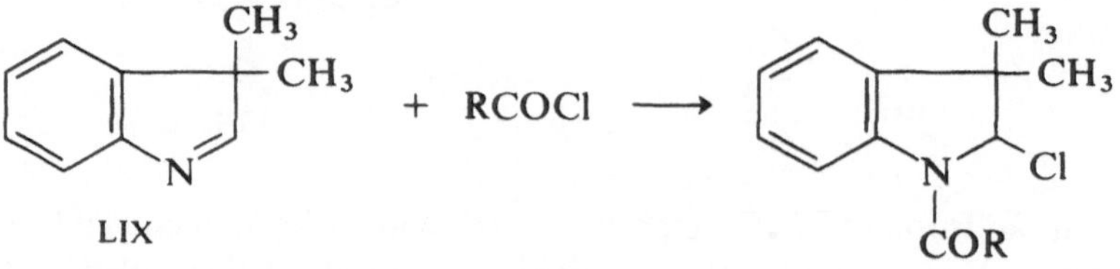

LIX   + RCOCl ⟶   LX

Another special method of synthesis of iminium halides involves the addition of dichlorocarbene to azomethines ([158]).

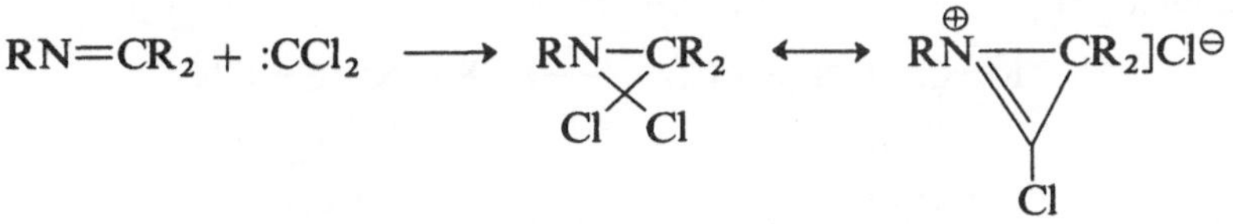

$$RN{=}CR_2 + :CCl_2 \longrightarrow \underset{\underset{Cl\quad Cl}{\diagdown\diagup}}{RN{-}CR_2} \longleftrightarrow \overset{\oplus}{RN} \underset{\underset{Cl}{|}}{\diagup\diagdown} CR_2]Cl^{\ominus}$$

In the Beckmann rearrangement, using phosphorus pentachloride or carbonyl chloride as the reagent, imidoyl chlorides are formed as intermediates ([90,127,204,266,269]). For example, Coleman and Pyle ([90]) used the Beckmann rearrangement of ketoximes to synthesize aldehydes. Thus, 4-chlorobenzophenone oxime (LXI) was treated with phosphorus pentachloride and the obtained imidoyl chloride (LXII) was reduced to the corresponding azomethine (LXIII), which on hydrolysis afforded 4-chlorobenzaldehyde (LXIV) in high yield.

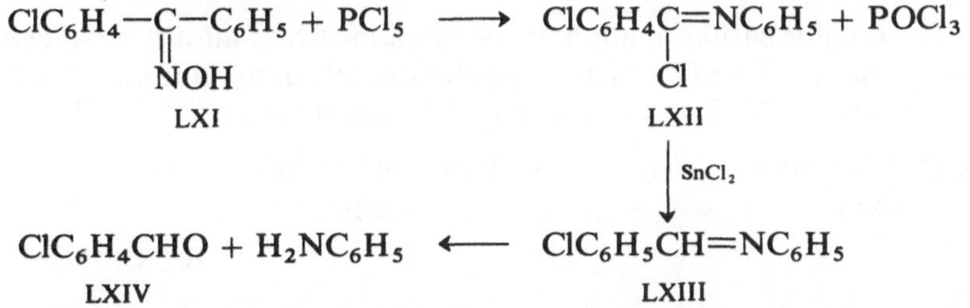

$$\underset{\underset{\text{LXI}}{}}{ClC_6H_4{-}\underset{\underset{NOH}{\|}}{C}{-}C_6H_5} + PCl_5 \longrightarrow \underset{\underset{\underset{\text{LXII}}{}}{\underset{Cl}{|}}}{ClC_6H_4\overset{}{C}{=}NC_6H_5} + POCl_3$$

$$\Big\downarrow SnCl_2$$

$$\underset{\text{LXIV}}{ClC_6H_4CHO + H_2NC_6H_5} \longleftarrow \underset{\text{LXIII}}{ClC_6H_5CH{=}NC_6H_5}$$

Stephen and Staskun ([269]) demonstrated that one mole of PCl_5 reacts with two moles of the ketoxime to yield approximately equimolecular amounts of amide and imidoyl chloride. The imidoyl chlorides can be scavenged by reaction with methyl anthranilate (LXV) to afford 4-quinazolones LXVI ([199,213,269]).

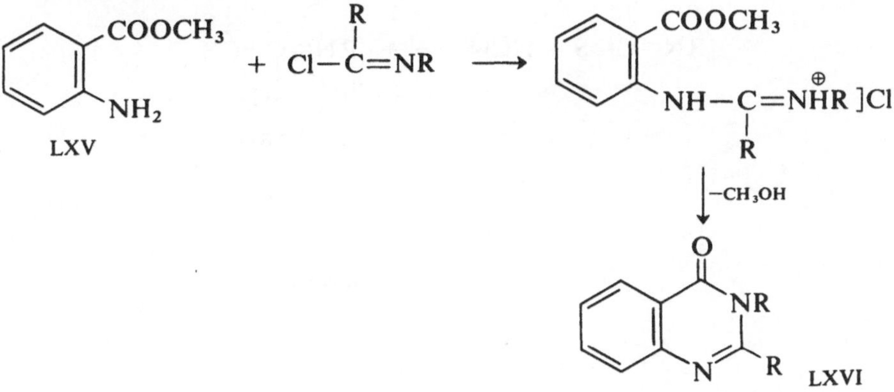

A low yield of imidoyl chlorides LXVII was obtained when aryl-glyoxylonitrile oximes LXVIII were treated with phosphorus penta-chloride (^{272}).

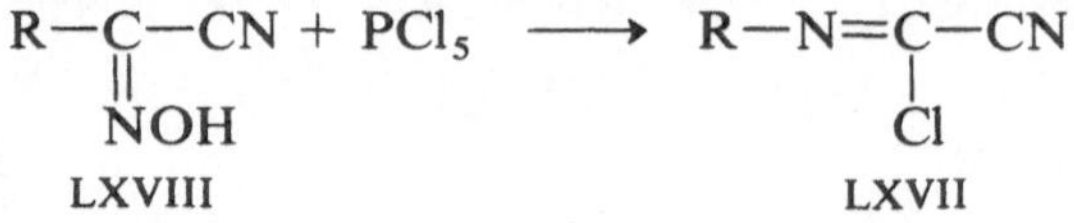

$$R-\underset{\underset{\text{NOH}}{\|}}{C}-CN + PCl_5 \longrightarrow R-N=\underset{\underset{\text{Cl}}{|}}{C}-CN$$

$$\text{LXVIII} \qquad\qquad\qquad\qquad \text{LXVII}$$

In addition to azomethines and oximes, various heterocumulenes, such as ketenimines, isocyanates and isothiocyanates, can be used to synthesize imidoyl chlorides. For example, addition of chlorine to ketenimines LXIX affords α-chloroimidoyl chlorides LXX (^{271}).

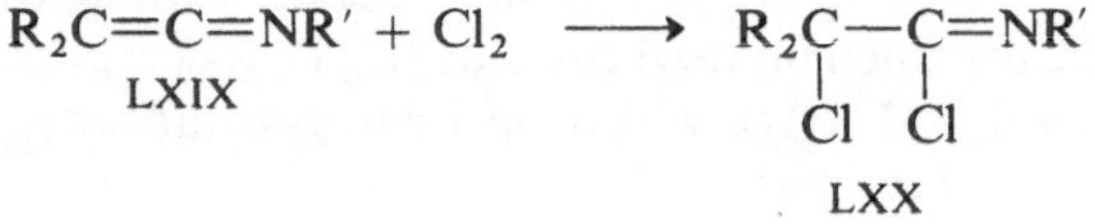

$$R_2C{=}C{=}NR' + Cl_2 \longrightarrow R_2\underset{\underset{\text{Cl}}{|}}{C}-\underset{\underset{\text{Cl}}{|}}{C}{=}NR'$$

$$\text{LXIX} \qquad\qquad\qquad\qquad\qquad \text{LXX}$$

The dichlorination of aliphatic isocyanates LXXI affords α,α-dichloro-alkyl isocyanates LXXII, which are predominantly in the tautomeric imidoyl chloride form LXXIII, as evidenced by infrared studies (^{97,99,100,244}).

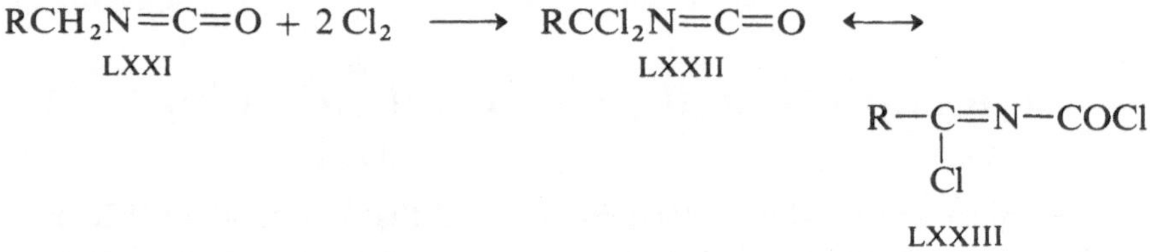

$$RCH_2N{=}C{=}O + 2\,Cl_2 \longrightarrow RCCl_2N{=}C{=}O \longleftrightarrow$$

$$\text{LXXI} \qquad\qquad\qquad \text{LXXII}$$

$$R-\underset{\underset{\text{Cl}}{|}}{C}{=}N-COCl$$

$$\text{LXXIII}$$

The addition of hydrocyanic acid to isothiocyanates yields the 1:1 adducts LXXIV, which on subsequent chlorination afford the imidoyl chlorides LXXV (^{98}).

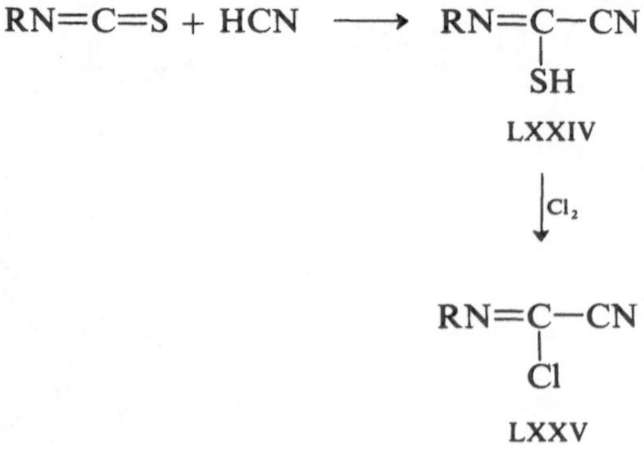

$$RN{=}C{=}S + HCN \longrightarrow RN{=}\underset{\underset{\text{SH}}{|}}{C}-CN$$

$$\text{LXXIV}$$

$$\downarrow^{Cl_2}$$

$$RN{=}\underset{\underset{\text{Cl}}{|}}{C}-CN$$

$$\text{LXXV}$$

E. Halogenation of Amines, Imines, and Carbamoyl Chlorides

The chlorination of certain cyclic amines and carbamoyl chlorides affords imidoyl chlorides ([151]); however, this method is of only limited synthetic value.

For example, high-temperature chlorination of the morpholine derivative (LXXVI) affords the heterocyclic imidoyl chloride (LXXVII) ([151]).

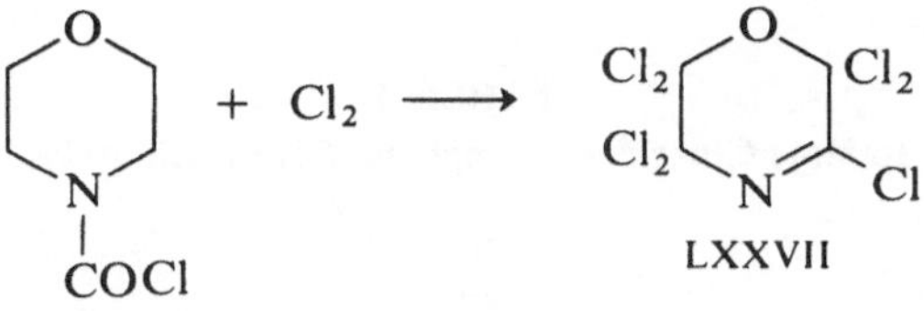

Likewise, the heterocyclic bis-imidoyl chloride (LXXVIII) has been obtained in the high-temperature chlorination of the quinone derivative LXXIX ([151]).

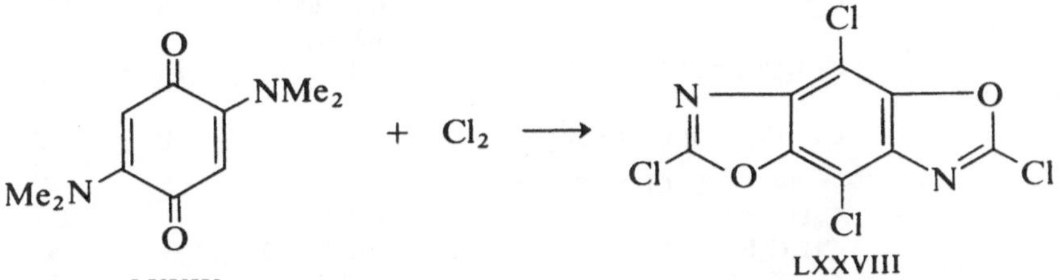

The chlorination of N-ethylbenzaldimine (LXXX) gives rise to the formation of the imidoyl chloride LXXXI ([152]).

$$C_6H_5CH{=}NC_2H_5 + Cl_2 \longrightarrow 4\text{-}ClC_6H_4CCl_2N{=}\underset{\underset{Cl}{|}}{C}{-}CCl_3$$

LXXX   LXXXI

F. Miscellaneous Methods

The reaction of trichlorovinylamines, such as LXXXII, which are obtained from N,N-disubstituted 2,2,2-trichloroacetamides and tertiary phosphines, with hydrogen chloride affords the iminium salts LXXXIII ([231]).

$$Cl_2C{=}C(Cl)NMe_2 + HCl \longrightarrow Cl_2CHCCl{=}\overset{\oplus}{N}Me_2]Cl^{\ominus}$$

LXXXII   LXXXIII

The formation of imidoyl chlorides in the reaction of carbonimidoyl dichlorides and benzene in the presence of aluminum chloride (Friedel–Crafts reaction) is described on page 48.

Perfluoro tertiary amines yield mixtures of fluorinated alkyl imidoyl chlorides on treatment with aluminum chloride ([207]).

The imidoyl chlorides and iminium chlorides, which have been characterized by either boiling point or melting point, are listed in Tables I–IV.

TABLE I

Imidoyl Chlorides of Aliphatic Carboxylic Acids

$$R-\underset{\underset{Cl}{|}}{C}=NR'$$

R	R'	Reagent*	B.p., °C/mm (M.p., °C)	Yield, %	Reference
CN	$(CH_2)_2C_6H_5$	A	91–93/3	—	214
CCl_3	C_2H_5	A	64–66/13	91	56, 152
	C_2Cl_5	—†	162–165/14 (69–71)	96	152
	$CCl_2CCl=CCl_2$	—†	124–136/0.3 (84)	86	152
	$CCl_2CCl_2N=\underset{\underset{Cl}{|}}{C}-CCl_3$	—†	(106)	10	152
	C_6H_{11}	A	129–131/12	64	281
	$CH_2C_6H_5$	A	148–153/11	—	56
	$CCl_2C_6H_5$	—†	(137–138)	86	152
	$CCl_2(2\text{-}ClC_6H_4)$	—†	(95)	97	152
	$CCl_2(3\text{-}ClC_6H_4)$	—†	(71)	90	152
	$CCl_2(4\text{-}ClC_6H_4)$	—†	(78)	87	152
	$CCl_2(2,5\text{-}Cl_2C_6H_3)$	—†	(120)	92	152
	$CCl_2(2,4\text{-}Cl_2C_6H_3)$	—†	180–190/0.25	68	152
	$(CH_2)_2C_6H_5$	A	168–170/12	—	56
	C_6H_5	A	135–138/14 (35–37)	—	56
	$4\text{-}CH_3C_6H_4$	A	(64–65)	—	56
$CHCl_2$	C_2H_5	A	72/3 128–129/14	—	265 56
	$(CH_2)_2C_6H_5$	A	155–160/17	—	56
	C_6H_5	A	83–86/0.5	—	265
	$4\text{-}CH_3C_6H_4$	A	145/15 (52)	—	56

* A = PCl_5, B = $COCl_2$, C = $SOCl_2$.
† Obtained by a chlorination procedure; see Reference 152.

Table I—*continued*

R	R'	Reagent*	B.p., °C/mm (M.p., °C)	Yield, %	Reference
CH_2Cl	C_6H_5	A	73–75/0.45	—	265
	$2\text{-}CH_3C_6H_4$	A	126–128/13	—	58
	$2\text{-}FC_6H_4$	A	83–85/0.2	80	61
	$3\text{-}FC_6H_4$	A	112–115/0.55	—	61
	$2\text{-}ClC_6H_4$	A	144–147/14	—	58
	$2\text{-}BrC_6H_4$	A	134–138/0.5	—	64
CH_3	C_6H_{11}	B	45–56/0.04	68	281
	C_6H_5	B	(118–120)	—	43
	$2\text{-}CH_3C_6H_4$	A	60/0.1	—	58
	$2\text{-}FC_6H_4$	A	70/0.25	45	61
	$2\text{-}ClC_6H_4$	A	111–114/14	—	58
	$2\text{-}BrC_6H_4$	A	142–143/12	—	58
CH_3CCl_2	C_2H_5	A	66–67/17	—	56
	C_6H_5	A	128–130/14	—	56
$CCl_2{=}CCl$	CH_3	A	75/17	77	152
C_2H_5	C_6H_{11}	B	43–44/0.02	66	281
$CH_3CH_2CCl_2$	C_2H_5	A	72–75/14	—	56
$C_6H_5(CH_2)_2CCl_2$	C_2H_5	A	170–172/16	—	56
$(CH_3)_2CCl$	C_6H_5	A	115–117/15	—	56
	$4\text{-}CH_3C_6H_4$	A	93–95/0.5	87	271
$(C_6H_5)_2CCl$	CH_3	A	147–152/0.4 (70–72)	92	271
	$n\text{-}C_4H_9$	A	160–163/0.2	88.5	271
	$4\text{-}CH_3C_6H_4$	A	(92.5–94)	87	270
$(CH_3)_2CH$	C_6H_{11}	B	40–41/0.001	67	281
	$2\text{-}CH_3C_6H_4$	A	67/0.3	77.5	237
	$4\text{-}CH_3C_6H_4$	A	80–85/0.8	82.5	271
	$4\text{-}CH_3OC_6H_4$	A	93–94/0.25	70	237
$n\text{-}C_4H_9(C_2H_5)CH$	$n\text{-}C_4H_9$	A	72–76/0.7	92.5	271
$(CH_3)_3C$	C_6H_{11}	C	104–106/20	39	281
$C_6H_{10}Cl$	C_2H_5	A	102/3	100	56
$C_9H_{17}\ddagger$	CH_3	A	125–127/15	—	56
$C_9H_{17}\S$	C_6H_5	A	165/11	—	56
$C_6H_5C(Cl){=}CH$	C_2H_5	A	140/0.2	—	68
$C_6H_5C(Cl){=}CH$	C_6H_5	A	160–170/0.1	—	68

‡ Campholyl.

§ Fencholyl.

TABLE II
Imidoyl Chlorides of Aromatic Acids*

$$R-\underset{\underset{Cl}{|}}{C}=NR'$$

R	R'	Reagent†	B.p., °C/mm (M.p., °C)	Yield, %	Reference
C_6H_5	CH_3	C	46–47/2	67	281
		A	90–92/13	~100	64
	C_2H_5	C	47–48/1	75	281
	$i\text{-}C_3H_7$	C	52–54/1	74	281
	$n\text{-}C_4H_9$	C	85–86/1	81	281
	C_6H_{11}	C	110–112/1 (66–67)	86	281
	$CH_2C_6H_5$	C	128–130/1	84	281
	C_6H_5	C	175–176/12 (40–41)	—	281
		A		95	90
	$2,6\text{-}(CH_3)_2C_6H_3$	C	153–156/1	89	281
	$2\text{-}CH_3OC_6H_4$	A	188–190/6	90	199
	$4\text{-}CH_3OC_6H_4$	C	198–200/20 (61–63)	88	281
	$2,4\text{-}(O_2N)_2C_6H_3$	A	(122–124)	59	281
$2\text{-}CH_3C_6H_4$	C_6H_5	A	174–177/10	90	199
$4\text{-}CH_3C_6H_4$	C_6H_5	C	141–144/1 (40–41)	82	281
$4\text{-}ClC_6H_4$	C_6H_5	A	(66–67)	81	90
$4\text{-}BrC_6H_4$	$4\text{-}BrC_6H_4$	A	(93–94)	72	127
$4\text{-}CH_3OC_6H_4$	C_6H_5	C	183–185/3 (73–76)	91	281
$4\text{-}O_2NC_6H_4$	C_6H_{11}	C	(40–42)	46	281
	C_6H_5	A	(137–138)	88	281
	$4\text{-}O_2NC_6H_4$	A	(132–134)	90	127 281
$3,5\text{-}(O_2N)_2C_6H_3$	C_6H_{11}	A	(86–87)	66	281
$2,4,6\text{-}(CH_3)_3C_6H_2$	C_6H_5	C	164–165/1 (60–62)	79	281
$\alpha\text{-}C_{10}H_7$	C_6H_5	A	(95)	81	199

* A variety of heterocyclic 2-furimidoyl chlorides are listed in ([137]).
† A = PCl_5, B = $COCl_2$, C = $SOCl_2$.

TABLE III
Miscellaneous Imidoyl Chlorides

$$R'—\underset{\underset{Cl}{|}}{C}=NR$$

R	R'	B.P., °C/mm (M.p., °C)	Yield, %	Reference
CN	C_6H_5	76/0.01	88	98
	C_6Cl_5	170/0.3 (108–110)	—	98
	4-$CH_3C_6H_4$	106–112/0.8 (21–22)	73.9	98
	2,4,6-$Cl_3C_6H_2$	130–140/0.09	78	98
$COOC_2H_5$	$CHCl_2$	115–118/11	—	66
	C_6H_5	100–103/0.05	73	219
	4-$CH_3C_6H_4$	125–127/0.05	48	219
	2-$CH_3OC_6H_4$	115–117/0.02	58	219
	3-$CH_3OC_6H_4$	110–112/0.03	53	219
	4-ClC_6H_4	105–108/0.05	63	219
$SO_2C_6H_5$	CH_2Cl	160–164/0.5	—	66
	CH_3	130/0.5	—	66
$OPCl_2$	CCl_3	40–50/6–7 (40–41)	67	183
	$(C_6H_5)_2CCl$	(81–83)	87.7	184
	4-$O_2NC_6H_4$	(121–124)	76	184
	3,5-$(O_2N)_2C_6H_3$	(82–84)	~100	185
$OP(C_6H_5)_2$	C_6H_5	(74–76)	83	100
	4-ClC_6H_4	(55–57)	~100	100
	3-$O_2NC_6H_4$	(127–129)	81.7	100
	4-$O_2NC_6H_4$	(87–89)	83.5	100

TABLE IV
Iminium Chlorides*

$$R-C \overset{\oplus}{\underset{|}{}}NR^1R^2]Cl^{\ominus}$$
$$\overset{|}{Cl}$$

R	R¹	R²	M.p., °C	Yield,	Reference
H	CH_3	H	75–80	—	43
H	CH_3	CH_3	140–145	95	42, 109
H	$-(CH_2)_5-$		58–66	—	107
H	CH_3	C_6H_5	Oil	—	42
CH_3	CH_3	CH_3	115–120	—	43, 107
C_2H_5	CH_3	CH_3	68–70	—	107
C_2H_5	$-(CH_2)_5-$		82–85	90–95	109
n-C_3H_7	CH_3	CH_3	82–84	95	109
n-C_3H_7	C_2H_5	C_2H_5	∼20	—	107
n-C_4H_9	CH_3	CH_3	50–55	—	107
i-C_4H_9	C_2H_5	C_2H_5	∼20	—	107
C_6H_5	CH_3	H	93–95	96	107
C_6H_5	CH_3	CH_3	95–96	90	43, 109
			124–125	97	126
C_6H_5	$-(CH_2)_5-$		136–140	85–90	109
4-$CH_3OC_6H_4$	$-(CH_2)_5-$		85	85–90	109
4-$O_2NC_6H_4$	CH_3	CH_3	>90 (dec.)	85–90	109
4-$O_2NC_6H_4$	$-(CH_2)_5-$		117–119	85–90	109
$-(CH_2)_3-$		CH_3	75–79	95	109
$-(CH_2)_2CCl_2-$		CH_3	50–68	—	107

* A variety of heterocyclic iminium chlorides, obtained by reacting 2-furimidoyl chloride with hydrogen or deuterium halides, are listed in (^{126}).

III. PHYSICOCHEMICAL PROPERTIES

Imidoyl halides are colorless liquids or low-melting solids, which are usually sensitive to moisture and heat.

In many instances, however, it is possible to purify imidoyl halides by vacuum distillation. If the compounds are thermally sensitive, purification can be conducted by solution in nonpolar organic solvents, in which the polar by-products, as well as products formed by self-condensation, are less soluble, or completely insoluble. While it is often difficult to establish melting points, because recrystallization and thereby exposure to moisture results in partial conversion to the corresponding carboxylic acid amides, identification by infrared spectroscopy is a good analytical tool. The characteristic spectral feature of the imidoyl halides is the C=N double bond absorption which occurs at 1650–1689 cm^{-1} in imidoyl chlorides (see Table V).

Theoretically, imidoyl chlorides could be either in the syn or anti form; however, attempts to obtain both isomers were unsuccessful, and

TABLE V

Asymmetrical Vibration Stretching of the
C=N Group in Imidoyl Chlorides

Type of compound	$\nu_{C=N}$ (cm^{-1})	Reference
$H-\underset{\underset{Cl}{\mid}}{C}=NCH_3$	1689	43
$C_6H_5-\underset{\underset{Cl}{\mid}}{C}=NC_6H_5$	1672	43
$NC-\underset{\underset{Cl}{\mid}}{C}=NCH_2CH_2C_6H_5$	1650	214
$R-\underset{\underset{Cl}{\mid}}{C}=N-COOEt$	1660–1670	219

TABLE VI

Asymmetrical Vibration Stretching of the
C=N Group in Iminium Halides*

Type of compound	$\nu_{C=N}$ (cm^{-1})	Reference
$H-\underset{\underset{Cl}{\mid}}{C}=\overset{\oplus}{N}(CH_3)_2]Cl^{\ominus}$	1681	43
$R-\underset{\underset{Cl}{\mid}}{C}=\overset{\oplus}{N}R_2]Cl^{\ominus}\dagger$	1621–1681	43, 126
$C_6H_5-\underset{\underset{Cl}{\mid}}{C}=\overset{\oplus}{N}(CH_3)_2]Cl^{\ominus}$	1634	43
$CH_3-\underset{\underset{Br}{\mid}}{C}=\overset{\oplus}{N}H_2]Br^{\ominus}$	1664	8
$CH_3-\underset{\underset{Br}{\mid}}{C}=\overset{\oplus}{N}D_2]Br^{\ominus}$	1626	8
$CH_3S-\underset{\underset{Br}{\mid}}{C}=\overset{\oplus}{N}H_2]Br^{\ominus}$	1587	8
$CH_3S-\underset{\underset{Br}{\mid}}{C}=\overset{\oplus}{N}D_2]Br^{\ominus}$	1545	8

* For related compounds, see also References 5, 6, and 7.
† R = alkyl.

the dielectric data observed for diaryl imidoyl chlorides indicates that they have the anti configuration ([127]).

The iminium halides, with the exception of the fluoro compounds, have a polar structure, as evidenced by conductivity measurements ([187]), single-crystal neutron defraction studies ([228]), and NMR work ([206,215]). For example, chlorodimethylformiminium chloride in sulfur dioxide was shown to be completely in the polar form. Further evidence is provided by the salt-like character of the compound, and by its characteristic C=N absorption at 1621–1681 cm^{-1} (see Table VI).

The NMR spectrum of a 2:1 complex of DMF and phosphorus trichloride has been reported recently ([257]).

The iminium halides are in general low-melting hygroscopic solids, which are insoluble in nonpolar solvents.

IV. CHEMICAL BEHAVIOR

A. Reaction with Oxygen–Hydrogen Bonds

The reaction of imidoyl chlorides and iminium chlorides with water readily affords carboxylic acid amides. Perhaps all investigators, using imidoyl chlorides, have utilized this reaction as a diagnostic tool, because imidoyl chlorides and iminium chlorides are often difficult to isolate in pure form. In 1876 Hallmann ([140]) converted an imidoyl chloride to the corresponding amide. The only kinetic investigation of the reaction of imidoyl chlorides with water has been reported by Ugi and his co-workers ([281]), who found that a two-step reaction, involving the ion-pair LXXXIV as the intermediate, occurs.

$$R-\underset{\underset{Cl}{|}}{C}=NR' \; \rightleftarrows \; R-C{\equiv}NR']^{\oplus}Cl^{\ominus} \; \xrightarrow{\text{H}_2\text{O}} \; RCONHR' + HCl$$

LXXXIV

Aliphatic imidoyl chlorides react faster than the N-alkyl imidoyl chlorides of aromatic carboxylic acids, and arylimidoyl chlorides react the slowest. Electron-withdrawing groups attached to the aryl moieties retard the reaction while electron-donating groups attached to the aryl groups increase the rate of reaction. Likewise, a carbonyl group adjacent to the C=N bond decreases the rate of hydrolysis. Cyclic imidoyl chlorides react rather slowly with water, as evidence by the fact that 2-chloro-Δ^1-pyrroline can be recrystallized from aqueous acetone ([274]).

The reaction of imidoyl halides with alkoxide ion proceeds as expected. For example, iminium chlorides undergo reaction with two equivalents of sodium alkoxide to afford amide acetals LXXXV which on treatment with alcohols under weak acidic conditions afford ortho esters ([108,109]).

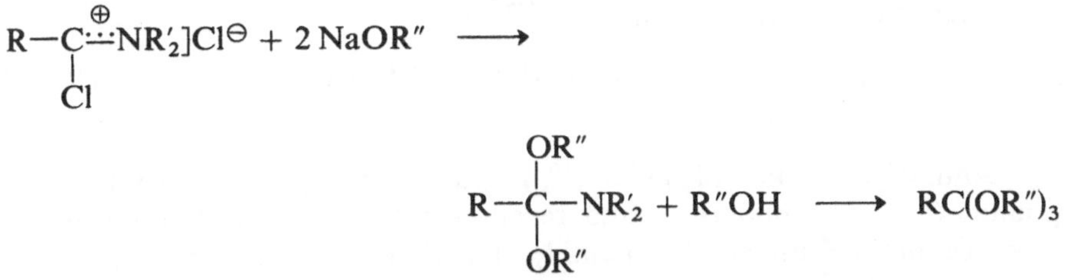

$$R-\overset{\oplus}{\underset{Cl}{C}}=NR_2']Cl^{\ominus} + 2\,NaOR'' \longrightarrow$$

$$R-\overset{OR''}{\underset{OR''}{C}}-NR_2' + R''OH \longrightarrow RC(OR'')_3$$

LXXXV

N,N-Dimethylformamide dimethylacetal ([109]). To a solution of 128 g (0.1 mole) of chlorodimethylformiminium chloride (see page 63) in 640 ml chloroform a solution of 48 g of sodium in 1000 ml of methanol is added dropwise at 0°C. After stirring for one hour at room temperature the solvent is evaporated, and vacuum distillation of the residue yields 65 g (55%) of N,N-dimethylformamide dimethylacetal, b.p. 104°C.

The reaction of iminium chloride with alcohols affords imidoyl ester hydrochlorides LXXXVI as intermediates, which eliminate alkyl halides upon heating, thereby regenerating the corresponding carboxylic acid amide ([107,109])

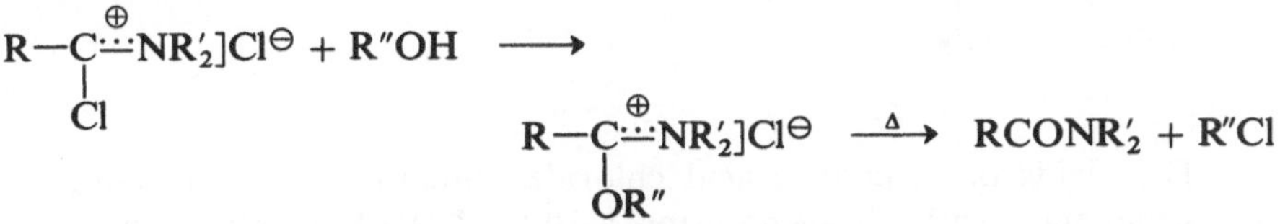

$$R-\overset{\oplus}{\underset{Cl}{C}}=NR_2']Cl^{\ominus} + R''OH \longrightarrow$$

$$R-\overset{\oplus}{\underset{OR''}{C}}=NR_2']Cl^{\ominus} \xrightarrow{\Delta} RCONR_2' + R''Cl$$

LXXXVI

Hydrolysis of the intermediate imidoyl ester hydrochlorides (LXXXVI) affords the corresponding carboxylic acid esters ([22,101,109,145]). For example, reaction of the imidoyl chloride LXXXVII with excess alcohol affords the carboxylic acid ester LXXXVIII in high yield ([101]).

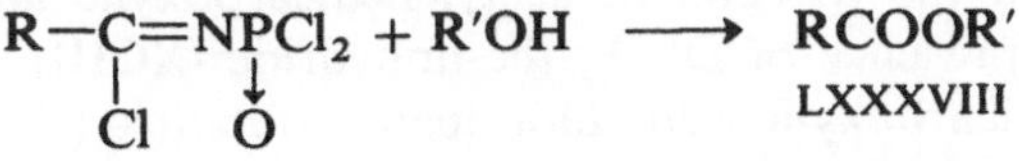

$$R-\underset{\underset{O}{\overset{|}{\downarrow}}}{\overset{|}{\underset{Cl}{C}}}=NPCl_2 + R'OH \longrightarrow RCOOR'$$

LXXXVIII

LXXXVII

The reaction of chlorodimethylformiminium chloride with alcohols has been utilized to synthesize the corresponding alkyl halides ([122,234,235]), and only a catalytic amount of N,N-dimethylformamide is necessary for conversion of an alcohol into the corresponding alkyl halide, using carbonyl chloride or thionyl chloride as the halogenating agent ([122,143]). For example, acetylene alcohols, such as butynediol (LXXXIX) can be converted to the halides XC and XCI ([122,203]).

$$HOCH_2C{\equiv}CCH_2OH + COCl_2 \xrightarrow{\text{DMF}}$$
LXXXIX

$$HOCH_2C{\equiv}CCH_2Cl + ClCH_2C{\equiv}CCH_2Cl$$
XC   XCI

Similarly, acidic phenols ([107]) and heterocyclic hydroxy compounds ([3,107,295]) undergo rapid reaction with carbonyl chloride in the presence of N,N-dimethylformamide (DMF) to yield the corresponding chlorides.

We have recently observed that the reaction of carbamates with carbonyl chloride to yield isocyanates is catalyzed by DMF ([283a]).

The reaction of imidoyl chlorides with carboxylic acids has been investigated ([42,91,93,107,150]). Again it was demonstrated that chlorodimethylformiminium chloride is an excellent catalyst to convert carboxylic acids to acid chlorides, and sulfonic acids to the corresponding sulfonyl chlorides ([42]). The reaction proceeds in the following manner:

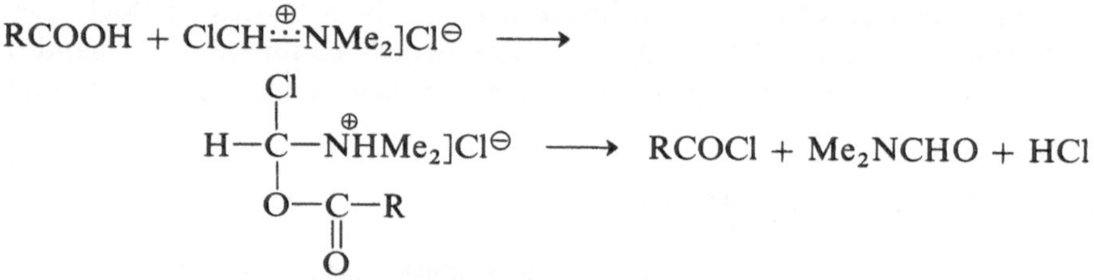

$$RCOOH + ClCH{\overset{\oplus}{\cdots}}NMe_2]Cl^{\ominus} \longrightarrow$$

$$
\begin{array}{c}
\text{Cl} \\
| \\
\text{H--C--} \overset{\oplus}{\text{N}}\text{HMe}_2]\text{Cl}^{\ominus} \longrightarrow RCOCl + Me_2NCHO + HCl\\
| \\
\text{O--C--R} \\
\| \\
\text{O}
\end{array}
$$

The yields of carboxylic acid chlorides, obtained in the reaction of carboxylic acids with phosgene using 1–10% of DMF as the catalyst, are between 83–93% ([107,211]). α,β-Unsaturated carboxylic acids afford β-chlorocarboxylic acid chlorides XCII ([107,247]).

$$R_2C{=}CHCOOH + COCl_2 \xrightarrow{\text{DMF}} R_2C(Cl)CH_2COCl$$
XCII

However, in the reaction of acetylenedicarboxylic acid with thionyl chloride in the presence of DMF, the anhydride (XCIII), rather than the corresponding dicarboxylic acid chloride, was obtained ([204]).

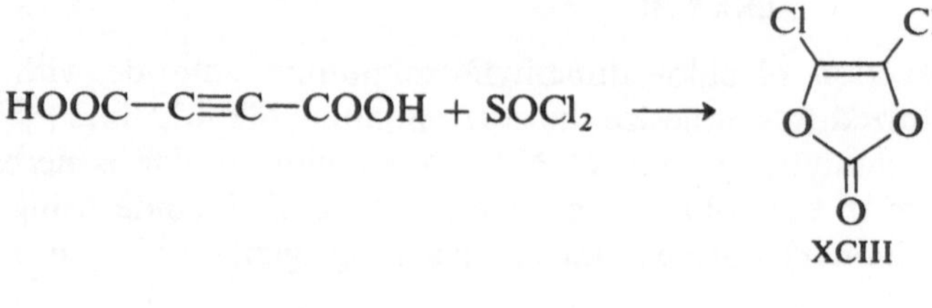

$$HOOC{-}C{\equiv}C{-}COOH + SOCl_2 \longrightarrow$$

XCIII

In the reaction of the imidoyl chloride XCIV [$R = C_6H_5$; $R' = 2,4(NO_2)_2C_6H_3$] with silver benzoate the intermediate mixed anhydride XCV has been isolated ([93]).

$$R = \underset{\underset{\text{Cl}}{|}}{C} - NR' + C_6H_5COOAg \longrightarrow R - \underset{\underset{NR'}{\|}}{C} - OCOC_6H_5$$

XCIV   XCV

The reaction of arylaminocarboxylic acids and arylaminosulfonic acids to afford isocyanatoarylcarboxylic acid chlorides and sulfonic acid chlorides, respectively, is also catalyzed by DMF ([284]).

Another interesting example of DMF catalysis involves the reaction of isatoic anhydride (XCVI) with phosgene to afford 2-isocyanatobenzoyl chloride (XCVII), which most likely occurs *via* o-isocyanatobenzoic acid (XCVIII) as the intermediate ([283b]).

XCVI   XCVIII   XCVII

B. Reaction with Sulfur–Hydrogen Bonds

The reaction of imidoyl chlorides with hydrogen sulfide to afford thioamides XCIX has been known since 1876 ([36,196,288]).

$$RC = \underset{\underset{\text{Cl}}{|}}{NR'} + H_2S \longrightarrow R - \underset{\underset{S}{\|}}{C} - NHR' + HCl$$

XCIX

This reaction has recently been used by Eilingsfeld and his colleagues ([107,109]) to synthesize thioamides from iminium chlorides. The reaction is quite simple, because it is not necessary to isolate the imidoyl chlorides. For example, one equivalent of carbonyl chloride is added to the solution of the corresponding N-substituted amide, followed by hydrogen sulfide until the evolution of hydrogen chloride ceases ([109]). Recently, Speziale and Smith ([263]) reacted iminium chlorides, obtained from enamines and hydrogen chloride, with hydrogen sulfide, and the corresponding thioamides were obtained in good yield. Likewise, addition of chlorine to trichlorovinylamines yields the trichloroiminium chlorides C, which are readily converted to the corresponding thioamide (CI) by means of H_2S ([263]).

$$Cl_2C = C(Cl)NR_2 + Cl_2 \longrightarrow$$

$$Cl_3C - \overset{\oplus}{\underset{\underset{Cl}{|}}{C} \colon NR_2}]Cl^{\ominus} \xrightarrow{H_2S} CCl_3 - \underset{\underset{S}{\parallel}}{C} - NR_2 + 2HCl$$

C   CI

N,N-Dimethylthioformamide ([109]). To a solution of 219 g (3 0 moles) of N,N-dimethylformamide in 250 ml of chloroform with stirring and cooling a solution of 330 g (3.3 moles) of phosgene in 300 ml of chloroform is added dropwise. After stirring for one hour hydrogen sulfide is added for approximately four hours with the termperature in the last hour ʃaised to 40°C. After removal of the solvent and vacuum distillation of the residue 227 g (85%) of N,N-dimethylthioformamide, b.p. 111–113°/10 mm is obtained.

Instead of hydrogen sulfide, the phosphorus compound

$$(EtO)_2P(O)SNH_4$$

can be used to convert imidoyl chlorides to thioamides ([243]).

The required imidoyl chlorides can also be prepared *in situ* from nitriles and hydrogen chloride. For example, Ishikawa ([159]) obtained thio amides in the reaction of nitriles with hydrogen chloride and thiocarboxylic S-acids CII.

$$RCN + HCl \longrightarrow [R - \underset{\underset{Cl}{|}}{C} = NH] + R' - \underset{\underset{O}{\parallel}}{C} - SH \longrightarrow$$

CII

$$R - \underset{\underset{NH}{\parallel}}{C} - S - \underset{\underset{O}{\parallel}}{C} - R' \xrightarrow{HCl} R - \underset{\underset{S}{\parallel}}{C} - NH_2 + R'COCl$$

CIII

When thioacetic S-acid is used in this reaction, N-acetylthioamide is obtained as a by-product. The rearrangement of the mixed anhydride intermediate CIII prior to cleavage can be avoided by conducting the reaction at room temperature ([33,34]). If isocyanides are treated with carboxylic acid chlorides and hydrogen sulfide, the thioamides CIV are obtained ([293]).

$$RNC + R'COCl \longrightarrow RN = \underset{\underset{Cl}{|}}{C} - \underset{\underset{O}{\parallel}}{C} - R' \xrightarrow{H_2S} RNH - \underset{\underset{S}{\parallel}}{C} - \underset{\underset{O}{\parallel}}{C} - R'$$

CIV

The reaction of iminium chlorides with mercaptans proceeds similarly to the alcohol reaction. For example, reaction of iminium chlorides with mercaptans affords the intermediate CV, which can be hydrolyzed to thiocarboxylic S-acid ester CVI, or treated with hydrogen sulfide to yield the dithiocarboxylic acid CVII ([109,296]).

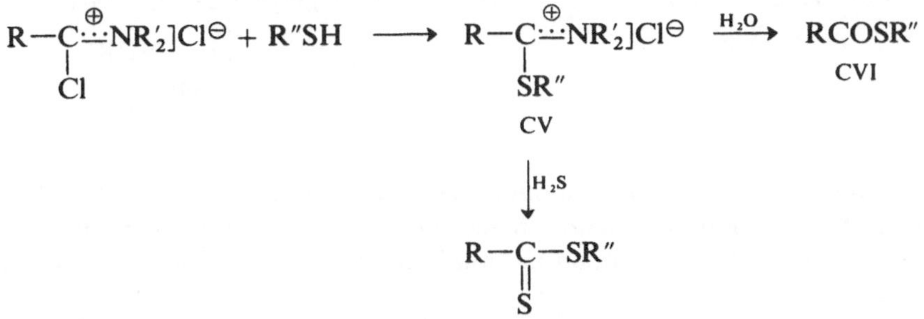

$$R-\overset{\overset{\oplus}{\underset{|}{C}}\cdots NR_2']Cl^{\ominus}}{Cl} + R''SH \longrightarrow R-\overset{\overset{\oplus}{\underset{|}{C}}\cdots NR_2']Cl^{\ominus}}{SR''} \xrightarrow{H_2O} RCOSR''$$

CV   CVI

$$\downarrow H_2S$$

$$R-\underset{\underset{S}{\|}}{C}-SR''$$

CVII

The reaction of phenylbenzimidoyl chloride with sodium thiophenolate affords the expected thioimidate CVIII ([11]).

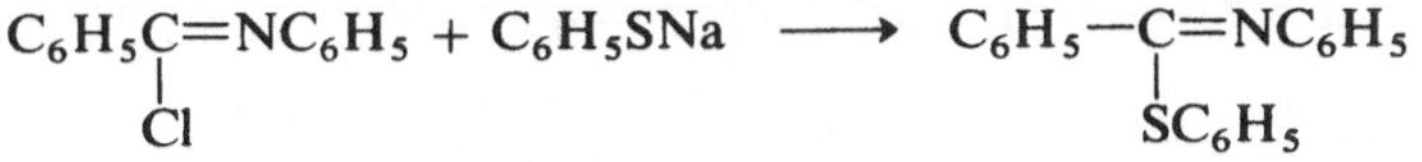

$$C_6H_5\underset{\underset{Cl}{|}}{C}{=}NC_6H_5 + C_6H_5SNa \longrightarrow C_6H_5-\underset{\underset{SC_6H_5}{|}}{C}{=}NC_6H_5$$

CVIII

Similarly, reaction of imidoyl chlorides with potassium xanthates yields the corresponding xanthates (CIX), which are aquatic herbicides ([194]).

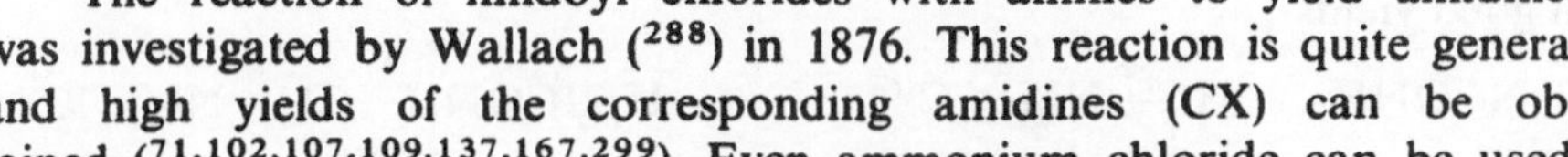

$$CH_3-\underset{\underset{Cl}{|}}{C}{=}NSO_2C_6H_5 + KS-\underset{\underset{S}{\|}}{C}-OEt \longrightarrow CH_3-\underset{\underset{SCSOEt}{|}}{C}{=}NSO_2C_6H_5$$

CIX

C. Reaction with Nitrogen–Hydrogen Bonds

The reaction of imidoyl chlorides with amines to yield amidines was investigated by Wallach ([288]) in 1876. This reaction is quite general and high yields of the corresponding amidines (CX) can be obtained ([71,102,107,109,137,167,299]). Even ammonium chloride can be used, provided the reaction is conducted at 150–180°C ([107]).

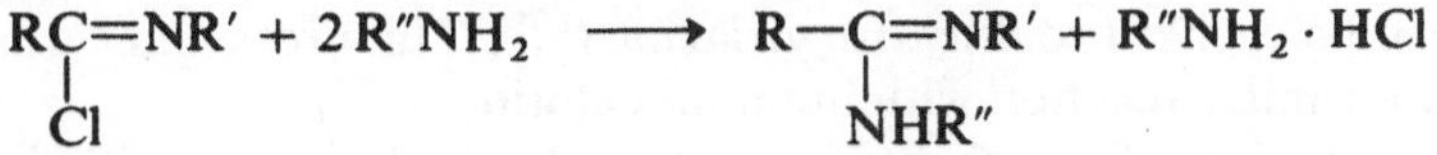

$$RC{=}NR' + 2R''NH_2 \longrightarrow R-\underset{\underset{NHR''}{|}}{C}{=}NR' + R''NH_2 \cdot HCl$$
$$\underset{Cl}{|}$$

CX

J. v. Braun and K. Weissbach ([62]) have investigated the rates of reaction of certain imidoyl chlorides with a variety of amines.

The reaction of the cyclic iminium chloride (CXI) with amines yields the linear amidines (CXII) ([158]).

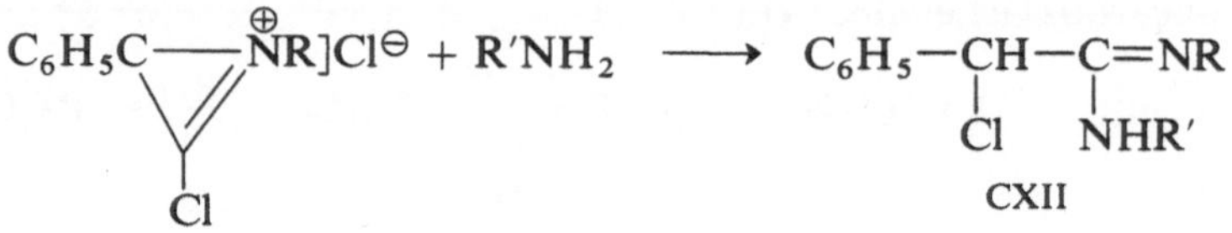

$$C_6H_5C{-}\overset{\oplus}{NR}]Cl^{\ominus} + R'NH_2 \longrightarrow C_6H_5{-}CH{-}C{=}NR$$

$$\underset{Cl}{|} \quad \underset{NHR'}{|}$$

CXI

CXII

Difunctional amino compounds, such as anthranilamide, o-phenylene-diamine, and o-aminophenol, react with imidoyl ester hydrochlorides, obtained from the corresponding iminium chlorides and alcohols, to yield heterocycles ([109]).

The reaction of amidines with imidoyl chloride N-carbonyl chlorides (CXIII) yields triazine derivatives CXIV ([97,99,244]).

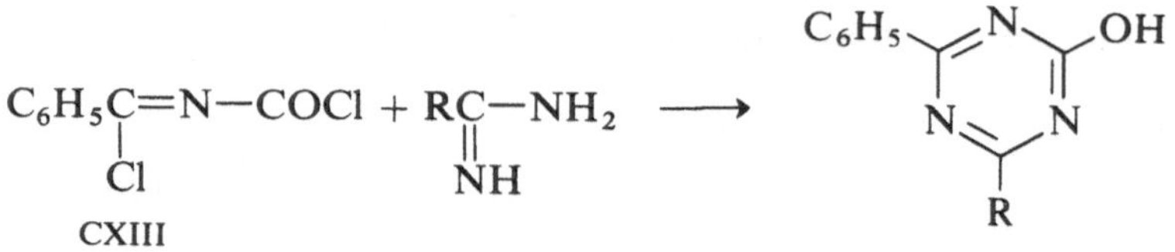

$$C_6H_5\underset{Cl}{\overset{|}{C}}{=}N{-}COCl + R\underset{NH}{\overset{||}{C}}{-}NH_2 \longrightarrow$$

CXIII

CXIV

If the imidoyl chlorides CXV are treated with primary amines, elimination with formation of the amide CXVI and isocyanide occurs ([281]).

$$CH_3CO\underset{Cl}{\overset{|}{C}}{=}NR + 2\,R'NH_2 \longrightarrow CH_3CONHR' + RNC + R'NH_2{\cdot}HCl$$

CXV

CXVI

The reaction of imidoyl chlorides with amines has been extended to ureas ([37,168,260]), and the expected amidinium chlorides CXVII are formed in good yields.

$$H_2NCONH_2 + HC\underset{Cl}{\overset{\oplus}{\cdots}NMe_2]Cl^{\ominus}} \longrightarrow H_2NCONH{-}CH\overset{\oplus}{\cdots}NMe_2]Cl^{\ominus}$$

CXVII

Hydrolysis of the amidinium salt CXVII affords N-formyl biuret, which can be cyclized to dihydroxytriazine ([109]). Thioureas and carbamates undergo a similar reaction with iminium chlorides ([107]).

The reaction of unsubstituted carboxylic acid amides with iminium chlorides affords the corresponding nitriles ([107]).

The expected products are obtained in the reaction of hydrazines ([109]) and hydroxylamines ([109,200,201,248]) with imidoyl chlorides. For example,

Ley (^{200}) reacted imidoyl chlorides with hydroxylamine in 1898, and he obtained the corresponding hydroxamic acid derivatives (CXVIII).

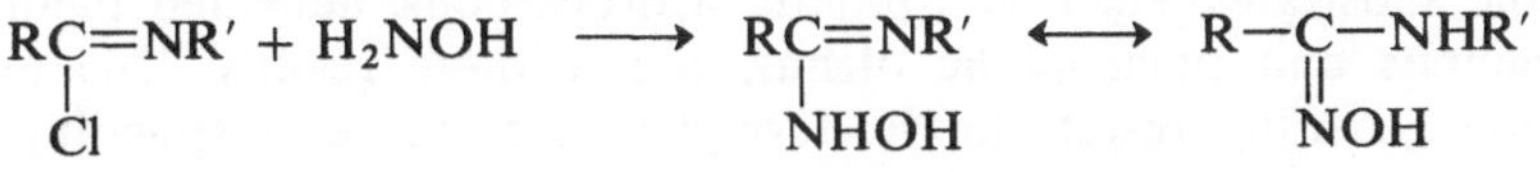

$$RC=NR' + H_2NOH \longrightarrow RC=NR' \longleftrightarrow R-C-NHR'$$
$$\underset{Cl}{|} \qquad\qquad \underset{NHOH}{|} \qquad\qquad \underset{NOH}{||}$$

CXVIII

Eilingsfeld and his colleagues have treated cyclic imidoyl chlorides with hydroxylamine and hydrazine (^{109}). For example, the azine CXX was obtained upon reaction of CXIX with hydrazine (^{109}).

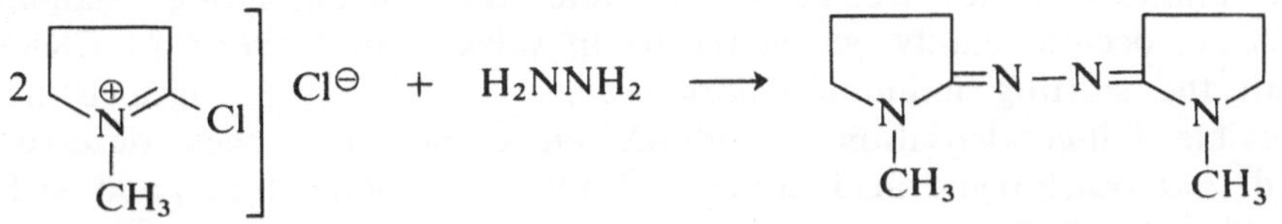

CXIX   CXX

The reaction of the difluoro compound CXXI with KOCN yields the isocyanate dimer CXXII and 1,2-bis(dialkylamino)-1,2-diisocyanatoethylene (CXXIII) (^{78}).

$$Me_2NCHF_2 + KOCN \longrightarrow$$

CXXI

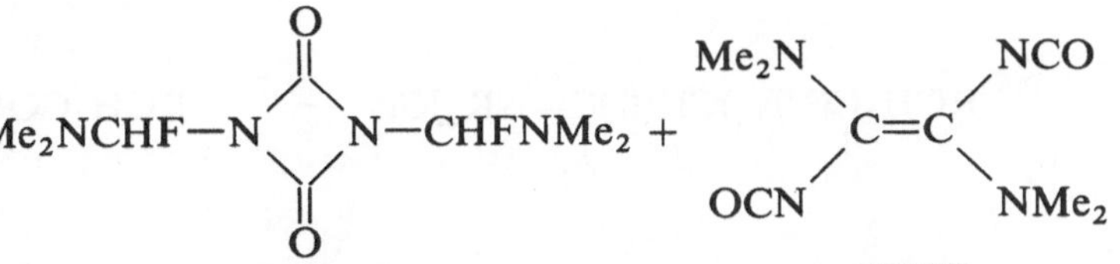

CXXII   CXXIII

Displacement of halogen in imidoyl chlorides by azide ion yields tetrazole derivatives (CXXIV) (^{60,70,245}).

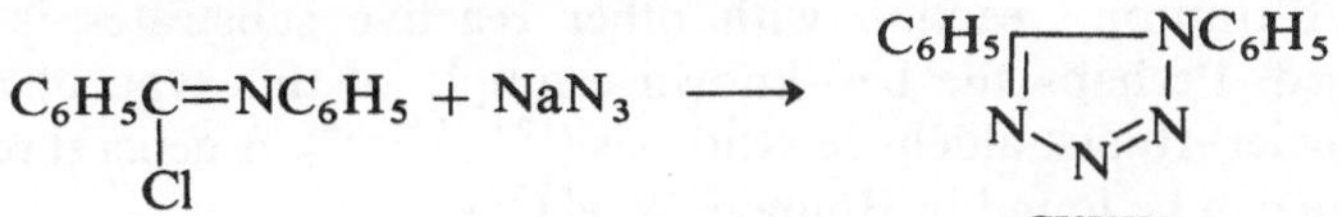

$$C_6H_5C=NC_6H_5 + NaN_3 \longrightarrow$$
$$\underset{Cl}{|}$$

CXXIV

D. Reaction with Carbon–Hydrogen Bonds

Iminium chlorides are strong electrophiles, because of the possibility of stabilization of the positive charge on the carbon atom.

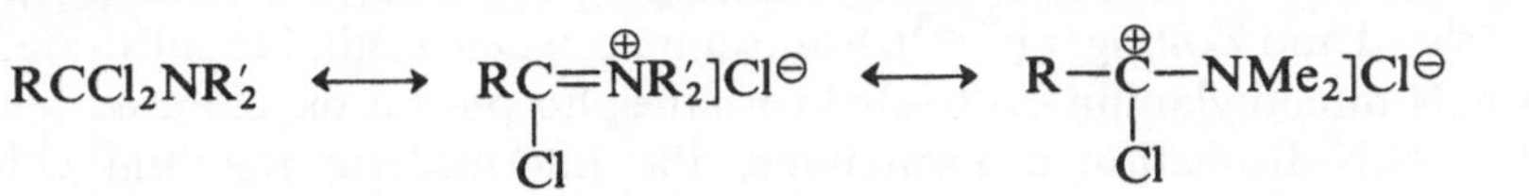

$$RCCl_2NR'_2 \longleftrightarrow RC=\overset{\oplus}{N}R'_2]Cl^{\ominus} \longleftrightarrow R-\overset{\oplus}{C}-NMe_2]Cl^{\ominus}$$
$$\underset{Cl}{|} \qquad\qquad\qquad \underset{Cl}{|}$$

Therefore, reaction of iminium chlorides, or their complexes with phosphorus oxychloride, thionyl chloride, carbonyl chloride, etc., with suitable substrates, such as aromatic hydrocarbons, activated methylene compounds and nucleophilic olefins, occurs quite readily. The overall reaction generally results in the formylation of the corresponding substrate.

An example of this type of reaction is the self-condensation of iminium chlorides, which is analogous to the self-condensation of imidoyl chlorides ([67]).

If, for example, an α-hydrogen is attached to the carbon atom of the iminium bond, transformation into the corresponding enamine (CXXV) occurs readily, which results in subsequent reaction of CXXV with the starting iminium chloride. Since attack on nitrogen is not possible self-condensation *via* attack on carbon has been observed, and β-ketocarboxylic acid amides CXXVI are obtained in good yield ([72,107,110]).

$$\text{RCH}_2\overset{\oplus}{\text{C}}\text{$\cdots$NR}'_2]\text{Cl}^{\ominus} + \text{RCH}=\underset{\underset{\text{CXXV}}{\overset{|}{\text{Cl}}}}{\text{C}}-\text{NR}'_2 \longrightarrow$$

$$\text{RCH}_2\underset{\overset{|}{\text{Cl}}}{\text{C}}(\text{NR}'_2)\text{CH(R)}\underset{\overset{|}{\text{Cl}}}{\overset{\oplus}{\text{C}}}\text{$\cdots$NR}'_2]\text{Cl}^{\ominus} \overset{\text{H}_2\text{O}}{\longrightarrow} \underset{\text{CXXVI}}{\text{RCH}_2\text{COC(R)HCONR}'_2}$$

Reaction of two different iminium chlorides is also possible, provided that one of the reactants has no α-hydrogens ([109]).

If the self-condensation of the iminium chlorides is minimized by suitable substitution, reaction with other reactive substrates is readily accomplished. Perhaps the best-known example of this type of reaction is the Vilsmeier–Haack aldehyde synthesis ([121,285,286]). A general review of this reaction can be found in Houben–Weyl ([32]).

Although Vilsmeier and Haack ([285]) in 1927 were the first who recognized the general applicability of this method, Dimroth and Zoeppritz ([105]) in 1902 had obtained dihydroxybenzaldehyde from resorcinol, formanilide, and phosphorus oxychloride. The modern mechanistic interpretation of this reaction was advanced by Arnold and Sorm ([12]), Jutz ([174]), and Bosshard and Zollinger ([35,43]). For example, when a suitable substrate, such as N,N-dimethylaniline, is treated with the phosphorus oxychloride complex of a N,N-disubstituted formamide, the intermediate iminium chloride

CXXVII is formed, which is hydrolyzed to the aldehyde in the final step of the reaction:

$$Me_2N-\langle\bigcirc\rangle + Cl_2POCH\overset{\oplus}{\cdots}NR_2']Cl^{\ominus} \longrightarrow Me_2N-\langle\bigcirc\rangle-CH\overset{\oplus}{\cdots}NR_2']Cl^{\ominus}$$

$$\underset{O}{\downarrow}$$

CXXVII

$$\downarrow$$

$$Me_2N-\langle\bigcirc\rangle-CHO$$

It is not necessary to use phosphorus oxychloride complexes of formamides, because Roh and Kochendörfer in 1937 ([239]) demonstrated that the iminium chlorides themselves react similarly and Bosshard and Zollinger ([43]) have shown that comparable yields can be obtained using chlorodimethylformiminium chloride as the reagent. However, in the synthesis of diarylketones good yields are only obtained when the phosphorus oxychloride complex of the corresponding arylcarboxylic acid amide has been used ([43]). The synthesis of ketones from carboxylic acid amides, phosphorus oxychloride and suitable substrates predates the aldehyde synthesis by approximately 40 years ([153]), but it is of only limited usefulness because side reactions are very often encountered ([107]).

Recently Bredereck and his students have compared the reactivity of a variety of amide complexes in their reaction with 4-amino-1,3-dimethyluracil.([75]) and they came to the conclusion that a mixed carbonyl chloride, phosphorus oxychloride amide complex is the most reactive reagent.

In general, one can conclude that the phosphorus oxychloride complex is the most useful reagent, but often the carbonyl chloride complex and/or the iminium chloride reacts equally well. A comparative study with regard to the use of N,N-dimethylformamides versus N-alkyl-N-arylformamides in the synthesis of alkylmercaptobenzaldehydes has been conducted by Dallacker and Eschelbach ([94]). These authors found that maximum yields at 50°C are obtained when the phosphorus oxychloride complex of N-4-methoxyphenyl-N-methylformanilide is used.

In addition to N,N-dimethylformamide and N-methylformanilides, N-formylpiperidine ([2]) and formamide itself ([179]) have been used in the Vilsmeier–Haack reaction.

Numerous aromatic and heterocyclic compounds can be utilized as substrates. While benzene does not yield benzaldehyde, benzene derivatives, such as N-substituted anilines ([10,89,129,173,189,285,286,303,304]), phenols ([179,258]), and thioether ([94]), formylate quite readily. Similarly naphthalene derivatives have been converted to the corresponding aldehydes. For example, naphthols, naphthol ethers, and naphthol thioethers react quite well ([1,79,85,179]). In reactive hydrocarbons, such as indanes ([30]) anthracenes ([83,117,118,307]), and pyrenes ([116,287]) substitution is not necessary to achieve rapid conversion.

If an excess of the formylating agent is used polyaldehydes can be obtained ([129,146,303,304]). In some instances even more than one aldehyde group can be introduced into a benzene ring ([129]).

The Vilsmeier–Haack aldehyde synthesis is especially useful for the synthesis of heterocyclic aldehydes. Thus, formylation of pyrroles ([112,220,240,246]), tetrahydroquinolines ([238]), indoles ([250,306]), naphthostyryl ([179]), carbazoles ([82,84,238]), benzazepines ([229]), thiophenes ([86,180, 181,236,301,302]), 2,2'-bithenyl ([31]), thionaphthenes ([179]), thiazines ([84]), dithiadienes ([225]), sydnones ([95]), and other heterocycles ([230]) are reported. In the pyrrole series ketones are sometimes obtained in good yield ([112]). In some instances the reaction occurs on the side chain of the heterocycle, rather than on the aromatic nucleus ([26,74,141,188]). For example, 2-methylpyrazine (CXXVIII) upon treatment with three moles of phosphorus oxychloride in excess N,N-dimethylformamide yields the enamine CXXIX, which can be hydrolyzed to the dialdehyde CXXX ([141]).

This reaction is perhaps best explained by the assumption that 2-methylpyrazine is partially in the enamine configuration, because enamines undergo reaction with iminium chlorides to afford the corresponding aldehyde arising from attack on the nucleophilic olefin ([19,314]). For example, 1-formylcyclohexanone (CXXXI) was obtained in the reaction of the enamine CXXXII with chlorodimethylformiminium chloride ([314]).

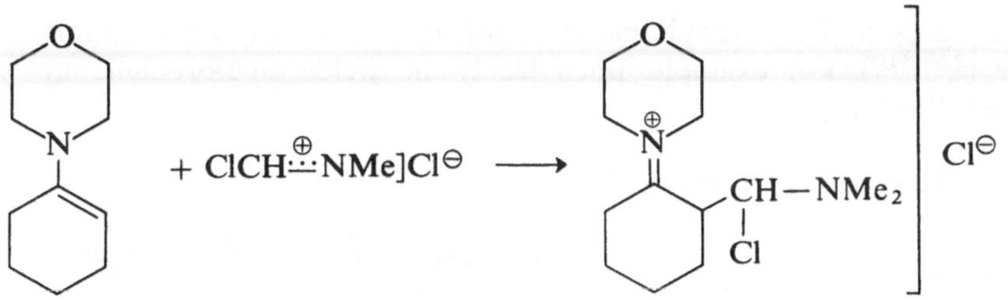

CXXXII

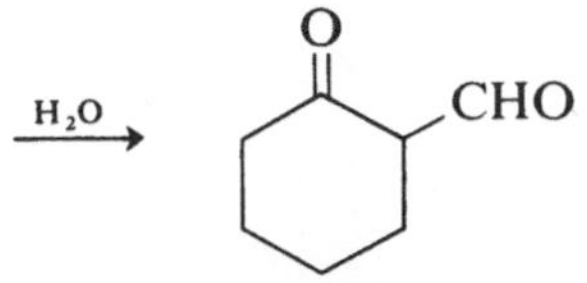

CXXXI

In addition to enamines, other reactive olefins, such as vinyl ethers ([12,111]), cyclopentadiene ([19,130,131,133,134]), fulvenes ([130,132,133,175,276]), styrenes ([202]), stilbenes ([240,249]), 1,1-diphenylethylene ([202]), and even isobutylene ([176]), undergo formylation with the Vilsmeier–Haack reagent.

For example, 2,2-diphenylacroleine (CXXXIII) is obtained in 50–60% yield in the reaction of 1,1-diphenylethylene with phosphorus oxychloride and N-methylformanilide ([202]).

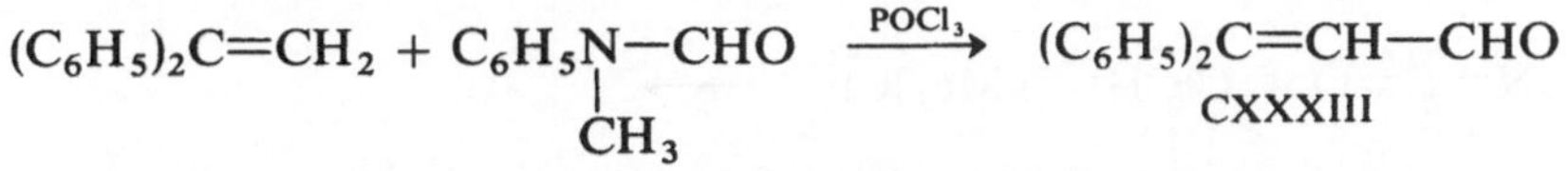

$$(C_6H_5)_2C{=}CH_2 + C_6H_5\underset{\underset{\textstyle CH_3}{|}}{N}{-}CHO \xrightarrow{\ POCl_3\ } (C_6H_5)_2C{=}CH{-}CHO$$

CXXXIII

Methylenecyclohexan (CXXXIV) affords the 1:2 adduct CXXXV upon treatment with chlorodimethylformiminium chloride ([178]).

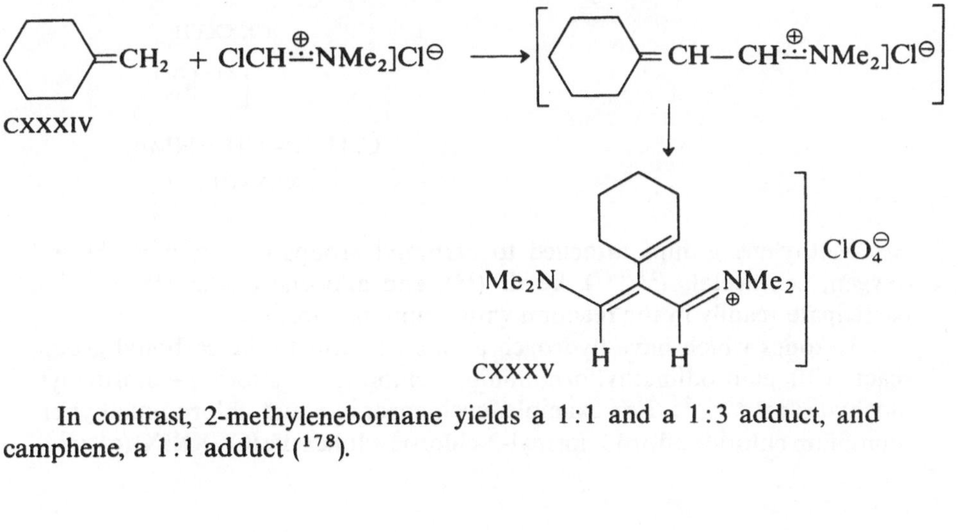

CXXXIV

CXXXV

In contrast, 2-methylenebornane yields a 1:1 and a 1:3 adduct, and camphene, a 1:1 adduct ([178]).

Acetylenes undergo rapid reaction with chlorodimethylformiminium chloride ([233,310]). For example, phenylacetylene yields α-chlorocinnamic aldehyde (CXXXVI) ([310]).

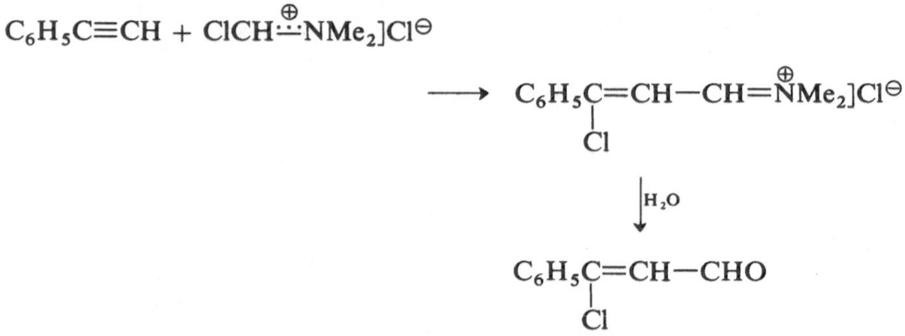

$$C_6H_5C\equiv CH + ClCH\overset{\oplus}{=\!\!=}NMe_2]Cl^{\ominus}$$

$$\longrightarrow \underset{\underset{Cl}{|}}{C_6H_5C}=CH-CH\overset{\oplus}{=\!\!=}\overset{\oplus}{N}Me_2]Cl^{\ominus}$$

$$\downarrow H_2O$$

$$\underset{\underset{Cl}{|}}{C_6H_5C}=CH-CHO$$

CXXXVI

Similar reactions occur on treatment of benzalaniline, carbodiimides, phenyl isocyanate and N-sulfinylaniline with iminium chlorides ([162]). For example, from chlorodimethylformiminium chloride and phenyl isocyanate the 1:1 adduct CXXXVII is obtained, which on thermolysis (with elimination of carbonyl chloride) or hydrolysis yields the formamidine CXXXVIII ([162]).

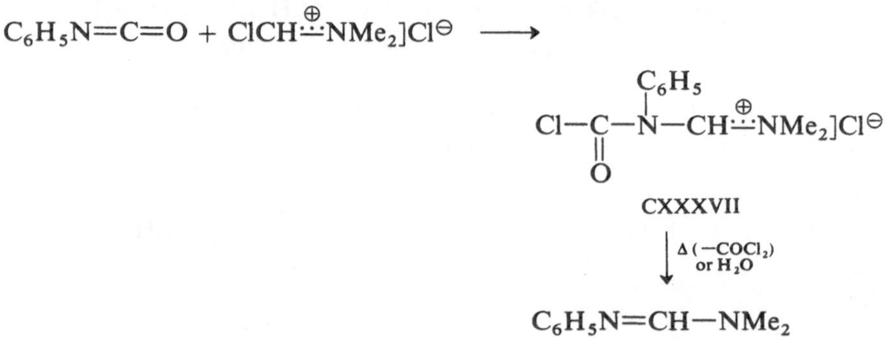

$$C_6H_5N=C=O + ClCH\overset{\oplus}{=\!\!=}NMe_2]Cl^{\ominus} \longrightarrow$$

$$\overset{\overset{C_6H_5}{|}}{\underset{\underset{O}{||}}{Cl-C}-N-CH\overset{\oplus}{=\!\!=}NMe_2]Cl^{\ominus}}$$

CXXXVII

$$\downarrow \begin{array}{c}\Delta(-COCl_2)\\ \text{or } H_2O\end{array}$$

$$C_6H_5N=CH-NMe_2$$

CXXXVIII

Methylene groups attached to carbonyl groups ([15,17,18,113,230]) and oxygen, i.e., acetals ([12,297]), ketals ([13]), and alkylene oxides ([160,310,311]), participate readily in the reaction with iminium chlorides.

Ketones which have hydrogen atoms adjacent to the carbonyl group, react with chlorodimethylformiminium chloride to afford β-chlorovinyl-aldehydes ([14,15,312]). For example, cyclohexanone and chlorodimethylformiminium chloride afford 1-formyl-2-chlorocyclohexene (CXXXIX) ([224,312]), which can be oxidized with oxygen to yield the corresponding dicarboxylic acid (CXL) ([312]).

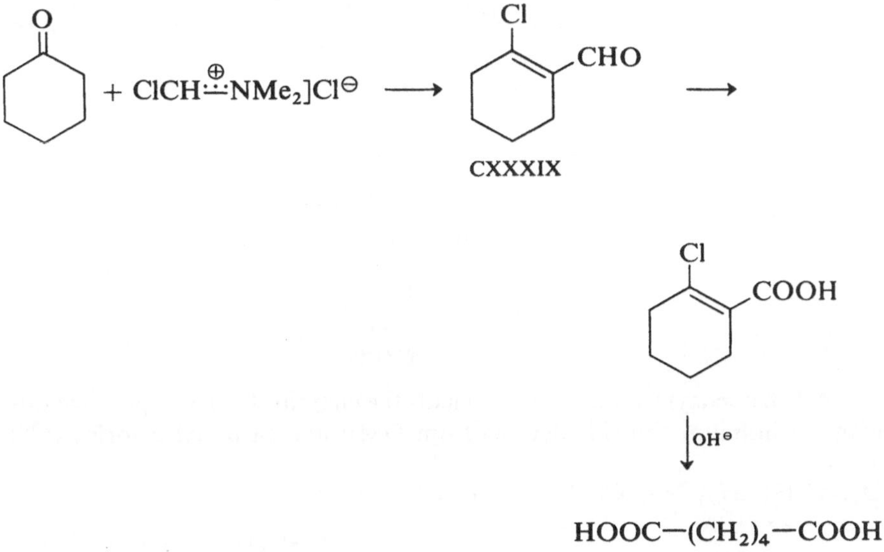

CXXXIX

HOOC—(CH$_2$)$_4$—COOH
CXL

A variety of β-chlorovinylaldehydes are obtained in the reaction of acetophenones, p-acetylbiphenyl, deoxybenzoin, acetonaphthenone, 4-chromanone, 4-thiochromanone, 5-tetrahydrobenzoxepinone, 5-tetrahydrobenzothiepinone (Weissenfels *et al.* report yields of 24–80%), and p-diacetylbenzene ([313]), with the DMF–POCl$_3$ complex ([300]).

In the reaction of dihydroxypyrimidines with the Vilsmeier–Haack reagent either ring cleavage or formylation has been observed ([76,77]).

Sodium malonate (CXLI) is easily converted to dimethylaminomethylene malonate (CXLII), using chlorodimethylformiminium chloride ([113]).

$$^{\ominus}CH(COOEt)_2]Na^{\oplus} + ClCH \overset{\oplus}{\overset{..}{=}} NMe_2]Cl^{\ominus} \longrightarrow$$
CXLI

$$Me_2NCH = C(COOEt)_2$$
CXLII

Malonodinitrile, on treatment with the DMF–POCl$_3$ complex yields the intermediate salt CXLIII, which cyclizes to the heterocycle CXLIV on heating with ammonium hydroxide ([177]).

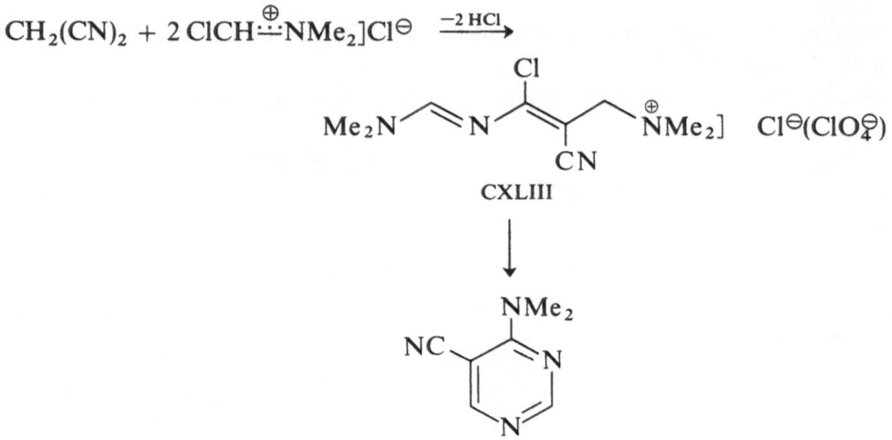

$$CH_2(CN)_2 + 2\ ClCH\overset{\oplus}{=\!\!=}NMe_2]Cl^{\ominus} \xrightarrow{-2\ HCl}$$

$$Me_2N\diagup{=}N\diagdown\underset{CN}{\overset{Cl}{C}}{=}C\diagdown\overset{\oplus}{NMe_2}]\quad Cl^{\ominus}(ClO_4^{\ominus})$$

CXLIII

CXLIV

4-Nitrobenzylcyanide (CXLV) yields the enamine CXLVI upon reaction with the iminium chloride derived from DMF and carbonyl chloride ([230]).

$$O_2NC_6H_4CH_2CN + ClCH\overset{\oplus}{=\!\!=}NMe_2]Cl^{\ominus} \longrightarrow$$

CXLV

$$O_2NC_6H_4\underset{CN}{\overset{}{C}}{=}CH{-}NMe_2$$

CXLVI

In the reaction of acetals with iminium chlorides dicarbonyl compounds CXLVII are obtained ([12,18,29]).

$$RCH(OEt)_2 \xrightarrow{DMF,\ COCl_2} Me_2NCH{=}\underset{R}{\overset{}{C}}{-}CHO \xrightarrow{NaOH} RCH(CHO)_2$$

CXLVII

The reaction of acetophenonediethylacetal (CXLVIII) with chlorodimethylformiminium chloride yields the iminium chloride CXLIX and the β-dimethylaminocinnamic aldehyde CL ([13]).

$$C_6H_5\underset{OEt}{\overset{OEt}{C}}{-}CH_3 + ClCH\overset{\oplus}{=\!\!=}NMe_2]Cl^{\ominus} \longrightarrow$$

CXLVIII

$$C_6H_5\underset{}{\overset{NMe_2}{C}}{=}CH{-}CH\overset{\oplus}{=\!\!=}NMe_2]Cl^{\ominus} + C_6H_5\underset{}{\overset{NMe_2}{C}}{=}CH{-}CHO$$

CXLIX   CL

In contrast, cyclic ethers afford the corresponding α-chloroalkyl-formates on treatment with chlorodimethylformiminium chloride ([160,310]). For example, reaction of ethylene oxide and chlorodimethylformiminium chloride gives rise to the formation of the formate CLI ([310]).

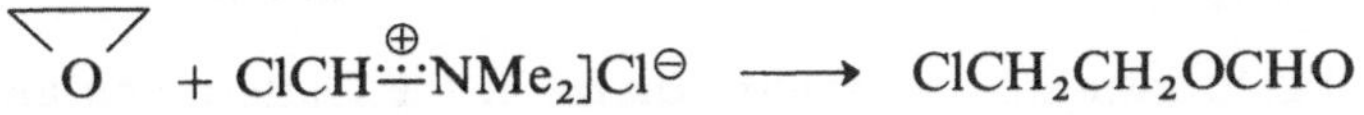

$$\overset{\triangle}{O} + ClCH\overset{\oplus}{=}NMe_2]Cl^{\ominus} \longrightarrow ClCH_2CH_2OCHO$$

CLI

The effect of inorganic salts on this reaction has been studied recently ([222]).

If the intermediate iminium salt is heated for several hours in an inert solvent, dichloroethane (CLII) is obtained ([311]).

$$ClCH_2CH_2OCH\overset{\oplus}{=}NMe_2]Cl^{\ominus} \overset{\triangle}{\longrightarrow} ClCH_2CH_2Cl + Me_2NCHO$$

CLII

From other cyclic ethers, such as oxetans and tetrahydrofuran, the corresponding 1,3- and 1,4-dichloroalkanes can be obtained similarly ([311]).

In some instances ring closure with formation of aromatic compounds occurs when unsaturated ketones are treated with chlorodimethylformimi-nium chloride ([28,154]). For example, the ketone CLIII affords the dialdehyde CLIV upon reaction with DMF and carbonyl chloride ([154]).

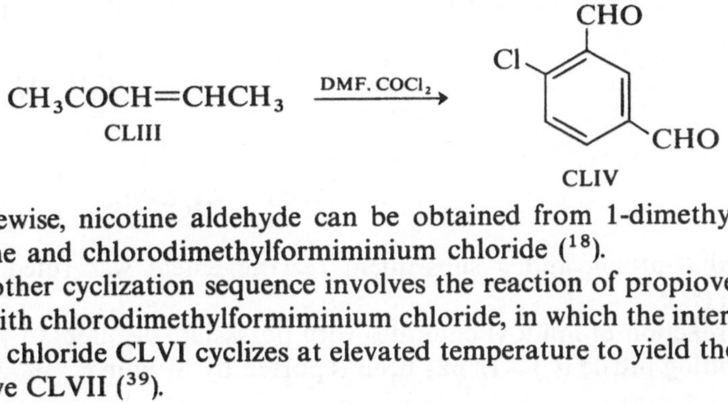

$$CH_3COCH{=}CHCH_3 \overset{DMF.\ COCl_2}{\longrightarrow}$$

CLIII

CLIV

Likewise, nicotine aldehyde can be obtained from 1-dimethylamino-butadiene and chlorodimethylformiminium chloride ([18]).

Another cyclization sequence involves the reaction of propioveratrone (CLV) with chlorodimethylformiminium chloride, in which the intermediate iminium chloride CLVI cyclizes at elevated temperature to yield the indene derivative CLVII ([39]).

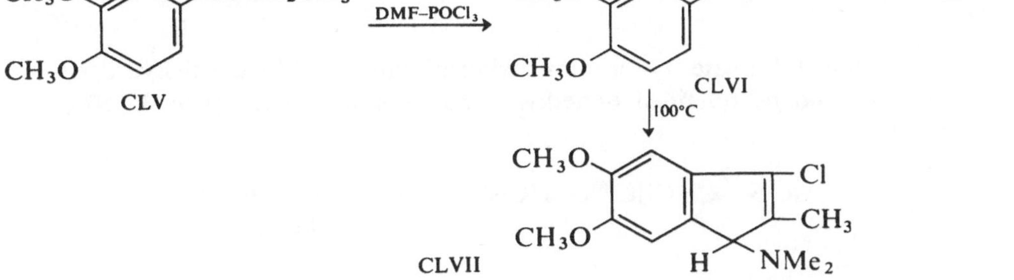

CLV $\overset{DMF-POCl_3}{\longrightarrow}$ CLVI   ↓ 100°C   CLVII

Tris-formylmethane (CLVIII) is obtained upon reaction of β-dimethylaminoacroleine with chlorodimethylformiminium chloride ([20]).

$$Me_2N{-}CH{=}CH{-}CHO \xrightarrow{DMF,\ COCl_2} \underset{\text{CLVIII}}{CH(CHO)_3}$$

Imidoyl chlorides are also intermediates in the Gattermann aldehyde synthesis and in the Houben–Hoesch ketone synthesis. The former reaction uses hydrocyanic acid, hydrogen chloride, and aluminum chloride as the catalyst, while in the latter reaction nitriles, hydrogen chloride, and zinc chloride are used. Hoesch ([149]) in 1915 assumed that the nitriles combine with hydrogen chloride to form imidoyl chlorides, which undergo electrophilic substitution reaction with a wide variety of substrates. Stephen ([268]) verified Hoesch's hypothesis by reacting phenylbenzimidoyl chloride with resorcinol, and he obtained the intermediate anil CLIX, which can be hydrolyzed to the corresponding ketone (CLX).

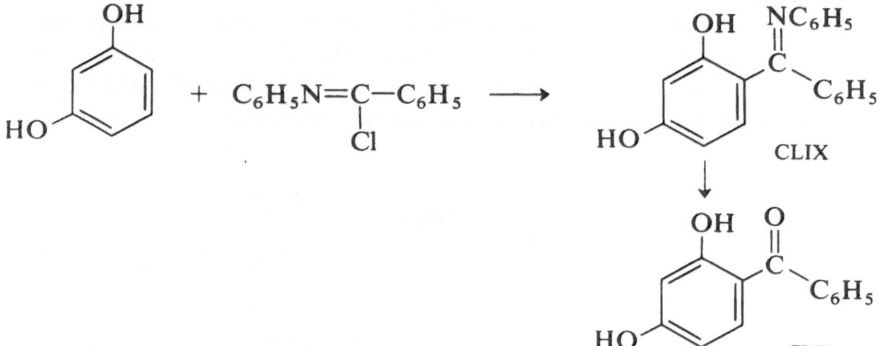

CLIX

CLX

Initial *O*-attack and a subsequent rearrangement was ruled out by Chapman ([87]).

The reaction of imidoyl chlorides with potassium cyanide, to yield the corresponding nitrile (CLXI), has been reported by Mumm ([212]).

$$C_6H_5{-}\underset{\displaystyle Cl}{\underset{|}{C}}{=}NC_6H_5 + KCN \longrightarrow C_6H_5{-}\underset{\displaystyle CN}{\underset{|}{C}}{=}NC_6H_5$$

CLXI

Arnold ([21]) treated chlorodimethylformiminium chloride with hydrogen cyanide, and he obtained dimethylaminomalonitrile (CLXII) in 45–50% yield.

$$Me_2N\overset{\oplus}{\cdots}CHCl]Cl^{\ominus} + HCN \longrightarrow \underset{\text{CLXII}}{Me_2NCH(CN)_2}$$

E. Addition and Elimination Reactions

The thermal elimination of alkyl halides to form nitriles is perhaps the best known reaction of imidoyl chlorides. J. v. Braun [47,48,49,60,61,67] investigated this reaction in detail over a period of 40 years, and he developed a host of useful new synthetic procedures for the synthesis of compounds, which are otherwise more difficult to obtain. Unfortunately, most of his work has been written up in detailed form in *Chemische Berichte* and it requires some effort to retrieve this information. However, he wrote one review article in 1934 [67] which is most informative. The elimination of alkyl halides on heating of imidoyl chlorides was recognized by Wallach in 1877 [290], and v. Pechmann [227] and Ley and Holzweissig [201] reported examples of this reaction prior to the work of J. v. Braun. The elimination reaction, in its most general terms, can be described by the following sequences:

$$RCOCl + HNR_2' \longrightarrow RCONR_2' + PCl_5 \longrightarrow RC[\overset{\oplus}{=}NR_2']Cl^{\ominus}$$
$$\underset{|}{\overset{}{Cl}}$$

$$RC[\overset{\oplus}{=}NR_2']Cl^{\ominus} \overset{\Delta}{\longrightarrow} R-\underset{|}{\overset{}{C}}=NR' + R'Cl$$
$$\underset{Cl}{} \qquad \qquad Cl$$

$$R-\underset{|}{\overset{}{C}}=NR' + H_2O \longrightarrow RCONHR' \longrightarrow RCOOH + R'NH_2$$
$$Cl$$

$$RCOCl + H_2NR' \longrightarrow RCONHR' + PCl_5 \longrightarrow R\underset{|}{\overset{}{C}}=NR'$$
$$\qquad \qquad Cl$$

$$R-\underset{|}{\overset{}{C}}=NR' \overset{\Delta}{\longrightarrow} RC\equiv N + R'Cl$$
$$Cl$$

$$RC\equiv N \overset{OH^{\ominus}}{\longrightarrow} RCOOH + NH_3$$

The overall reaction amounts to a stepwise conversion of a secondary amine *via* a primary amine to ammonia. Of course, both steps can be conducted simultaneously. For example, the reaction of N-benzoylpiperidine with phosphorus pentachloride at 120°C yields the imidoyl chloride CLXIII which is readily converted to the amide CLXIV. Hydrolysis of CLXIV yields ε-chloropentylamine (CLXV) [48].

$$C_6H_5CON\langle\rangle + PCl_5 \xrightarrow{120°C} C_6H_5\underset{\underset{Cl}{|}}{C}=N(CH_2)_5Cl \xrightarrow{H_2O}$$

$$\text{CLXIII}$$

$$C_6H_5CONH(CH_2)_5Cl \longrightarrow C_6H_5COOH + H_2N(CH_2)_5Cl$$

$$\underset{\text{CLXIV}}{} \qquad\qquad \underset{\text{CLXV}}{}$$

If the reaction is conducted at 200°C, double elimination with formation of benzonitrile and 1,5-dichloropentane (CLXVI) occurs ([48]).

$$C_6H_5CON\langle\rangle + PCl_5 \xrightarrow{200°C} C_6H_5CN + Cl(CH_2)_5Cl$$

$$\underset{\text{CLXVI}}{}$$

The reaction can equally well be conducted with phosphorus pentabromide, and 1,5-dibromopentane is obtained in good yield ([49]). The dibromoalkanes, of course, can also be obtained from linear diamines and phosphorus pentabromide. For example, 1,6-dibromohexane (CLXVII) has been obtained upon reaction of N,N'-dibenzoylhexamethylenediamine with phosphorus pentabromide ([50]).

$$C_6H_5CONH(CH_2)_6NHCOC_6H_5 + 2\,PBr_5 \longrightarrow Br(CH_2)_6Br$$

$$\underset{\text{CLXVII}}{}$$

The conversion of an alkyl halide into its next higher homolog can be accomplished by the following sequences ([61]):

$$CH_3(CH_2)_7I + KCN \longrightarrow CH_3(CH_2)_7CN \xrightarrow{H_2,\,NH_3} CH_3(CH_2)_8NH_2$$

$$CH_3(CH_2)_8NH_2 + C_6H_5COCl \longrightarrow$$

$$CH_3(CH_2)_8NHCOC_6H_5 + PCl_5 \longrightarrow CH_3(CH_2)_8Cl$$

The smooth transformation of a primary amine into the corresponding alkyl halide allows a simple conversion of a primary amine into the corresponding alcohol. Normally the stereoconfiguration of the leaving group remains intact, which allowed stereochemical assignments in the terpene series ([54]).

In the case of N,N'-disubstituted amides usually selective elimination of one alkyl halide occurs under mild conditions ([52,53]).

The elimination of hydrogen chloride from imidoyl chlorides is also the standard method of synthesis of ketenimines CLXVIII ([45,106,237,271]).

$$RCH_2\underset{\underset{Cl}{|}}{-}C=NR' \xrightarrow{Et_3N} RCH=C=NR'$$

$$\underset{\text{CLXVIII}}{}$$

This dehydrochlorination occurs readily in the presence of a tertiary amine, such as triethylamine, and the reaction is similar to the dehydrochlorination of carboxylic acid chlorides to ketenes ([271]). However, in the dehydrochlorination of the N-allylimidoyl chloride (CLXIX) with triethylamine, rearrangement to the nitrile CLXX occurs ([45]).

$$R_2CHC\underset{\underset{\text{CLXIX}}{\overset{|}{Cl}}}{=}NCH_2CH=CH_2 \xrightarrow{\text{Et}_3\text{N}} \underset{\text{CLXX}}{CH_2=CHCH_2C(CN)R_2}$$

N-(2-Propynyl)butyramide (CLXXI) was converted in a similar manner, although with poor yield, to 2-ethyl-3,4-pentadienenitrile (CLXXII) ([45]).

$$\underset{\text{CLXXI}}{CH_3(CH_2)_2CONHCH_2C\equiv CH} \xrightarrow[2)\text{ Et}_3\text{N}]{1)\text{ COCl}_2} \underset{\text{CLXXII}}{CH_2=C=CH-CH(C_2H_5)CN}$$

Dimethylketen-p-anisylimine ([237]). To the solution of 37 g (0.175 mole) of p-methoxyphenylisobutyroyl chloride in 200 ml of dry toluene 90 ml of triethylamine is added, and the solution is refluxed for two hours. After cooling and removal of the precipitated triethylamine hydrochloride by filtration, the solvent is removed under reduced pressure. Vacuum distillation of the residue yields 15 g (49 %) of dimethylketen-p-anisylimine, b.p. 92–93°C/0.25 mm.

The compound is a yellow liquid lachrymator and it can be stored at −20°C for several weeks.

Ketenimines are also obtained by dechlorination of α-chloroimidoyl chlorides, using sodium iodide ([270]).

$$R_2C\underset{\underset{Cl}{\overset{|}{}}}{-}C\underset{\underset{Cl}{\overset{|}{}}}{=}NR' \xrightarrow{\text{NaI}} R_2C=C=NR'$$

The elimination of two equivalents of hydrogen chloride from iminium chlorides, which can be accomplished by the use of lithium dialkylamides, yields ynamines CLXXIII ([80]).

$$RCH_2C\overset{\oplus}{\underset{\underset{Cl}{\overset{|}{}}}{-}}NR'_2]Cl^{\ominus} \xrightarrow{\text{LiNR}_2} \underset{\text{CLXXIII}}{RC\equiv C-NR'_2}$$

The iminium chlorides can be regenerated from the ynamines upon addition of two equivalents of hydrogen chloride ([80]).

Elimination of hydrogen chloride from arylimidoyl chlorides having α-hydrogens attached to the nitrogen is also possible. The generated nitrile ylides can be trapped with a variety of substrates, such as acrylonitrile and acetylenecarboxylic acid esters [156]. For example, if the nitrile ylides CLXXIV and CLXXV, generated from the corresponding imidoyl chlorides, are trapped with methyl acrylate, the same epimeric Δ-pyrrolines CLXXVI are obtained [157].

$$C_6H_5-\underset{\underset{Cl}{|}}{C}=NCH_2C_6H_4NO_2\,(4) \qquad\qquad (4)\,NO_2C_6H_4\underset{\underset{Cl}{|}}{C}=NCH_2C_6H_5$$

$$\left[\begin{array}{c} C_6H_5C\overset{\oplus}{\equiv}\overset{\ominus}{N}-\overset{\ominus}{C}HC_6H_4NO_2\,(4) \\ \updownarrow \\ C_6H_5\overset{\ominus}{C}=\overset{\oplus}{N}=CHC_6H_4NO_2\,(4) \end{array}\right] \longleftrightarrow \left[\begin{array}{c} (4)\,NO_2C_6H_4C\overset{\oplus}{\equiv}\overset{\ominus}{N}-\overset{\ominus}{C}HC_6H_5 \\ \updownarrow \\ (4)\,NO_2C_6H_4\overset{\ominus}{C}=\overset{\oplus}{N}=CHC_6H_5 \end{array}\right]$$

CLXXIV   CLXXV

$$\downarrow H_2C=CHCOOMe$$

CLXXVI

Apparently isomerization of the nitrile ylides accounts for this result [157].

If highly reactive dipolarophiles, such as benzaldehyde and nitroso-benzene, are used, equilibrium is not established and two different pair of diastereoisomers are formed [157].

The addition of hydrogen halides to imidoyl chlorides affords the corresponding iminium halides. For example, Grdinic and Hahn [126] added hydrogen chloride, hydrogen bromide, deuterium chloride, and deuterium bromide to N-alkylimidoyl chlorides, and the authors isolated the corresponding iminium salts. Likewise, hydrogen iodide has been added to imidoyl chlorides [195]. However, reaction of iminium chlorides with hydrogen fluoride results in the displacement of chlorine by fluorine. For example, chlorodimethylformiminium chloride reacts with hydrogen fluoride to afford α,α-difluorotrimethylamine [27]. The latter compound can be obtained directly from N,N-dimethylformamide and carbonyl fluoride [114, 115].

$$ClCH\overset{\oplus}{\cdots}NMe_2]Cl^{\ominus} \xrightarrow{\ HF\ } HCF_2NMe_2 \xleftarrow{\ COF_2\ } OHCNMe_2$$

The reaction of iminium chlorides with diazonium salts occurs *via* addition to the diazo group. For example, reaction of an iminium chloride CLXXVII with *p*-nitrophenyldiazonium fluoroborate yields the aryl-hydrazone of an α-ketocarboxylic acid amide CLXXVIII ([110]).

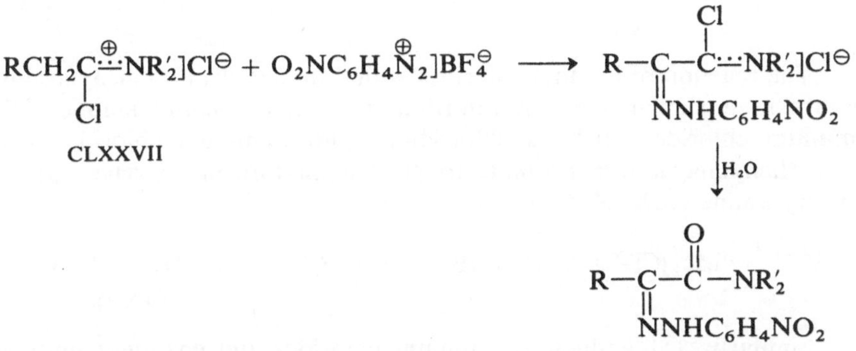

$$RCH_2\overset{\oplus}{C}\text{---}NR_2']Cl^{\ominus} + O_2NC_6H_4\overset{\oplus}{N}_2]BF_4^{\ominus} \longrightarrow R\text{---}\underset{\underset{NNHC_6H_4NO_2}{\|}}{\overset{\overset{Cl}{|}}{C}}\text{---}\overset{\oplus}{C}\text{---}NR_2']Cl^{\ominus}$$

$$\underset{Cl}{\overset{\ }{|}}$$

CLXXVII

$$\downarrow H_2O$$

$$R\text{---}\underset{\underset{NNHC_6H_4NO_2}{\|}}{\overset{\overset{O}{\|}}{C}}\text{---}\overset{\ }{C}\text{---}NR_2'$$

CLXXVIII

However, the reaction of diazonium chlorides with N,N-dimethyl-formamide proceeds in a different manner, giving rise to an intramolecular aldehyde formation. For example, from benzenediazonium chloride and N,N-dimethylformamide, salicylaldehyde CLXXIX was obtained in low yield ([308]).

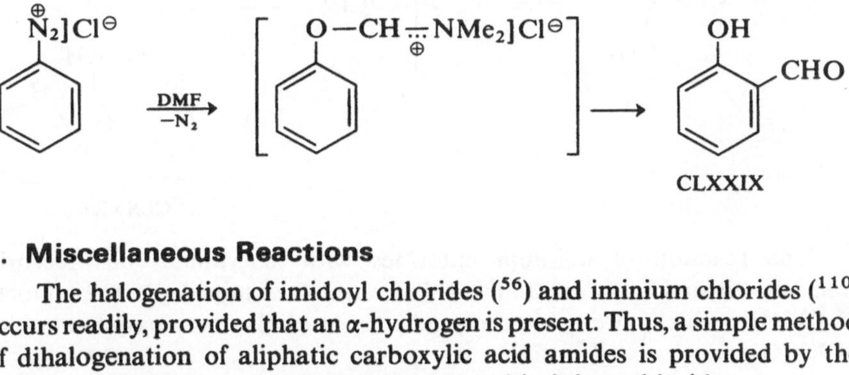

CLXXIX

F. Miscellaneous Reactions

The halogenation of imidoyl chlorides ([56]) and iminium chlorides ([110]) occurs readily, provided that an α-hydrogen is present. Thus, a simple method of dihalogenation of aliphatic carboxylic acid amides is provided by the addition of halogen to imidoyl chlorides and iminium chlorides.

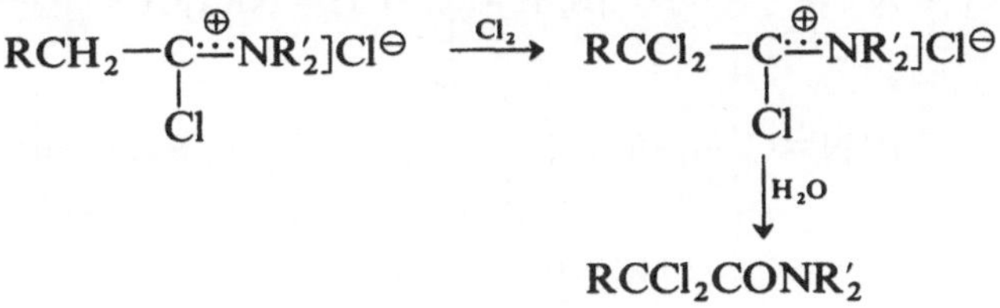

$$RCH_2\text{---}\overset{\oplus}{C}\text{---}NR_2']Cl^{\ominus} \xrightarrow{Cl_2} RCCl_2\text{---}\overset{\oplus}{C}\text{---}NR_2']Cl^{\ominus}$$

$$\underset{Cl}{\overset{\ }{|}} \qquad\qquad\qquad \underset{Cl}{\overset{\ }{|}}$$

$$\downarrow H_2O$$

$$RCCl_2CONR_2'$$

The Grignard reaction of imidoyl chlorides with alkyl and arylmagnesium halides has been investigated ([81,103]), and the corresponding ketones have been obtained.

$$C_6H_5\underset{\underset{Cl}{|}}{C}{=}NC_6H_5 + RMgX \longrightarrow C_6H_5COR$$

The reaction of imidoyl chlorides with 4-dimethylaminophenyllithium occurs in a similar manner, affording the corresponding ketone ([255]). Iminium chlorides, such as chlorodimethylformiminium chloride, react with phenylmagnesium bromide to yield a mixture of benzene and the tertiary amine CLXXX ([221]).

$$ClCH\overset{\oplus}{\cdots}NMe_2]Cl^\ominus + C_6H_5MgBr \longrightarrow C_6H_6 + (C_6H_5)_2CHNMe_2$$
$$\text{CLXXX}$$

Similarly, 1:1 adducts of iminium chlorides and enamines undergo reaction with phenylmagnesium bromide to afford benzhydrylamines ([162]).

Ring closure of the 1:1 adducts of iminium chloride and enamines with zinc–copper complex is feasible; however, the yield of diaminocyclopropanes CLXXXI is low ([162]).

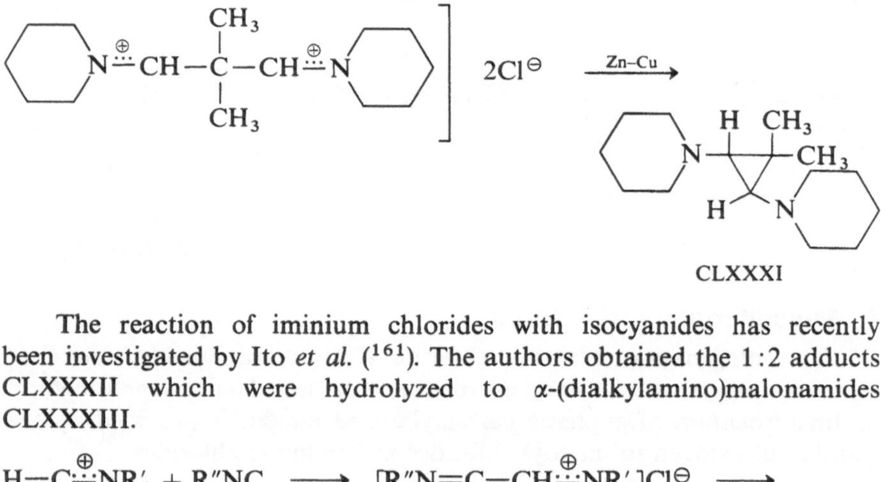

CLXXXI

The reaction of iminium chlorides with isocyanides has recently been investigated by Ito *et al.* ([161]). The authors obtained the 1:2 adducts CLXXXII which were hydrolyzed to α-(dialkylamino)malonamides CLXXXIII.

$$H{-}\underset{\underset{Cl}{|}}{C}\overset{\oplus}{\cdots}NR_2' + R''NC \longrightarrow [R''N{=}\underset{\underset{Cl}{|}}{C}{-}CH\overset{\oplus}{\cdots}NR_2']Cl^\ominus \longrightarrow$$

$$R''N{=}\underset{\underset{Cl}{|}}{C}{-}CH(NR_2')\underset{\underset{Cl}{|}}{C}{=}NR'' \xrightarrow{H_2O} R_2'NCH(CONHR'')_2$$
$$\text{CLXXXII} \qquad\qquad\qquad \text{CLXXXIII}$$

The Michalis–Arbuzov reaction, involving imidoyl chlorides and phosphites, occurs quite readily ([147,191,315]). For example, imidoyl chlorides upon treatment with trialkyl phosphites yield the phosphonates CLXXXIV, which can be hydrolyzed to the corresponding aroylphosphonates (CLXXXV) ([315]).

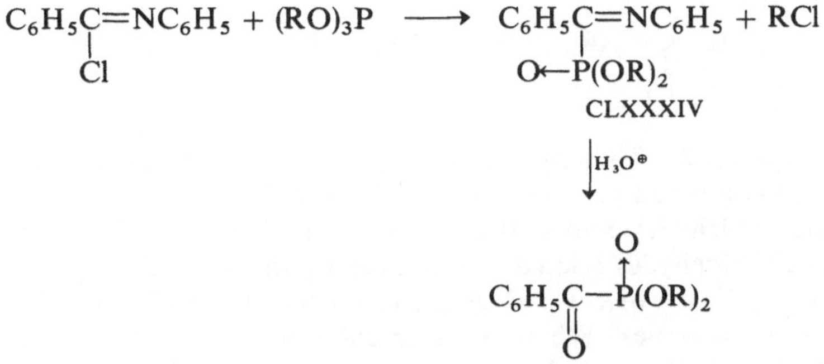

$$
\underset{\overset{|}{Cl}}{C_6H_5C}{=}NC_6H_5 + (RO)_3P \longrightarrow \underset{\overset{|}{O{\leftarrow}P(OR)_2}}{C_6H_5C}{=}NC_6H_5 + RCl
$$

CLXXXIV

$$\downarrow H_3O^{\oplus}$$

$$
C_6H_5\underset{\overset{||}{O}}{C}{-}\overset{\overset{O}{\uparrow}}{P}(OR)_2
$$

CLXXXV

In the reaction of N-chloro amides with trialkyl phosphites the corresponding trialkyl phosphates and imidoyl chlorides are obtained ([210]).

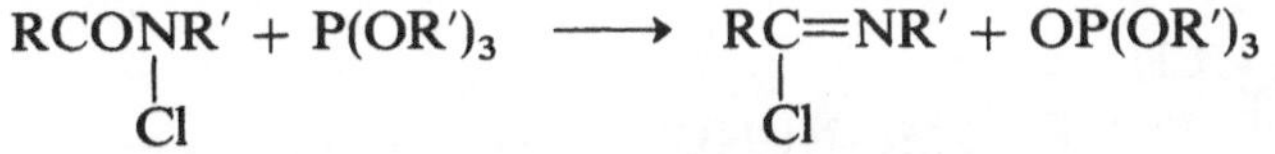

$$
\underset{\overset{|}{Cl}}{RCONR'} + P(OR')_3 \longrightarrow \underset{\overset{|}{Cl}}{RC}{=}NR' + OP(OR')_3
$$

If the reaction is conducted in the presence of a secondary amine, the corresponding amidines are isolated ([210]).

The reaction of phenylbenzimidoyl chloride with dimethyl sulfoxide yields benzanilide (CLXXXVI) and chloromethylmethylsulfide (CLXXXVII) ([205]).

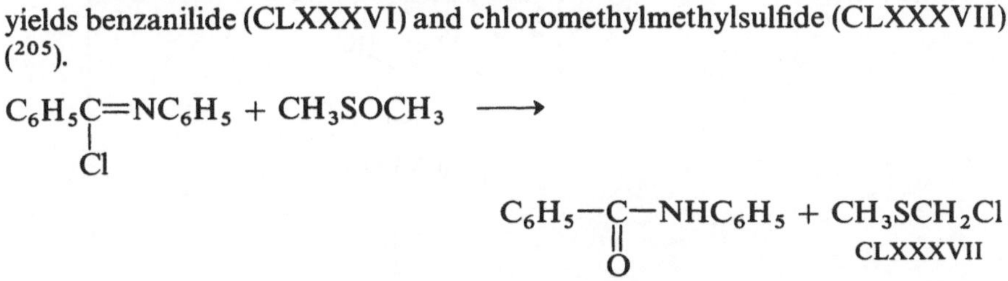

$$
\underset{\overset{|}{Cl}}{C_6H_5C}{=}NC_6H_5 + CH_3SOCH_3 \longrightarrow
$$

$$
\underset{\overset{||}{O}}{C_6H_5{-}C}{-}NHC_6H_5 + CH_3SCH_2Cl
$$

CLXXXVII

CLXXXVI

An interesting reaction of iminium chlorides involves the synthesis of peptide bonds. Zaoral and Arnold ([309]), for example, have treated the protected amino acid with one equivalent of the iminium chloride and an aminoester hydrochloride, thereby forming the peptide bond. The iminium chloride converts the acid to the acid chloride, as shown by Bosshard and Zollinger ([43]), which in turn reacts with the amino group to form the peptide bond. Cramer and his students ([92]) utilized chlorodimethylformiminium chloride for the synthesis of diesters of orthophosphoric and pyrophosphoric acid, as well as for the synthesis of nucleotide esters.

The synthesis of aldehydes from imidoyl chlorides *via* reduction and hydrolysis of the azomethine has been demonstrated by Sonn and Müller ([259]), who used stannous chloride as the reducing agent.

$$R-\underset{\underset{Cl}{|}}{C}=NR' \longrightarrow RCH=NR' \longrightarrow RCHO$$

J. v. Braun ([63,69]) improved this method by using $CrCl_2$ as the reducing agent. This method can be used equally well for the synthesis of α,β-unsaturated aldehydes, and v. Braun and Kurtz ([69]) synthesized a variety of long-chain aldehydes related to perfumes by this method.

Gold ([125]) has discovered an interesting reaction leading to a Vilsmeier–Haack-type complex. When cyanuric chloride (CLXXXVIII) is dissolved in N,N-dimethylformamide at 20°C reaction occurs with formation of the 1:2 complex CLXXXIX. If the reaction is conducted at slightly higher temperature an exothermic reaction occurs with evolution of carbon dioxide affording the iminium chloride CXC ([125]).

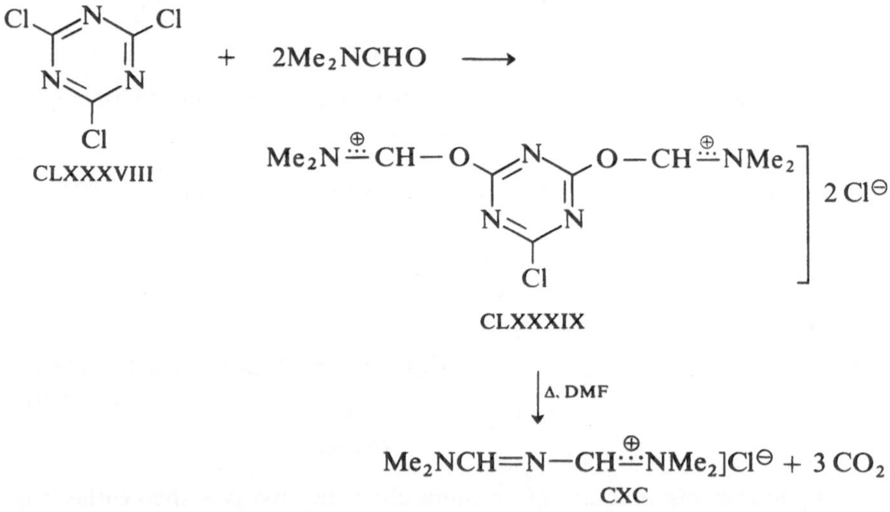

CLXXXVIII

CLXXXIX

$\downarrow$ Δ, DMF

$$Me_2NCH=N-CH\overset{\oplus}{\ldots}NMe_2]Cl^{\ominus} + 3\,CO_2$$
CXC

The iminium chloride CXC is a building block for the synthesis of a variety of heterocyclic systems ([125]).

Phosphonitrile chloride $(PNCl_2)_3$ undergoes a similar reaction with DMF to afford the salt CXCI in high yield ([267]).

$$\begin{array}{c} Cl_2P{\overset{N}{\underset{N}{\diagdown}}}PCl_2 \\ N{\underset{P}{\diagdown}}N \\ Cl_2 \end{array} + 4\,Me_2NCHO \longrightarrow$$

$$\left. \begin{array}{c} (Me_2N{\cdots}CHO)_2P{\overset{N}{\underset{N}{\diagdown}}}P(OCH{\cdots}NMe)_2 \\ N{\underset{P}{\diagdown}}N \\ Cl_2 \\ \text{CXCI} \end{array} \right] \quad 4\,Cl^\ominus$$

Reaction of CXCI with 4-aminoazobenzene results in the formation of N,N-dimethyl-N'-(4-phenylazophenyl)formamidine hydrochloride ([267]).

V. REFERENCES

1. Adams, R., and Wicks, Z. W., *J. Am. Chem. Soc.* **66**, 1315 (1944).
2. Akabori, S., and Senoh, J., *Bull. Chem. Soc. Japan* **14**, 166 (1939).
3. Albers, H., Oster, R., and Schroeder, H., German Pat. 1,178,052 (1964); *Chem. Abstr.* **61**, 16080 (1964).
4. Albright, J. D., Benz, E., Lanzilotti, A. E., and Goldman, L., *Chem. Commun.* 413 (1965).
5. Allenstein, E., and Schmidt, A., *Ber.* **97**, 1286 (1964).
6. Allenstein, E., and Quis, P., *Ber.* **97**, 1857 (1964).
7. Allenstein, E., and Schmidt, A., *Ber.* **97**, 1863 (1964).
8. Allenstein, E., and Quis, P., *Ber.* **97**, 3162 (1964).
9. D'Amico, J. J., Webster, S. T., Campbell, R. H., and Twine, C. E., *J. Org. Chem.* **30**, 3618 (1965).
10. Anker, R. M., and Cook, A. H., *J. Chem. Soc.* 489 (1944).
11. Arbuzov, A. E., Shishkin, V. E., and Tyulenev, S. S., *Zh. Organ. Khim.* **1**, 1442 (1965).
12. Arnold, Z., and Sorm, F., *Coll. Czech. Chem. Commun.* **23**, 452 (1958).
13. Arnold, Z., and Zemlicka, J., *Coll. Czech. Chem. Commun.* **24**, 786 (1959).
14. Arnold, Z., and Zemlicka, J., *Coll. Czech. Chem. Commun.* **24**, 2378 (1959).
15. Arnold, Z., and Zemlicka, J., *Coll. Czech. Chem. Commun.* **24**, 2385 (1959).
16. Arnold, Z., *Coll. Czech. Chem. Commun.* **24**, 4048 (1959).
17. Arnold, Z., and Zemlicka, J., *Coll. Czech. Chem. Commun.* **25**, 1302 (1960).
18. Arnold, Z., *Coll. Czech. Chem. Commun.* **25**, 1308 (1960).
19. Arnold, Z., *Coll. Czech. Chem. Commun.* **25**, 1313 (1960).
20. Arnold, Z., and Zemlicka, J., *Coll. Czech. Chem. Commun.* **25**, 1318 (1960).
21. Arnold, Z., *Coll. Czech. Chem. Commun.* **26**, 1113 (1961).
22. Arnold, Z., *Coll. Czech. Chem. Commun.* **26**, 1723 (1961).
23. Arnold, Z., and Holy, A., *Coll. Czech. Chem. Commun.* **26**, 3059 (1961).
24. Arnold, Z., *Angew. Chem.* **73**, 176 (1961).
25. Arnold, Z., and Holy, A., *Coll. Czech. Chem. Commun.* **27**, 2886 (1962).
26. Arnold, Z., *Coll. Czech. Chem. Commun.* **28**, 863 (1962).
27. Arnold, Z., *Coll. Czech. Chem. Commun.* **28**, 2047 (1963).
28. Arnold, Z., and Kornilov, M., *Coll. Czech. Chem. Commun.* **29**, 645 (1964).
29. Arnold, Z., and Holy, A., *Coll. Czech. Chem. Commun.* **30**, 47 (1965).

30. Arnold, Z., *Coll. Czech. Chem. Commun.* **30**, 2783 (1965).
31. Badger, G. M., Elix, J. A., and Lewis, G. E., *Austr. J. Chem.* **19**, 1477 (1966).
32. Bayer, O., *Houben-Weyl, Methoden der Organischen Chemie*, Fourth Edition, Vol. 7, Part 1, G. Thieme, Stuttgart (1954), p. 29.
33. Behringer, H., Hauser, L., and Kohl, K., *Ber.* **92**, 910 (1959).
34. Behringer, H., and Weber, D., *Ann.* **682**, 197 (1965).
35. Bellefontaine, A., Lorenz, H., and Wizinger, R., *Helv. Chim. Acta* **28**, 600 (1945).
36. Bernthsen, A., *Ber.* **10**, 1238 (1877).
37. Beyer, K. H., and Eilingsfeld, H., German Pat. 1,110,625 (1961).
38. Biltz, H., *Ber.* **25**, 2533 (1892).
39. Bodendorf, K., and Mayer, R., *Ber.* **98**, 3565 (1965).
40. Böhme, H., Hartke, K., and Müller, A., *Ber.* **96**, 607 (1963).
41. Böshagen, H., *Ber.* **99**, 2566 (1966).
42. Bosshard, H. H., Mory, R., Schmid, M., and Zollinger, H., *Helv. Chim. Acta* **42**, 1653 (1959).
43. Bosshard, H. H., and Zollinger, H., *Helv. Chim. Acta* **42**, 1659 (1959).
44. Brace, N. O., *J. Org. Chem.* **28**, 3093 (1963).
45. Brannock, K. C., and Burpitt, R. D., *J. Org. Chem.* **30**, 2564 (1965).
46. Braun, J. v., *Ber.* **37**, 2678 (1904).
47. Braun, J. v., *Ber.* **37**, 2812 (1904).
48. Braun, J. v., *Ber.* **37**, 2915 (1904).
49. Braun, J. v., *Ber.* **37**, 3210 (1904).
50. Braun, J. v. and Müller, C., *Ber.* **39**, 2018 (1906).
51. Braun, J. v., and Sobecki, W., *Ber.* **44**, 1464 (1911).
52. Braun, J. v., *Ber.* **55**, 3165 (1922).
53. Braun, J. v., *Ann.* **449**, 249 (1926).
54. Braun, J. v., *Ber.* **59**, 1444 (1926).
55. Braun, J. v., Jostes, F., and Heymons, A., *Ber.* **60**, 92 (1927).
56. Braun, J. v., Jostes, F., and Münch, W., *Ann.* **453**, 113 (1927).
57. Braun, J. v., and Heymons, A., *Ber.* **62**, 409 (1929).
58. Braun, J. v., and Silvermann, H., *Ber.* **63**, 498 (1930).
59. Braun, J. v., and Heymons, A., *Ber.* **63**, 502 (1930).
60. Braun, J. v., and Heymons, A., *Ber.* **63**, 3191 (1930).
61. Braun, J. v., and Rudolph, W., *Ber.* **64**, 2265 (1931).
62. Braun, J. v., and Weissbach, K., *Ber.* **65**, 1574 (1932).
63. Braun, J. v., and Rudolph, W., *Ber.* **67**, 269 (1934).
64. Braun, J. v., and Pinkernelle, W., *Ber.* **67**, 1218 (1934).
65. Braun, J. v., and Rudolph, W., *Ber.* **67**, 1735 (1934).
66. Braun, J. v., and Rudolph, W., *Ber.* **67**, 1762 (1934).
67. Braun, J. v., *Angew. Chem.* **47**, 611 (1934).
68. Braun, J. v., and Ostermayer, H., *Ber.* **70**, 1002 (1937).
69. Braun, J. v., and Kurtz, P., *Ber.* **70**, 1009 (1937).
70. Braun, J. v., and Rudolph, W., *Ber.* **74**, 264 (1941).
71. Bredereck, H., Gompper, R., Klemm, K., and Rempfer, H., *Ber.* **92**, 837 (1959).
72. Bredereck, H., Gompper, R., and Klemm, K., *Ber.* **92**, 1456 (1959).
73. Bredereck, H., and Bredereck, K., *Ber.* **94**, 2278 (1961).
74. Bredereck, H., and Simchen, G., *Angew. Chem. Intern. Ed.* **2**, 738 (1963).
75. Bredereck, H., Effenberger, F., and Simchen, G., *Ber.* **97**, 1403 (1964).
76. Bredereck, K., and Richter, R., *Ber.* **98**, 131 (1965).
77. Bredereck, H., Simchen, G., and Santos, A. A., *Ber.* **100**, 1344 (1967).

78. Brown, M., United States Pat. 3,214,412 (1965); *Chem. Abstr.* **64**, 3542 (1966).
79. Brownell, W., and Weston, A., *J. Am. Chem. Soc.* **73**, 4971 (1951).
80. Buijle, R., Halleux, A., and Viehe, H. G., *Angew. Chem. Intern. Ed.* **5**, 584 (1966).
81. Busch, M., and Fleischmann, M., *Ber.* **43**, 2553 (1910).
82. Buu-Hoi, N. P., and Hoan, N., *J. Am. Chem. Soc.* **73**, 98 (1951).
83. Buu-Hoi, N. P., and Hoan, N., *J. Org. Chem.* **16**, 874 (1951).
84. Buu-Hoi, N. P., and Hoan, N., *J. Org. Chem.* **16**, 1327 (1951).
85. Buu-Hoi, N. P., and Hoan, N., *J. Chem. Soc.* 1834 (1951).
86. Campaigne, E., and Archer, L., *J. Am. Chem. Soc.* **75**, 989 (1953).
87. Chapman, A. W., *J. Chem. Soc.* **121**, 1676 (1922).
88. Ciba, British Pat. 870,454 (1961); *Chem. Abstr.* **56**, 11421 (1962).
89. Ciba, French Pat. 1,181,605 (1959).
90. Coleman, G. H., and Pyle, R. E., *J. Am. Chem. Soc.* **68**, 2007 (1947).
91. Cramer, F., and Baer, K., *Ber.* **93**, 1233 (1960).
92. Cramer, F., Rittner, S., Reinhard, W., and Desai, P., *Ber.* **99**, 2252 (1966).
93. Curtin, D. Y., and Miller, L. L., *Tetrahedron Letters* 1869 (1965).
94. Dallacker, F., and Eschelbach, F. E., *Ann.* **689**, 171 (1965).
95. Dallacker, F., and Kern, J., *Ber.* **99**, 3830 (1966).
96. Decombe, J., and Verry, C., *Compte Rend.* **256**, 5156 (1963).
97. Degener, E., Holtschmidt, H., and Schmelzer, H. G., French Pat. 1,379,156 (1964); *Chem. Abstr.* **63**, 619 (1965).
98. Degener, E., and Holtschmidt, H., German Pat. 1,224,305 (1966); *Chem. Abstr.* **65**, 20071 (1966).
99. Degener, E., Schmelzer, H. G., and Holtschmidt, H., *Angew. Chem. Intern. Ed.* **5**, 960 (1966).
100. Derkach, G. I., *Zh. Obshch. Khim.* **29**, 241 (1959); *Chem. Abstr.* **53**, 21750 (1959).
101. Derkach, G. I., Lepesa, A. M., and Kirsanov, A. V., *Zh. Obshch. Khim.* **31**, 3424 (1961); *Chem. Abstr.* **57**, 3353 (1962).
102. Derkach, G. I., Gubnitskaya, E. S., and Kirsanov, A. V., *Zh. Obshch. Khim.* **31**, 3746 (1961); *Chem. Abstr.* **57**, 9735 (1962).
103. Derkach, G. I., Gubnitskaya, E. S., and Kirsanov, A. V., *Zh. Obshch. Khim.* **31**, 3679 (1961); *Chem. Abstr.* **57**, 9876 (1962).
104. Derkach, G. I., and Rudavskii, V. P., *Zh. Obshch. Khim.* **35**, 1202 (1965); *Chem. Abstr.* **63**, 11398 (1965).
105. Dimroth, O., and Zoeppritz, *Ber.* **35**, 995 (1902).
106. Dimroth, P., German Pat. 1,166,771 (1964).
107. Eilingsfeld, H., Seefelder, M., and Weidinger, H., *Angew. Chem.* **72**, 836 (1960).
108. Eilingsfeld, H., Seefelder, M., and Weidinger, H., German Pat. 1,119,872 (1960).
109. Eilingsfeld, H., Seefelder, M., and Weidinger, H., *Ber.* **96**, 2671 (1963).
110. Eilingsfeld, H., Seefelder, M., and Weidinger, H., *Ber.* **96**, 2899 (1963).
111. Eistert, B., and Haupter, F., *Ber.* **92**, 1921 (1959).
112. Ermili, A., Castro, A. J., and Westfall, P. A., *J. Org. Chem.* **30**, 339 (1965).
113. Farbenfabriken Bayer, A.G., British Pat. 917,436 (1963); *Chem. Abstr.* **58**, 12425 (1963).
114. Fawcett, F. S., Tullock, C. W., and Coffman, D. D., *J. Am. Chem. Soc.* **84**, 22 (1962).
115. Fawcett, F. S., Tullock, C. W., and Coffman, D. D., *J. Am. Chem. Soc.* **84**, 4275 (1962).
116. Fieser, L. F., and Hershberg, E. B., *J. Am. Chem. Soc.* **60**, 2547 (1938).
117. Fieser, L. F., and Hartwell, J. L., *J. Am. Chem. Soc.* **60**, 2555 (1938).
118. Fieser, L. F., and Jones, J. E., *J. Am. Chem. Soc.* **64**, 1666 (1945).
119. Fieser, L. F., Hartwell, J. L., and Jones, J. E., *in* "Organic Synthesis Collective Volumes," Vol. 3, J. Wiley and Sons, New York (1955), p. 98.

120. Finkbeiner, H., *J. Org. Chem.* **30**, 2861 (1965).

121. Fischer, O., Müller, A., and Vilsmeier, A., *J. Prakt. Chem.* [2] **198**, 60 (1925).

122. Fischer, R., and Pasedach, H., German Pat. 1,133,717 (1962).

123. Friedel, M. C., *Bull. Soc. Chim.* [3]**61**, 1027 (1894).

124. Godefroy, E. F., van der Eycken, C. A. M., and Janssen, P. A., *J. Org. Chem.* **32**, 1259 (1967).

125. Gold, H., *Angew. Chem.* **72**, 956 (1960).

126. Grdinic, M., and Hahn, V., *J. Org. Chem.* **30**, 2381 (1965).

127. Greenberg, B., and Aston, J. G., *J. Org. Chem.* **25**, 1894 (1960).

128. Grundmann, C., Weisse, G., and Seide, S., *Ann.* **577**, 77 (1952).

129. Grundmann, C., and Dean, J. M., *Angew. Chem. Intern. Ed.* **4**, 955 (1965).

130. Hafner, K., and Bernhard, C., *Angew. Chem.* **64**, 533 (1957).

131. Hafner, K., and Vöpel, K. H., *Angew. Chem.* **71**, 672 (1959).

132. Hafner, K., *Angew. Chem.* **72**, 574 (1960).

133. Hafner, K., Vöpel, K. H., Ploss, G., and König, C., *Ann.* **666**, 52 (1963).

134. Hafner, K., German Pat. 1,105,411 (1961).

135. Hagedorn, I., and Tönges, H., *Pharmazie* **12**, 570 (1957).

136. Hagedorn, I., Etling, H., and Lichtel, K. E., *Ber.* **99**, 520 (1966).

137. Hahn, V., and Grdinic, M., *J. Chem. Eng. Data* **11**, 211 (1966).

138. Hall, H. K., Jr., *J. Am. Chem. Soc.* **77**, 5993 (1955).

139. Hall, H. K., Jr., *J. Am. Chem. Soc.* **78**, 2717 (1956).

140. Hallmann, F., *Ber.* **9**, 846 (1876).

141. Hansen, H. V., Caputo, J. A., and Meltzer, R. I., *J. Org. Chem.* **31**, 3845 (1966).

142. Hantzsch, A., *Ber.* **64**, 667 (1931).

143. Held, P., German Pat. (East) 32,562 (1964); *Chem. Abstr.* **63**, 13107 (1965).

144. Hertler, W. R., and Corey, E. J., *J. Org. Chem.* **23**, 1221 (1958).

145. Hettler, H., *Tetrahedron Letters* 4049 (1966).

146. Hickmott, P. W., *J. Chem. Soc.* 666 (1966).

147. Hilgetag, G., Zieloff, K., and Paul, H., *Angew. Chem.* **77**, 261 (1965).

148. Hinkel, L. E., and Treharne, G. J., *J. Chem. Soc.* 866 (1945).

149. Hoesch, K., *Ber.* **48**, 1122 (1915).

150. Holsten, J. R., and Lilyquist, M. R., United States Pat. 3,210,422 (1965); *Chem. Abstr.* **63**, 17990 (1965).

151. Holtschmidt, H., *Angew. Chem. Intern. Ed.* **1**, 632 (1962).

152. Holtschmidt, H., Degener, E., and Schmelzer, H. G., *Ann.* **701**, 107 (1967).

153. Hoechster Farbwerke, German Pat. 41,751; 44,077 (1887).

154. Holy, A., and Arnold, Z., *Coll. Czech. Chem. Commun.* **30**, 53 (1964).

155. Howard, Jr., E. G., United States Pat. 2,810,726 (1957).

156. Huisgen, R., Stangl, H., Sturm, H. J., and Wagenhofer, H., *Angew. Chem. Intern. Ed.* **1**, 50 (1962).

157. Huisgen, R., and Raab, R., *Tetrahedron Letters* 649 (1966).

158. Ichimura, K., and Ohta, M., *Tetrahedron Letters* 807 (1966).

159. Ishikawa, S., *Sci. Papers Inst. Phys. Chem. Res. Tokyo* **7**, 293 (1927).

160. Ito, Y., Katsuragawa, S., Okano, M., and Oda, R., *Tetrahedron Letters* 1321 (1965).

161. Ito, Y., Okano, M., and Oda, R., *Tetrahedron* **22**, 447 (1966).

162. Ito, Y., Katsuragawa, S., Okano, M., and Oda, R., *Tetrahedron* **23**, 2159 (1967).

163. Janz, G. J., and Danyluk, S. S., *J. Am. Chem. Soc.* **81**, 3846 (1959).

164. Janz, G. J., and Danyluk, S. S., *J. Am. Chem. Soc.* **81**, 3850 (1959).

165. Janz, G. J., and Danyluk, S. S., *J. Am. Chem. Soc.* **81**, 3854 (1959).

166. Jentzsch, W., *Ber.* **97**, 1361 (1964).

167. Jentzsch, W., and Seefelder, M., German Pat. 1,172,269 (1964).
168. Jentzsch, W., and Seefelder, M., German Pat. 1,175,223 (1964).
169. Jentzsch, W., *Ber.* **97**, 2755 (1964).
170. Jentzsch, W., and Seefelder, M., *Ber.* **98**, 274 (1965).
171. Johnson, F., and Nasutavicus, W. A., *J. Heterocyclic Chem.* **2**, 26 (1965).
172. Johnson, F., and Madronero, R., *Advances in Heterocyclic Chemistry*, Vol. 6, Academic Press, New York (1966), p. 131.
173. Jostes, F., German Pat. 547,108 (1929).
174. Jutz, C., *Ber.* **91**, 850 (1958).
175. Jutz, C., *Ber.* **97**, 2050 (1964).
176. Jutz, C., Müller, E., *Ber.* **99**, 2479 (1966).
177. Jutz, C., and Müller, W., *Angew. Chem. Intern. Ed.* **5**, 1042 (1966).
178. Jutz, C., and Müller, W., *Ber.* **100**, 1536 (1967).
179. Kalischer, G., Scheyer, H., and Keller, K., German Pat. 514,415; 519,444; 519,806 (1927).
180. King, W. J., and Nord, F. F., *J. Org. Chem.* **13**, 635 (1948).
181. King, W. J., and Nord, F. F., *J. Org. Chem.* **14**, 405 (1949).
182. Kirsanov, A. V., *Izv. Akad. Nauk SSSR, Otd. Khim. Nauk*, 646 (1954); *Chem. Abstr.* **49**, 13161 (1955).
183. Kirsanov, A. V., and Derkach, G. I., *Zh. Obshch. Khim.* **26**, 2009 (1956); *Chem. Abstr.* **51**, 1821 (1957).
184. Kirsanov, A. V., and Derkach, G. I., *Zh. Obshch. Khim.* **27**, 3248 (1957); *Chem. Abstr.* **52**, 8997 (1958).
185. Kirsanov, A. K., and Derkach, G. I., *Zh. Obshch. Khim.* **28**, 1187 (1958); *Chem. Abstr.* **53**, 1208 (1959).
186. Kirsanov, A. V., Bodnarchuk, M. D., and Shevchenko, V. I., *Dopovidi Akad. Nauk Ukr. RSR*, 221, (1963); *Chem. Abstr.* **59**, 12666 (1963).
187. Klages, F., and Grill, W., *Ann.* **594**, 21 (1955).
188. Klutchko, S., Hansen, H. K., and Melzer, R. I., *J. Org. Chem.* **30**, 3454 (1965).
189. Koike, E., Obayashi, K., Oka, R., Taura, S., and Takada, H., Japanese Pat. 19,100 (1965); *Chem. Abstr.* **63**, 16263 (1965).
190. Kotjscheff, T., Wolf, F., and Wolter, G., *Z. Chem.* **6**, 148 (1966); *Chem. Abstr.* **65**, 7051 (1966).
191. Kreutzkamp, N., and Cordes, G., *Ann.* **623**, 103 (1959).
192. Krieble, V. K., and Peiker, A. L., *J. Am. Chem. Soc.* **51**, 3368 (1929).
193. Kühle, E., *Angew. Chem. Intern. Ed.* **1**, 647 (1962).
194. Kühle, E., Belgian Pat. 660,171 (1965); *Chem. Abstr.* **64**, 603 (1966).
195. Lander, G. C., and Laws, H. E., *J. Chem. Soc.* **85**, 1695 (1904).
196. Leo, H., *Ber.* **10**, 2133 (1877).
197. Leuchs, H., Wulkow, G., and Gerland, H., *Ber.* **65**, 1586 (1932).
198. Leuchs, H., and Schlötzer, A., *Ber.* **67**, 1572 (1934).
199. Levy, P. R., and Stephen, H., *J. Chem. Soc.* 985 (1956).
200. Ley, H., *Ber.* **31**, 240 (1898).
201. Ley, H., and Holzweissig, E., *Ber.* **36**, 18 (1903).
202. Lorenz, H., and Wizinger, R., *Helv. Chim. Acta* **28**, 6000 (1945).
203. Ludsteck, D., Neubauer, G., Pasedach, H., and Seefelder, M., German Pat. 1,133,716 (1962).
204. McDonald, R. N., and Krueger, R. A., *J. Org. Chem.* **28**, 2542 (1963).
205. Martin, D., and Weise, A., *Ann.* **702**, 86 (1967).
206. Martin, G., and Martin, M., *Bull. Soc. Chim. France* 637 (1963); *Chem. Abstr.* **60**, 391 (1964).

207. Mazalov, S. A., and Sokolov, S. V., *Zh. Obshch. Khim.* **36**, 1330 (1966); *Chem. Abstr.* **65**, 16851 (1966).
208. Meerwein, H., Laasch, P., Mersch, R., and Spille, J., *Ber.* **89**, 209 (1956).
209. Michael, A., and Wing, J. F., *Am. Chem. J.* **7**, 71 (1885).
210. Mitin, Y. V., and Vlasov, G. P., *Probl. Organ. Sinteza, Akad. Nauk SSSR*, 297 (1965); *Chem. Abstr.* **64**, 11122 (1966).
211. Mory, R., Stöcklin, E., and Schmid, M., German Pat. 1,026,750 (1958); *Chem. Abstr.* **55**, 5428 (1961).
212. Mumm, O., *Ber.* **43**, 886 (1910).
213. Mumm, O., and Hesse, H., *Ber.* **43**, 2508 (1910).
214. Murakami, M., Akagi, K., and Takahashi, K., *J. Am. Chem. Soc.* **83**, 2002 (1961).
215. Namanworth, E., Dissertation, Yale University, 1965.
216. Nef, J. U., *Ann.* **270**, 267 (1892).
217. Nef, J. U., *Ann.* **280**, 298 (1894).
218. Nef, J. U., *Ann.* **287**, 301 (1895).
219. Neidlein, R., and Bottler, R., *Tetrahedron Letters* 1069 (1966).
220. Nenitzescu, C. D., and Isacescu, D., *Bull. Soc. Chim. (Rumania)* **11**, 135 (1929).
221. Oda, R., Katsuragawa, S., Ito, Y., and Okano, M., *Nippon Kagaku Zasshi* **87**, 1236 (1966).
222. Oda, R., Katsuragawa, S., Ito, Y., and Okano, M., *Nippon Kagaku Zasshi* **87**, 490 (1966).
223. Okawa, K., Abe, J., Watanabe, T., and Fujimoto, K., Japanese Pat. 13,726 (1965); *Chem. Abstr.* **63**, 14774 (1965).
224. Paquette, L. A., Johnson, B. A., and Hinga, F. M., *Org. Syn.* **46**, 118 (1966).
225. Parham, W. E., Roder, T. M., and Hasek, W. R., *J. Am. Chem. Soc.* **75**, 1647 (1953).
226. Paul, H., Weise, A., and Dettmer, R., *Ber.* **98**, 1450 (1965).
227. Pechmann, H. v., *Ber.* **33**, 611 (1900).
228. Peterson, S. W., and Williams, J. M., *J. Am. Chem. Soc.* **88**, 2866 (1966).
229. Porai-Koshits, B. A., Kvitko, I. Y., and Favorskii, O. V., *Zh. Organ. Khim.* **1**, 1516 (1965); *Chem. Abstr.* **64**, 698 (1966).
230. Porai-Koshits, B. A., Kvitko, I. Y., and Shutkova, E. A., *Latvijas PSR Zinatnu Akad. Vestis, Khim. Ser.* 587 (1965); *Chem. Abstr.* **64**, 8168 (1966).
231. Prajsnar, B., *Zeszyty Nauk Politech. Slask. Chem.* No. 20 (1963); *Chem. Abstr.* **62**, 16019 (1965).
232. Prajsnar, B., *Zeszyty Nauk Politech. Slask. Chem.* No. 24, 241 (1964); *Chem. Abstr.* **63**, 11289 (1965).
233. Quang, Y. V., Cadiot, P., Willemart, A., *Compt. Rend.* **248**, 2356 (1959).
234. Rätz, R., and Sweeting, O. J., *J. Org. Chem.* **28**, 1608 (1963).
235. Rätz, R., and Sweeting, O. J., *J. Org. Chem.* **30**, 438 (1965).
236. Ricci, A., Balucani, D., and Buu-Hoi, N. P., *J. Chem. Soc.* 779 (1967).
237. Ried, W., and Junker, P., *Ann.* **700**, 32 (1966).
238. Roh, N., German Pat. 645,880 (1935); 660,693 (1935).
239. Roh, N., and Kochendörfer, G., German Pat. 677,207 (1937).
240. Rogers, M. A. T., *J. Chem. Soc.* 596 (1943).
241. Sayigh, A. A. R., and Ulrich, H., *J. Chem. Soc.* 3146 (1963).
242. Schand, E. W., United States Pat. 2,411,064 (1946); *Chem. Abstr.* **41**, 1236 (1947).
243. Schicke, H. G., Belgian Pat. 664,091 (1965); *Chem. Abstr.* **65**, 659 (1966).
244. Schmelzer, H. G., Degener, E., and Holtschmidt, A., *Angew. Chem. Intern. Ed.* **5**, 960 (1966).
245. Schroeter, G., *Ber.* **42**, 3360 (1909).
246. Schulte, K. E., Reisch, J., and Stoess, U., *Angew. Chem. Intern. Ed.* **4**, 1081 (1965).

247. Seefelder, M., German Pat. 1,167.819 (1964); *Chem. Abstr.* **61**, 1761 (1964).
248. Seefelder, M., German Pat. 1,204,663 (1965); *Chem. Abstr.* **64**, 2025 (1966).
249. Seus, E. J., *J. Org. Chem.* **30**, 2818 (1965).
250. Shabica, A. C., Howe, E. E., Ziegler, J. B., and Tishler, M., *J. Am. Chem. Soc.* **68**, 1156 (1946).
251. Shevchenko, V. I., Bodnarchuk, N. D., and Kirsanov, A. V., *Zh. Obshch. Khim.* **33**, 1342 (1963); *Chem. Abstr.* **59**, 11239 (1963).
252. Shevchenko, V. I., and Kukhar, V. P., *Zh. Obshch. Khim.* **36**, 1260 (1966); *Chem. Abstr.* **65**, 16845 (1966).
253. Silverstein, R. M., Ryskiewicz, E. E., Willard, C., and Koehler, C., *J. Org. Chem.* **20**, 668 (1955).
254. Simchen, G., *Angew. Chem. Intern. Ed.* **5**, 663 (1966).
255. Smissman, E. E., and Diebold, J. L., *J. Org. Chem.* **30**, 4002 (1965).
256. Smith, G. F., *J. Chem. Soc.* 3842 (1954).
257. Smith, T. D., *J. Chem. Soc.* A, 841 (1966).
258. Sommers, A. H., Michaels, R. J., Jr., and Weston, A. W., *J. Am. Chem. Soc.* **74**, 5546 (1951).
259. Sonn, A., and Müller, E., *Ber.* **52**, 1927 (1919).
260. Spaenig, H., and Schoenleben, W., Belgian Pat. 635,435 (1964); *Chem. Abstr.* **61**, 13334 (1964).
261. Speziale, A. J., and Freeman, R. C., *J. Am. Chem. Soc.* **82**, 903 (1960).
262. Speziale, A. J., and Smith, L. R., *J. Org. Chem.* **27**, 4361 (1962).
263. Speziale, A. J., and Smith, L. R., *J. Org. Chem.* **28**, 3492 (1963).
264. Speziale, A. J., Smith, L. R., and Fedder, J. E., *J. Org. Chem.* **30**, 4303 (1965).
265. Speziale, A. J., and Smith, L. R., United States Pat. 3,230,255 (1966); *Chem. Abstr.* **64**, 8044 (1966).
266. Stamicorn, N. V., British Pat. 1,007,413 (1965); *Chem. Abstr.* **64**, 8158 (1966).
267. Stepanov, B. I., and Migachev, G. I., *Zh. Vses. Khim. Obshchestva im. D.I. Mendeleeva* **11**, 472 (1966).
268. Stephen, H. J., *J. Chem. Soc.* **117**, 1529 (1920).
269. Stephen, H., and Staskun, B., *J. Chem. Soc.* 980 (1956).
270. Stevens, C. L., and French, J. C., *J. Am. Chem. Soc.* **75**, 657 (1953).
271. Stevens, C. L., and French, J. C., *J. Am. Chem. Soc.* **76**, 4398 (1954).
272. Stevens, T. E., *J. Org. Chem.* **32**, 670 (1967).
273. Stewart, R., and Clark, R. H., *J. Am. Chem. Soc.* **69**, 713 (1949).
274. Tafel, J., and Wassmuth, O., *Ber.* **40**, 2841 (1907).
275. Thurman, J. C., *Chem. & Ind.* (*London*) 752 (1964).
276. Treibs, W., Hiebsch, A., and Neupert, H. J., *Naturwiss.* **44**, 352 (1957).
277. Tullock, C. W., Coffman, D. D., and Mutterties, E. L., *J. Am. Chem. Soc.* **86**, 357 (1964).
278. Ugi, I., and Meyr, R., *Angew. Chem.* **70**, 702 (1958).
279. Ugi, I., and Meyr, R., *Ber.* **93**, 239 (1960).
280. Ugi, I., Betz, W., Fetzer, U., and Offermann, K., *Ber.* **94**, 2814 (1961).
281. Ugi, I., Beck, F., and Fetzer, U., *Ber.* **95**, 126 (1962).
282. Ulrich, H., Tilley, J. N., and Sayigh, A. A. R., *J. Org. Chem.* **29**, 2401 (1965).
283. Ulrich, H., Tucker, B., and Sayigh, A. A. R., *J. Org. Chem.* **30**, 2779 (1965).
283a. Ulrich, H., Tucker, B., and Sayigh, A. A. R., *Angew. Chem. Intern. Ed.* **6**, 636 (1967).
283b. Ulrich, H., Tucker, B., and Sayigh, A. A. R., *J. Org. Chem.* **32**, 4052 (1967).
284. Ulrich, H., Tucker, B., and Sayigh, A. A. R., unpublished results.
285. Vilsmeier, A., and Haack, A., *Ber.* **60**, 119 (1927).

286. Vilsmeier, A., *Chemiker Z.* **75**, 133 (1951).
287. Vollmann, H., Becker, H., Corell, M., Streeck, H., and Langbein, G., *Ann.* **531**, 1 (1937).
288. Wallach, O., *Ber.* **9**, 1212 (1876).
289. Wallach, O., and Hoffmann, M., *Ann.* **184**, 75 (1877).
290. Wallach, O., *Ann.* **184**, 1 (1877).
291. Wallach, O., *Ann.* **214**, 193, 226 (1882).
292. Wallach, O., and Kamensky, H., *Ann.* **214**, 234 (1882).
293. Walter, W., and Bode, K. D., *Angew. Chem. Intern. Ed.* **2**, 510 (1962).
294. Warren, W. H., and Wilson, F. E., *Ber.* **68**, 957 (1935).
295. Weidinger, H., and Wellenreuther, G., British Pat. 927,974 (1963); *Chem. Abstr.* **60**, 2987 (1964).
296. Weidinger, H., and Eilingsfeld, H., Belgian Pat. 631,291 (1963); *Chem. Abstr.* **61**, 4277 (1964).
297. Weidinger, H., and Eilingsfeld, H., German Pat. 1,192,180 (1965); *Chem. Abstr.* **63**, 6922 (1965).
298. Weidinger, H., and Kranz, J., *Ber.* **96**, 2070 (1963).
299. Weise, W., Scheurer, G., Aumann, A., and Pommer, H., German Pat. 1,104,949 (1961).
300. Weissenfels, M., Schurig, H., and Huehsam, G., *Z. Chem.* **6**, 471 (1966); *Chem. Abstr.* **66**, 55177f (1967).
301. Weston, A. W., and Michaels, I. R., Jr., *J. Am. Chem. Soc.* **72**, 1422 (1950).
302. Weston, A. W., and Michaels, I. R., Jr., *Org. Syn.* **31**, 108 (1951).
303. Wilson, C. D., United States Pat. 2,437,370 (1945).
304. Wilson, C. D., United States Pat. 2,558,285 (1947).
305. Wolf, W., Degener, E., and Petersen, S., *Angew. Chem.* **72**, 963 (1960).
306. Wolff, P., German Pat. 614,325; 615,130 (1933).
307. Wood, J. H., and Bost, R. W., *J. Am. Chem. Soc.* **59**, 1722 (1937).
308. Yaroslavsky, S., *Tetrahedron Letters* 1503 (1965).
309. Zaoral, M., and Arnold, Z., *Tetrahedron Letters* 9 (1960).
310. Ziegenbein, W., and Franke, W., *Angew. Chem.* **71**, 573 (1959).
311. Ziegenbein, W., and Hornung, K. H., *Ber.* **93**, 2976 (1962).
312. Ziegenbein, W., Lang, W., and Schmitt, J., German Pat. 1,125,902 (1962).
313. Ziegenbein, W., German Pat. 1,213,830 (1966); *Chem. Abstr.* **64**, 19497 (1966).
314. Ziegenbein, W., *Angew. Chem. Intern. Ed.* **4**, 358 (1965).
315. Zieloff, K., Paul, H., and Hilgetag, G., *Ber.* **99**, 357 (1966).

Chapter 4

HALOFORMAMIDINES

I. INTRODUCTION

Haloformamidines are derivatives of formic acid; the name haloformamidines is used by *Chemical Abstracts*, which numbers the center carbon atom 1, and lists the nitrogens as N and N'. However, this class of compounds is more closely related to carbamic acid, as evidenced by its synthesis from urea (carbamic acid amide) and cyanamide (carbamic acid nitrile). In a recent article ([15]) concerned with the synthesis and reactions of substituted chloroformamidine hydrochlorides, the term carbamido chlorides was used. Since chloroformamidine hydrochlorides I are the amido chlorides of carbamide, this generic name is not unreasonable. There is no evidence for the isomeric geminate chloride structure II, and the polar character of I is evidenced by the relatively high melting points of chloroformamidine hydrochlorides, and by their insolubility in nonpolar solvents ([15]).

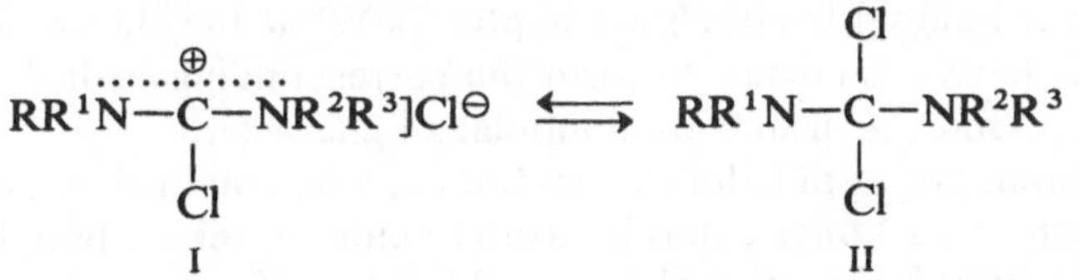

$$
RR^1\overset{\oplus}{N}-\underset{\underset{Cl}{|}}{C}-NR^2R^3]Cl^{\ominus} \rightleftharpoons RR^1N-\underset{\underset{Cl}{|}}{\overset{\overset{Cl}{|}}{C}}-NR^2R^3
$$

I   II

The parent haloformamidines are not known; however, their hydrohalides were isolated as early as 1875, when Drechsel ([14]) treated cyanamide with hydrogen halides. However, its true structure was not recognized until recently ([39,41]).

The first substituted chloroformamidine hydrochloride was synthesized in 1895 by Lengfeld and Stieglitz ([44]), who added hydrogen chloride to diphenylcarbodiimide. Several years later, Steindorff ([55]) synthesized the first free chloroformamidine by reacting 1,1,3-triphenylurea with phosphorus pentachloride.

In general, chloroformamidine hydrochlorides are formed in the reaction of thiourea and urea derivatives with either carbonyl chloride or phosphorus pentachloride. If one of the substituents R in haloformamidine hydrohalides is hydrogen, elimination of hydrogen halide can be achieved readily by either heat or by using a base as the hydrogen halide scavenger. The free haloformamidines III are thus obtained, usually in good yield.

$$\overset{\oplus}{\text{RNH}}\text{—}\underset{\underset{\text{X}}{|}}{\text{C}}\text{—NR}^1\text{R}^2]\text{X}^\ominus \xrightarrow{-\text{HX}} \text{RN}=\underset{\underset{\text{X}}{|}}{\text{C}}\text{—NR}^1\text{R}^2$$

III

When two of the substituents R in II are hydrogen, two moles of hydrogen halide can be eliminated, and the resulting product is the corresponding carbodiimide (IV). The reversal of this reaction, i.e., the addition of hydrogen halide to carbodiimides, is a useful method of synthesis of haloformamidine hydrohalides.

$$\overset{\oplus}{\text{RNH}}\text{—}\underset{\underset{\text{X}}{|}}{\text{C}}\text{—NHR}^1]\text{X}^- \underset{+2\,\text{HX}}{\overset{-2\,\text{HX}}{\rightleftarrows}} \underset{\text{IV}}{\text{RN}=\text{C}=\text{NR}^1}$$

The dehydrohalogenation to the corresponding haloformamidine, carbodiimide, or cyanamide is prevented in tetraalkylhaloformamidinium halides (I, R = alkyl), which in this respect resemble quaternary ammonium halides.

Depending upon the basicity of the nitrogens, either haloformamidines or their hydrohalides are formed in the general synthetic procedures. The haloformamidines are in this, and in many other respects, quite closely related to the imidoyl halides (see Chapter 3). While the chloro and bromo compounds have a tendency to form the corresponding hydrohalides, the fluoro compounds, as usual, are nonpolar in character.

A different group of haloformamidines can be obtained by adding acid halides, such as carbonyl chloride, oxalyl chloride, thiocarbonyl chloride, phosphorus oxychloride, phosphorus trichloride, sulfur chlorides, and acetyl chloride to carbodiimides.

In view of their facile reaction with nucleophiles, haloformamidines are exceedingly useful in organic synthesis, and a wide variety of carbonic acid derivatives have been synthesized from haloformamidines, especially in recent years.

II. METHODS OF SYNTHESIS

A. Halogenation of Thioureas

The reaction of substituted thioureas with carbonyl chloride is perhaps the most useful synthetic method for chloroformamidines. Thus, alkyl

[15,16,52,59,60,62,63,66], aryl ([15,16,66]), and arenesulfonylthioureas ([4,61,65,66]) have been converted to chloroformamidines, or its hydrochlorides. The reaction proceeds well with di-, tri- and tetrasubstituted thioureas, and in one instance a monosubstituted thiourea, t-butylthiourea, has been converted to the corresponding chloroformamidine hydrochloride ([15]).

$$RR^1N-\underset{\underset{S}{\|}}{C}-NR^2R^3 + COCl_2 \longrightarrow RR^1\overset{\oplus}{N}-\underset{\underset{Cl}{|}}{C}-NR^2R^3]Cl^{\ominus} + COS$$

In the case of 1-arenesulfonyl-3-alkylthioureas (V) elimination of hydrogen chloride occurs simultaneously with the generation of carbonyl sulfide and the free chloroformamidines VI are obtained at room temperature ([4,61,65, 66]). 1-Chloro-N-arenesulfonyl-N'-alkyl formamidines (VI) in refluxing chlorobenzene (130°C) eliminate the second mole of hydrogen chloride, and the sulfonylcarbodiimides VII are obtained ([61,65,66]). Sulfonylcarbodiimides are the precursors of hypoglycemic drugs, because addition of water produces sulfonylureas VIII, which are oral antidiabetics.

$$RSO_2NH-\underset{\underset{\underset{V}{S}}{\|}}{C}-NHR' + COCl_2 \longrightarrow$$

$$RSO_2N=\underset{\underset{\underset{VI}{Cl}}{|}}{C}-NHR' + HCl + COS$$

$$\downarrow 130°C$$

$$RSO_2NH-\underset{\underset{\underset{VIII}{O}}{\|}}{C}-NHR' \xleftarrow{H_2O} \underset{VII}{RSO_2N=C=NR'} + HCl$$

N-p-Toluenesulfonyl-N'-n-butylchloroformamidine ([66]). To a solution of 20 g (0.07 mole) of 1-*p*-toluenesulfonyl-3-*n*-butylthiourea in 150 ml of carbon tetrachloride dropwise and with stirring a solution of 7 g (0.07 mole) of phosgene in carbon tetrachloride is added. After stirring for 90 min at room temperature the precipitated N-*p*-toluenesulfonyl-N'-*n*-butylchloroformamidine (17.9 g; 88.6%), m.p. 90–95°C is collected by filtration.

In contrast, the reaction of 1-arenesulfonyl-3-arylthioureas (V, R = R' = aryl), with carbonyl chloride affords the four-membered ring 1,3-thiazetidine-2-ones (IX) almost exclusively ([65]). The compounds (IX) are also formed as minor by-products in the reaction of aliphatic and aromatic 1,3-disubstituted thioureas with carbonyl chloride ([15,70]).

$$\text{RSO}_2\text{NH}-\underset{\underset{\text{S}}{\|}}{\text{C}}-\text{NHR'} + \text{COCl}_2 \longrightarrow$$

(structure IX)

IX

The reaction of thioureas with carbonyl chloride can be conducted at room temperature in an inert diluent, preferentially in a solvent, in which the starting thiourea is soluble. Suitable solvents are benzene, chlorobenzene, ethylene dichloride, dioxane, etc. If by-products are formed, they are easily separated with diethyl ether, in which the salt-like chloroformamidine hydrochlorides are completely insoluble. Sometimes it is possible to isolate the initial reaction product X, especially if a nonpolar solvent is being used, which retards the elimination of carbonyl sulfide ([15]).

$$\text{RNH}-\underset{\underset{\text{S}}{\|}}{\text{C}}-\text{NHR} + \text{COCl}_2 \longrightarrow \text{R\overset{..}{N}H}\overset{\oplus}{\overset{..}{\cdots}\overset{..}{\text{C}}\cdots}\overset{..}{\text{N}}\text{HR}]\text{Cl}^{\ominus}$$
$$\underset{\text{SCOCl}}{|}$$

X

$$\Big\downarrow -\cos$$

$$\text{R\overset{..}{N}H}\overset{\oplus}{\overset{..}{\cdots}\overset{..}{\text{C}}\cdots}\overset{..}{\text{N}}\text{HR}]\text{Cl}^{\ominus}$$
$$\underset{\text{Cl}}{|}$$

The use of an excess of carbonyl chloride in the reaction with di-substituted thioureas has to be avoided, because the chloroformamidine hydrochlorides eliminate hydrogen chloride and the generated carbodiimides add carbonyl chloride to afford chloroformamidine-N-carbonyl chlorides XI ([59,66]).

$$\text{R\overset{..}{N}H}\overset{\oplus}{\overset{..}{\cdots}\overset{..}{\text{C}}\cdots}\overset{..}{\text{N}}\text{HR}]\text{Cl}^{\ominus} + \text{COCl}_2 \longrightarrow \underset{\underset{\text{Cl}}{|}}{\text{RN}}-\underset{}{\text{C}}=\text{NR} + 2\,\text{HCl}$$
$$\overset{\overset{\text{COCl}}{|}}{}$$

XI

The reaction of thioureas with carbonyl chloride can be extended to cyclic derivatives. Thus, chloroimidazolidine hydrochloride (XII) is obtained in the reaction of ethylene thiourea with carbonyl chloride (see Chapter 8).

$$\underset{\underset{\text{S}}{\|}}{\text{HN}}\diagdown\diagup\text{NH} + \text{COCl}_2 \longrightarrow \Big[\overset{\oplus}{\text{HN}}\diagdown\underset{\underset{\text{Cl}}{|}}{}\diagup\text{NH}\Big]\text{Cl}^{\ominus}$$

XII

The use of carbonyl chloride in the conversion of thioureas to chloroformamidines offers the advantage of gaseous by-product. Phosphorus pentachloride can be used equally well, but complete separation from the liquid trichlorophosphorus sulfide is sometimes more difficult to accomplish. In one instance, chlorine has been used to convert a pseudothiourea derivative to the corresponding chloroformamidine hydrochloride. Thus, addition of chlorine to a solution of the pseudothiourea derivative XIII in carbon tetrachloride afforded a quantitative yield of trimethylchloroformamidine hydrochloride (XIV) ([42]).

$$(CH_3)_2N-\underset{\underset{\text{XIII}}{SCH_3}}{C}=NCH_3 + Cl_2 \longrightarrow (CH_3)_2\overset{\oplus}{N}-\underset{\underset{\text{XIV}}{Cl}}{C}-NHCH_3]Cl^{\ominus}$$

Chloroformamidines are not produced in the reaction of thioureas with oxalyl chloride, because ring closure occurs and 2-imino-1,3-thiazolidin-4,5-diones (XV) are obtained in high yield ([64,66]). The heterocyclic derivatives XV isomerize readily to form thioparabanic acids XVI ([57,64,66]).

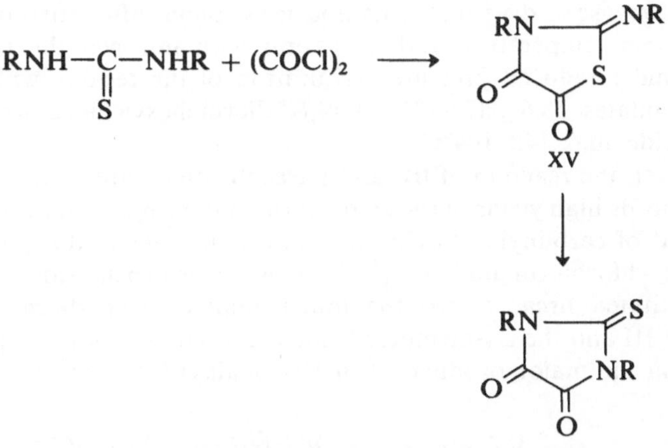

$$RNH-\underset{\underset{S}{\|}}{C}-NHR + (COCl)_2 \longrightarrow$$

XV

XVI

B. Halogenation of Ureas

The reaction of substituted ureas with carbonyl chloride provides another useful method of synthesis of chloroformamidines ([15,16,60,62,63,66]). However, oxygen is not as nucleophilic as sulfur and consequently attack on nitrogen often occurs. In the case of 1,3-dialkylureas good yields of chloroformamidine hydrochlorides are obtained only when the substituents are secondary and tertiary alkyl groups ([60,62,63,66]). The N-attack products, N,N'-dialkylallophanoyl chlorides XVII, are the major reaction products, when one of the substituents is an *n*-alkyl group ([62,66]).

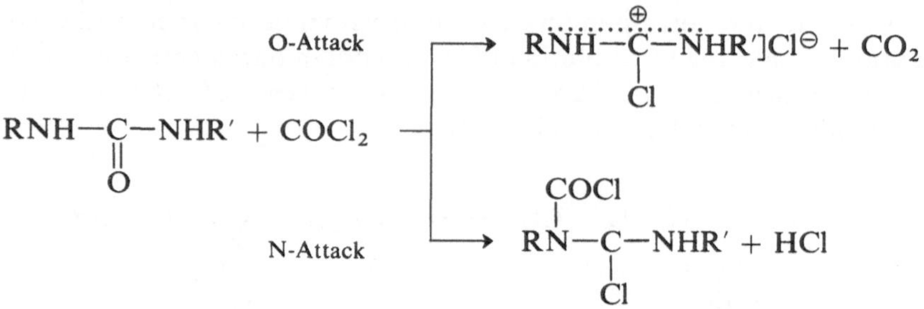

Similarly, N,N-dimethylurea reacts with carbonyl chloride *via* N-attack ([6]).

N,N'-Dicyclohexylchloroformamidine hydrochloride ([62]). To a solution of 60.4 g (0.27 mole) of 1,3-dicyclohexylurea in 400 ml of ethylene dichloride a solution of 27.4 g (0.28 mole) of phosgene in 200 ml of ethylene dichloride is added dropwise and with stirring and ice-cooling. After stirring for one hour at room temperature and purgation with nitrogen the solvent is removed under reduced pressure. Trituration of the residue with diethyl ether precipitates 58.6 g (77.8%) of N,N'-dicyclohexylchloroformamidine hydrochloride, m.p. 143–144°C.

However, the reaction of tri- and tetrasubstituted ureas with carbonyl chloride affords high yields of the chloroformamidine hydrochlorides ([15,16]).

Instead of carbonyl chloride, phosphorus pentachloride can be used to prepare chloroformamidines ([8,20,55]) or its hydrochlorides ([60,63,66]). 1,3-Disubstituted ureas afford the four-membered ring diazaphosphetidinones XVIII and the six-membered ring compounds XIX as by-products, XVIII being the major product when R = *n*-alkyl ([40,60,63,66]).

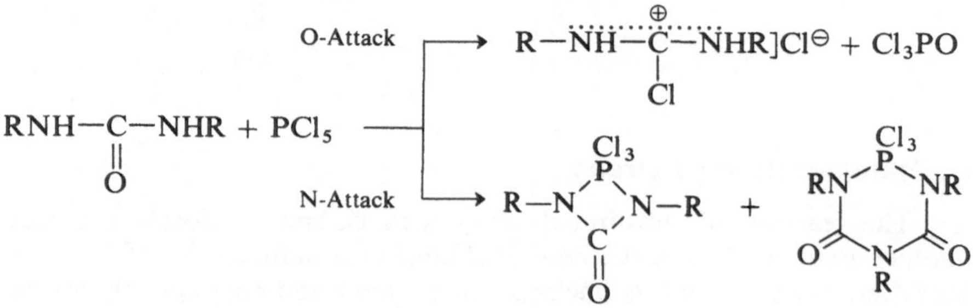

The reaction of thionyl chloride with 1,3-disubstituted ureas affords either N-sulfinylamines and isocyanates (R = n-alkyl), or chloroformamidine hydrochlorides (R = sec-alkyl, t-alkyl) [66]. However, the reaction of oxalyl chloride with ureas does not produce chloroformamidines, parabanic acids being the sole reaction product [51,65,66].

C. Addition of Halides to Cyanamides and Carbodiimides

The addition of hydrogen halides to cyanamide which in its tautomeric form can be written as carbodiimide (XX) was reported by Drechsel [14] in 1875. The author obtained both the dihydrochloride XXI (X = Cl), and the dihydrobromide XXI (X = Br) of cyanamide.

$$H_2N-CN \longleftrightarrow \underset{XX}{HN=C=NH} + 2\,HX \longrightarrow H_2\overset{\oplus}{N}-\underset{X}{\overset{|}{C}}-NH_2]X^{\ominus}$$
$$XXI$$

The dihydrohalides of cyanamide have been obtained by other investigators [24,46,49,56], but its true structure was recognized only recently [39,41]. Upon reaction of chloroformamidine hydrochloride with nitric acid, the corresponding chloroformamidine nitrate, an explosive, was obtained [39,50].

Chloroformamidine hydrochlorides were also obtained by addition of hydrogen chloride to dicyano imide [1,2], and to mono- [39,48,71] and disubstituted [39] cyanamides.

By addition of hydrogen chloride to diphenylcarbodiimide, Lengfeld and Stieglitz [44] synthesized 1-chloro-N,N'-diphenylformamidine hydrochloride (XXII) in 1895. The reaction is not limited to diarylcarbodiimides, and from dialkylcarbodiimides the corresponding chloroformamidine hydrochlorides can be obtained in quantitative yield [59,62].

$$C_6H_5N=C=NC_6H_5 + 2\,HCl \longrightarrow C_6H_5\overset{\oplus}{N}H-\underset{Cl}{\overset{|}{C}}-NHC_6H_5]Cl^{\ominus}$$
$$XXII$$

N,N'-Di-n-butylchloroformamidine hydrochloride [62]. To a solution of 15.9 g (0.1 mole) of di-n-butylcarbodiimide in 100 ml of chloroform gaseous hydrogen chloride is added until the exothermic reaction ceases. Excess hydrogen chloride is removed with nitrogen and evaporation of the solvent affords N,N'-di-n-butylchloroformamidine hydrochloride as a colorless oil.

If the addition of hydrogen chloride to cyanamide is conducted in alcohol, isourea hydrochlorides are obtained [56].

The reaction of dialkylcyanamides with carboxylic acid chlorides at 150°C affords 1:1 adducts, which are formulated as chloroformamidines XXIII because they react with ammonia to give the corresponding guanidines (XXIV) (^{5}).

$$R_2N\text{—}CN + C_6H_5COCl \longrightarrow R_2\overset{\cdot}{N}\text{—}\underset{\underset{Cl}{|}}{C}\text{=}N\text{—}COC_6H_5$$

XXIII

$$\downarrow \text{NH}_3$$

$$R_2N\text{—}\underset{\underset{NH_2}{|}}{C}\text{=}N\text{—}COC_6H_5$$

XXIV

The yields of XXIII are not high, because diazapyrylium salts are formed as the main products; however, from p-nitrobenzoyl chloride and dimethylcyanamide at room temperature a 52 % yield of the corresponding 1:1 adduct is obtained (^{5}).

In the reaction of dialkyl cyanamides with carbonyl chloride 2:1 adducts XXV are initially formed, as evidenced by the isolated oligomeric chloroformamidinium chloride (XXVI), which is obtained in high yield (^{5}).

$$2\,R_2N\text{—}CN + COCl_2 \longrightarrow [R_2N\text{—}\underset{\underset{Cl}{|}}{C}\text{=}N\text{—}CO\text{—}N\text{=}\underset{\underset{Cl}{|}}{C}\text{—}NR_2]$$

XXV

$$\downarrow \text{COCl}_2$$

$$R_2N\text{—}\underset{\underset{Cl}{|}}{C}\text{=}N\text{—}\underset{\underset{Cl}{|}}{C}\text{=}N\text{—}\underset{\underset{Cl}{|}}{C}\text{=}\overset{\oplus}{N}R_2]Cl^{\ominus}$$

XXVI

Carboxylic acid chlorides add in a like manner to carbodiimides to form the corresponding 1:1 adducts, which have a chloroformamidine structure. For example, acetyl chloride was added to carbodiimides, and the structure of the labile 1:1 adducts (XXVII) was established by subsequent reactions with anthranilamide, leading to 4-quinazolones XXVIII (^{25,27}).

$$RN\text{=}C\text{=}NR + CH_3COCl \longrightarrow RN\text{=}\underset{\underset{Cl}{|}}{C}\text{—}N(COCH_3)R$$

XXVII

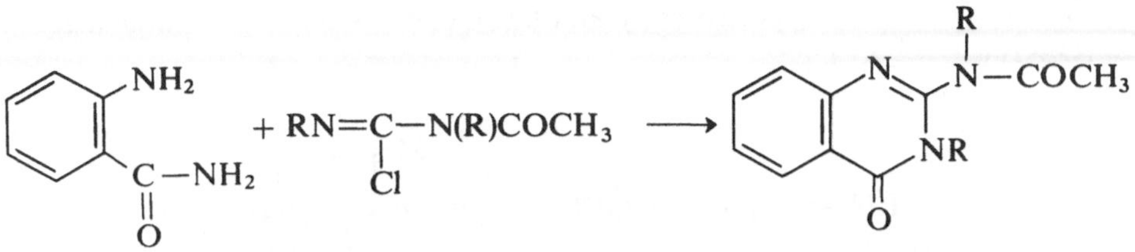

XXVIII

In contrast, benzoyl chloride and ethyl chloroformate react considerably more slowly than the aliphatic carboxylic acid chlorides ([27]).

N,N'-Diethyl-N-acetylchloroformamidine ([27]). To 4.9 g (0.05 mole) of diethylcarbodiimide a solution of 3.93 g (0.05 mole) of acetyl chloride in 100 ml of methylene chloride is added at 0°C. After standing for one hour at room temperature the solvent is evaporated and vacuum distillation of the residue yields 8.1 g (92%) of N,N'-diethyl-N-acetylchloroformamidine, b.p. 33–34°C/0.01 mm.

The 1:1 adducts of carbodiimides and carbonyl chloride ([21,27,59,66]) and thiocarbonyl chloride ([21]) are more stable. The structure of the carbonyl chloride adducts XXIX was confirmed by infrared spectroscopy ($v_{C=O}$ 1745 cm^{-1}; $v_{C=N}$ 1670 cm^{-1}) and reactions with hydrazine, which produces 1,2,4-triazoles XXX ([59]).

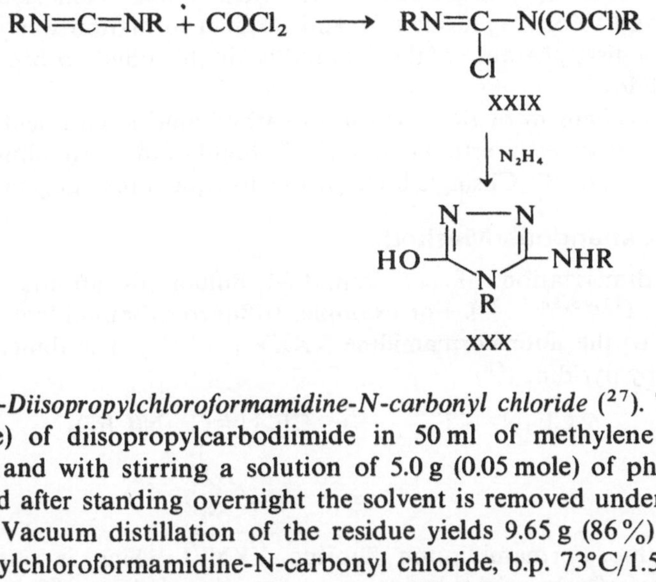

XXIX

XXX

N,N'-Diisopropylchloroformamidine-N-carbonyl chloride ([27]). To 6.31 g (0.05 mole) of diisopropylcarbodiimide in 50 ml of methylene chloride dropwise and with stirring a solution of 5.0 g (0.05 mole) of phosgene is added and after standing overnight the solvent is removed under reduced pressure. Vacuum distillation of the residue yields 9.65 g (86%) of N,N'-diisopropylchloroformamidine-N-carbonyl chloride, b.p. 73°C/1.5 mm.

Likewise, oxalyl chloride adds to carbodiimides to yield 1:1 adducts, for which the geminate dichloride structure XXXI has been postulated on

the basis of infrared spectroscopy and facile conversion to parabanic acids
XXXII ([54,64,65,66]).

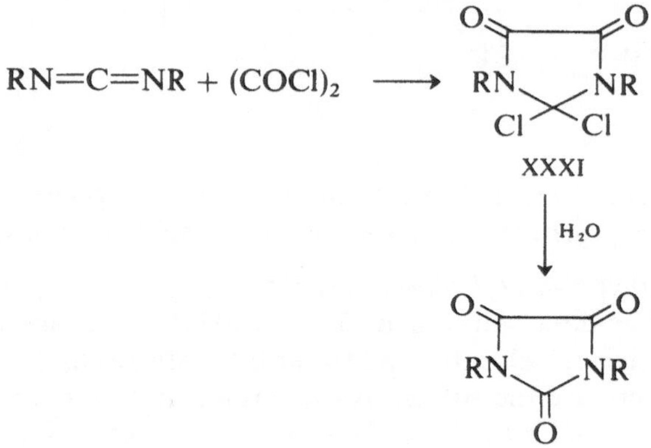

$$RN{=}C{=}NR + (COCl)_2 \longrightarrow$$

XXXI

$$\downarrow H_2O$$

XXXII

The reaction of malonyl chloride with carbodiimides proceeds in a
different manner ([31]).

Other halides which have been added to carbodiimides include phosphorus trichloride, phosphorus oxychloride, sulfur dichloride, sulfuryl
chloride, and thionyl chloride ([21]), and product formation was judged on
the basis of disappearance of the cumulative double bond absorption of the
carbodiimides.

The mechanism of the reaction of carbodiimides with acid chlorides
most likely involves insertion of one C=N bond of the carbodiimide into a
C—Cl, P—Cl, or S—Cl single bond, with subsequent rearrangement ([67]).

D. Miscellaneous Methods

The dimerization of carbonimidoyl difluorides affords fluoroformamidines ([17,45,58,72,73]). For example, trifluorocarbonimidoyl difluoride
dimerizes to the fluoroformamidine XXXV ([17,45,72]). The dimerization is
catalyzed by pyridine ([58]).

$$2\,CF_3N{=}CF_2 \longrightarrow CF_3N{=}\underset{\underset{F}{|}}{C}{-}N(CF_3)_2$$

XXXV

The fluoroformamidinium fluoride XXXVI which is obtained by
fluorination of cyanuric chloride or aminoiminomethanesulfinic acid has
the nonpolar bis(difluoroamino)difluoromethane structure XXXVII ([34]).

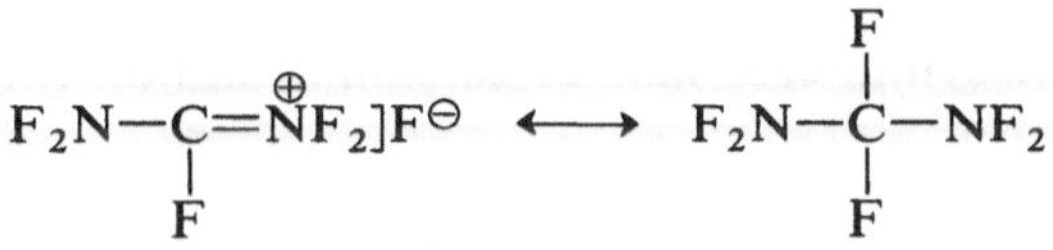

$$F_2N-\overset{|}{\underset{F}{C}}=\overset{\oplus}{N}F_2]F^{\ominus} \longleftrightarrow F_2N-\overset{F}{\underset{F}{\overset{|}{C}}}-NF_2$$

XXXVI   XXXVII

In the reaction of the cyanamide derivative XXXVIII with hydrogen chloride, the chloroformamidine hydrochloride derivative XXI (see page 119) is obtained ([30]).

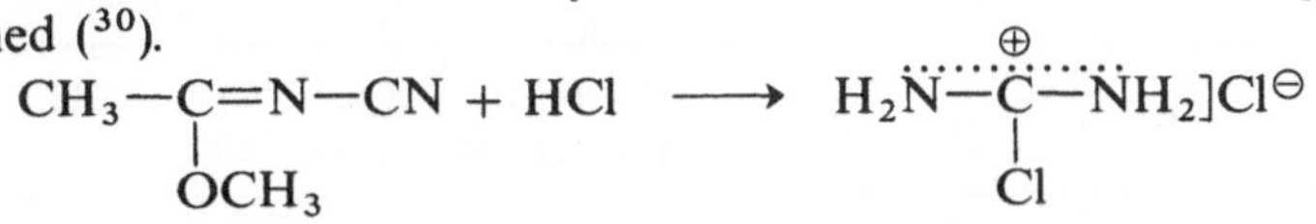

$$CH_3-\overset{|}{\underset{OCH_3}{C}}=N-CN + HCl \longrightarrow H_2\overset{\cdots}{N}\overset{\oplus}{-}\overset{|}{\underset{Cl}{C}}\overset{\cdots}{-}\overset{\cdots}{N}H_2]Cl^{\ominus}$$

XXXVIII   XXI

The reaction of the uretone imine XXXIX (R = ethyl) with carbonyl chloride proceeds *via* ring opening with formation of the chloroformamidine XL ([19]).

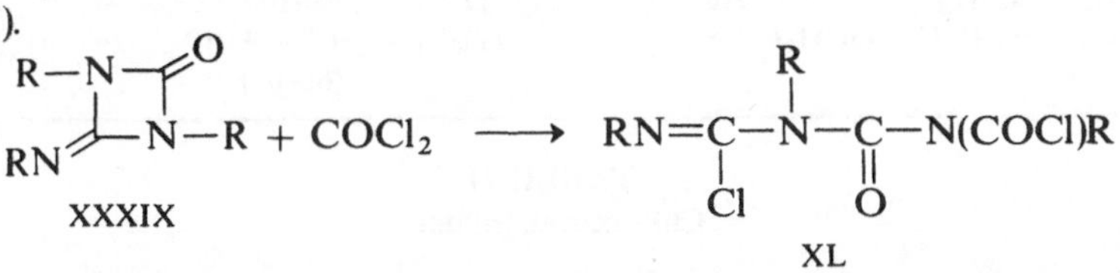

$$\underset{RN}{\overset{R-N}{\diagdown}}\overset{O}{\underset{N-R}{\diagup}} + COCl_2 \longrightarrow RN=\overset{R}{\underset{Cl}{\overset{|}{C}}-N-\overset{|}{\underset{O}{\overset{||}{C}}}-N(COCl)R$$

XXXIX

XL

The intermediacy of chloroformamidines in the reaction of chloramine T (sodium *p*-toluenesulfonchloramide) with isocyanides has been postulated ([3,4]).

The chloroformamidine XLI, which is obtained from the carbonimidoyl dichloride XLII and dimethylamine, yields on high-temperature chlorination the unusual chloroformamidine XLIII ([29]). See also Chapter 2.

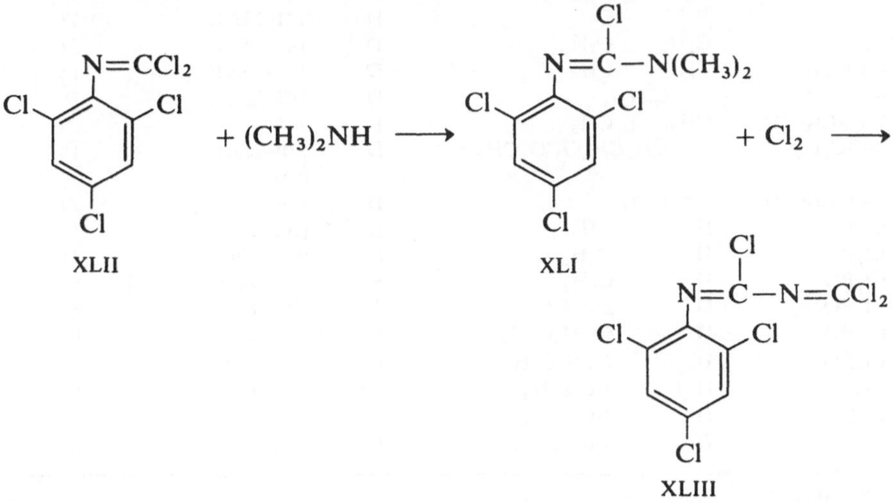

XLII   + (CH$_3$)$_2$NH $\longrightarrow$ XLI   + Cl$_2$ $\longrightarrow$

XLIII

The hitherto-synthesized haloformamidines and their hydrohalides are listed in Tables I–V.

TABLE I
Fluoroformamidines

$$R-N=C-NR^1R^2$$
$$|$$
$$F$$

R	R^1	R^2	Method of preparation	B.p., °C/mm (M.p., °C)	Yield, %	Reference
CF_3	CF_3	CF_3	D	38	—	72, 73
SF_5	CF_3	SF_5	D	88–92	—	58
C_6H_5	CH_3	CH_3	D	71–72/0.1	—	36
$4\text{-}ClC_6H_4$	CH_3	CH_3	D	91–93/0.25	—	36
$4\text{-}ClC_6H_4$	C_2H_5	C_2H_5	D	98–100/0.1	—	36
$4\text{-}ClC_6H_4$	$-CH_2CH_2OCH_2CH_2-$		D	193–194/19 (66–69)	—	36

TABLE II
Chloroformamidines

$$RN=C-NR^1R^2$$
$$|$$
$$Cl$$

R	R^1	R^2	Method of preparation	B.p., °C/mm (M.p., °C)	Yield, %	Reference
C_6H_5	CH_3	CH_3	B	96–97/1.0	—	20
			D	131–133/12	—	18
C_6H_5	C_2H_5	C_2H_5	D	145–150/10	—	37
$4\text{-}ClC_6H_4$	CH_3	CH_3	D	157–160/10	—	18
$4\text{-}ClC_6H_4$	$-(CH_2)_4-$		D	202–207/15	—	18
$2,4,6\text{-}Cl_3C_6H_2$	CH_3	CH_3	D	—*	—	29
$4\text{-}ClC_6H_4$	$-CH_2CH_2OCH_2CH_2-$		D	210–212/15 (85)	—	18
$2,4,6\text{-}Cl_3C_6H_2$	$=CCl_2$		D	136–138/0.1	60	29
Cl_2PO	H	CH_3	B	(103–105)	—	11
Cl_2PO	H	C_2H_5	B	(98–100)	—	11
Cl_2PO	H	C_6H_5	B	(138–141)	—†	8
Cl_2PO	H	$2\text{-}CH_3C_6H_4$	B	(105–107)	—	8
Cl_2PO	H	$4\text{-}CH_3C_6H_4$	B	(152–154)	—	8
Cl_2PO	H	$4\text{-}CF_3C_6H_4$	B	(138–140)	—	8
Cl_2PO	H	$3\text{-}ClC_6H_4$	B	(135–136)	—	8
Cl_2PO	H	$4\text{-}ClC_6H_4$	B	(153–154)	—	8
Cl_2PO	H	$4\text{-}BrC_6H_4$	B	(147–149)	—	8

* Not reported.
† Reference 8 reports yields of 67–98 %.

Table II—*continued*

R	R^1	R^2	Method of preparation	B.p., °C/mm (M.p., °C)	Yield, %	Reference
Cl$_2$PO	H	4-IC$_6$H$_4$	B	(139–141)	—	8
Cl$_2$PO	H	2-O$_2$NC$_6$H$_4$	B	(98–100)	—	8
Cl$_2$PO	H	3-O$_2$NC$_6$H$_4$	B	(147–149)	—	8
Cl$_2$PO	H	4-O$_2$NC$_6$H$_4$	B	(164–165)	—	8
Cl$_2$PO	H	4,3(Me)O$_2$NC$_6$H$_3$	B	(136–137)	—	8
(CH$_3$O)$_2$PO	H	CH$_3$	B	100–103/0.6	—	11
(CH$_3$O)$_2$PO	H	C$_2$H$_5$	B	110–112/0.5	—	11
(C$_2$H$_5$O)$_2$PO	H	C$_2$H$_5$	B	120–122/0.6	—	11
(C$_6$H$_5$O)$_2$PO	H	CH$_3$	B	(86–87)	—	11
(C$_6$H$_5$O)$_2$PO	H	C$_2$H$_5$	B	(98–100)	—	11
(C$_6$H$_5$O)$_2$PO	H	C$_6$H$_5$	B	(96–98)	55	12
(C$_6$H$_5$O)$_2$PO	H	4-ClC$_6$H$_4$	B	(126–127)	67	12
(C$_6$H$_5$O)$_2$PO	H	4-BrC$_6$H$_4$	B	(129–132)	70	12
(C$_6$H$_5$O)$_2$PO	H	2,4-Cl$_2$C$_6$H$_3$	B	(95–98)	53	12
(C$_6$H$_5$O)$_2$PO	H	β-C$_{10}$H$_7$	B	(107–109)	55	12
(4-ClC$_6$H$_4$O)$_2$PO	H	CH$_3$	B	(98–106)	—	11
(4-ClC$_6$H$_4$O)$_2$PO	H	C$_2$H$_5$	B	(88–90)	—	11
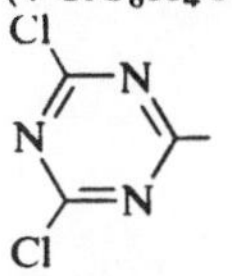 (dichlorotriazinyl)	CH$_3$	CH$_3$	D	141–144	—	32
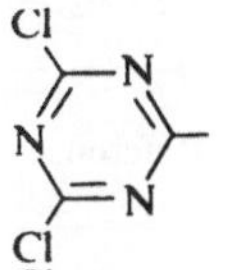 (dichlorotriazinyl)	C$_2$H$_5$	C$_2$H$_5$	D	(78–80)	94	53, 23
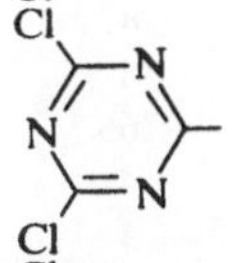 (dichlorotriazinyl)	i-C$_3$H$_7$	i-C$_3$H$_7$	D	(94–98)	—	23
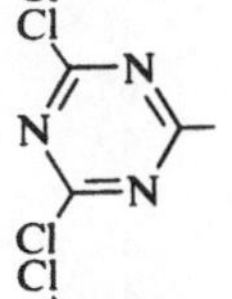 (dichlorotriazinyl)	—CH$_2$CH$_2$OCH$_2$CH$_2$—		D	(107–110)	—	32
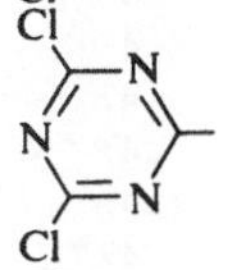 (dichlorotriazinyl)	CH$_3$	C$_6$H$_5$	D	—	—	23
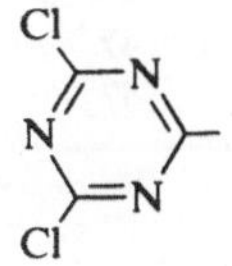 (chlorotriazinyl)	C$_6$H$_5$	C$_6$H$_5$	D	153.5–154.5	—	32

Table II—*continued*

R	R^1	R^2	Method of preparation	B.p., °C/mm (M.p., °C)	Yield, %	Reference
CH_3	COCl	t-C_4H_9	C	74/3.0	85	29
C_2H_5	COCl	C_2H_5	C	52/0.5	84	21
i-C_3H_7	COCl	i-C_3H_7	C	73/1.5	86	29
n-C_4H_9	COCl	n-C_4H_9	C	86/9.5	98	59
			C	84–86/0.1	—	21
C_6H_{11}	COCl	C_6H_{11}	C	140–142/0.8	100	59
C_6H_5	COCl	C_6H_5	C	Oil	—	21
4-$CH_3C_6H_4$	COCl	4-$CH_3C_6H_4$	C	Oil	98.8	59
4-ClC_6H_4	COCl	4-ClC_6H_4	C	Oil	—	21
3,4-$Cl_2C_6H_3$	COCl	3,4-$Cl_2C_6H_3$	C	Oil	—	21
2-$O_2NC_6H_4$	COCl	2-$O_2NC_6H_4$	C	Oil	—	21
n-C_4H_9	CSCl$_2$	n-C_4H_9	C	80–84/0.1	—	21
C_6H_5CO	CH_3	CH_3	C	(147–150)	52	5
C_2H_5	$COCH_3$	C_2H_5	C	33–34/0.–1	92	27
C_3H_7	$COCH_3$	C_3H_7	C	48.5–49.5/0.01	95	27

TABLE III
Arenesulfonylchloroformamidines

$$RSO_2N{=}C{-}NR^1R^2$$
$$\mid$$
$$Cl$$

R	R^1	R^2	Method of preparation	M.p., °C	Yield, %	Reference
CH_3	CH_3	C_6H_5	D	121–122	82	47
C_6H_{11}	H	i-C_4H_9	A	68–70	—	4
4-$CH_3C_6H_4$	H	n-C_4H_9	A	98–99	88.6	65
			A	100–101	—	4
4-$CH_3C_6H_4$	H	i-C_4H_9	A	90–92	—	4
4-$CH_3C_6H_4$	H	C_6H_{11}	A	128–130	—	4
4-ClC_6H_4	H	n-C_3H_7	A	100–108	80	65
			A	113–115	—	4
4-ClC_6H_4	H	$C_6H_5CH_2$	A	114–116	—	4
4-$CH_3OC_6H_4$	H	$CH_3O(CH_2)_3$	A	88–90	—	4
4-$CH_3CONHC_6H_4$	H	i-C_4H_9	A	133–135	—	4
C_6H_5	CH_3	C_6H_5	D	111–112	94	47
C_6H_5	C_2H_5	C_6H_5	D	88–89	88	47
C_6H_5	C_6H_5	C_6H_5	D	193–195	80	47
4-$CH_3C_6H_4$	CH_3	C_6H_5	D	137–138	65	47
4-$CH_3C_6H_4$	C_2H_5	C_6H_5	D	132–133	93	47
4-$CH_3C_6H_4$	C_6H_5	C_6H_5	D	210–211	75	47

TABLE IV
2,2-Dichloroimidazolidine-4,5-Diones

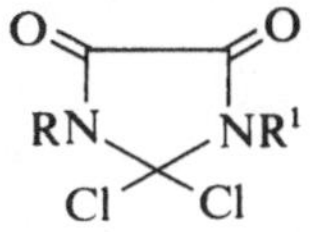

R	R^1	M.p., °C	Yield, %	Reference
i-C$_3$H$_7$	i-C$_3$H$_7$	—	—	54
2-CH$_3$C$_6$H$_4$	2-CH$_3$C$_6$H$_4$	167–168	86	64
CH$_3$SO$_2$	n-C$_3$H$_7$	128	83.3	65
C$_6$H$_5$SO$_2$	C$_2$H$_5$	131	77.3	65
4-CH$_3$C$_6$H$_4$SO$_2$	n-C$_4$H$_9$	125	74.2	65

TABLE V
Chloroformamidine Hydrochlorides

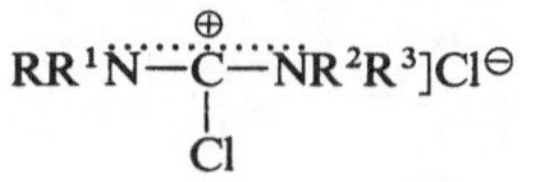

$$RR^1\overset{\oplus}{N}\!\!=\!\!\underset{\underset{\text{Cl}}{|}}{C}\!\!=\!\!NR^2R^3]Cl^{\ominus}$$

R	R^1	R^2	R^3	Method of preparation	M.p., °C	Yield, %	Reference
H	H	H	H	C	177–179	—	14, 24, 30, 39, 41, 46, 49
CH$_3$	H	H	H	C	—	—	39
CH$_3$	CH$_3$	H	H	C	—	—	39
t-C$_4$H$_9$	H	H	H	C	110–113	—	15
(C$_6$H$_5$)$_2$CH	H	H	H	C	180–181	92	71
CH$_3$	H	H	CH$_3$	A	162–164	86	16
				A	138–143	—	15
				B	138–140	28	62
i-C$_3$H$_7$	H	H	i-C$_3$H$_7$	A	100–105	88	16
				B	97–100	75.1	62
n-C$_4$H$_9$	H	H	n-C$_4$H$_9$	C	Oil	97.7	59
n-C$_{18}$H$_{37}$	H	H	n-C$_{18}$H$_{37}$	B	104	28.1	62
C$_6$H$_{11}$	H	H	C$_6$H$_{11}$	A	139–141	95	16
				B	143–144	77.8	62
C$_6$H$_5$	H	H	i-C$_3$H$_7$	A	158–160	99	16
C$_6$H$_5$	H	H	C$_6$H$_{11}$	B	148–150	67	16
C$_6$H$_5$	H	H	C$_6$H$_5$	A	126–129	81	16

Table V—*continued*

R	R¹	R²	R³	Method of preparation	M.p., °C	Yield, %	Reference
3-ClC₆H₄	H	H	3-ClC₆H₄	A	120–122	83	16
4-CH₃OC₆H₄	H	H	4-CH₃OC₆H₄	A	116–118	93	16
β-C₁₀H₇	H	H	β-C₁₀H₇	A	>300	69	16
C₆H₁₁	H	—(CH₂)₄—		B	146–150	62	16
C₆H₅	H	CH₃	CH₃	B	155–158	25	16
4-ClC₆H₄	H	CH₃	CH₃	B	191–197	30	16
C₆H₅	H	—(CH₂)₄—		A	166–171	96	16
CH₃	CH₃	H	CH₃	A	69–71	100	42
CH₃	CH₃	CH₃	CH₃	B	110–112	96	16
CH₃	CH₃	—(CH₂)₅—		B	144–145	86	16
—(CH₂)₅—		—(CH₂)₅—		B	75–80	89	16
C₆H₅	CH₃	CH₃	CH₃	B	40–50	77	16
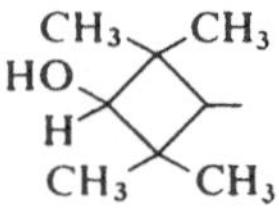 (structure: CH₃ CH₃ HO— H —CH₃ CH₃)	H	H	H	C	162–164	67	48

III. PHYSICOCHEMICAL PROPERTIES

The chloroformamidines are basic compounds, but their base strength has not been reported. However, it is indicated that the alkylchloroformamidines are considerably more basic than the aryl derivatives because moderate heating causes elimination of hydrogen chloride from disubstituted aryl-chloroformamidine hydrochlorides ([66]). N-Arenesulfonyl-N′-alkylchloroformamidines are almost neutral compounds because no hydrochloride formation was observed ([61,65,66]).

The free haloformamidines are either low-melting solids, or colorless liquids, which can be distilled under vacuum without decomposition. The halo group is readily displaced by a variety of nucleophiles, as shown in Section IV.

The hydrohalides of chloroformamidine and alkylchloroformamidines are hydroscopic solids. The infrared spectra of haloformamidines show the characteristic C=N absorption at approximately 1546–1725 cm⁻¹ (see Table VI).

NMR spectroscopy is useful for the determination of the position of the hydrogen in 1-chloro-N-arenesulfonyl-N′-alkylformamidines (XLIV),

TABLE VI
Asymmetrical Stretching Vibration of the
C=N Group in Chloroformamidines

Class of compounds	$\nu_{C=N}$ (cm^{-1})	Reference
$R_FN=\overset{\underset{\mid}{F}}{C}-N(R_F)_2$	1725	58
$RSO_2N=\overset{\underset{\mid}{Cl}}{C}-NHR$	1600–1605	65
$RSO_2N=\overset{\underset{\mid}{Cl}}{C}-NR_2$	1546–1590	47
$H_2\overset{\oplus}{N}-\overset{\underset{\mid}{Cl}}{C}-NH_2]Cl^{\ominus}$	1680	39
$R\overset{\oplus}{NH}-\overset{\underset{\mid}{Cl}}{C}-NH_2]Cl^{\ominus}$	1642	48
$R\overset{\oplus}{NH}-\overset{\underset{\mid}{Cl}}{C}-NHR]Cl^{\ominus}$	1661–1685	62
$RN=\overset{\underset{\mid}{Cl}}{C}-N(COCH_3)R$	1670	27
$RN=\overset{\underset{\mid}{Cl}}{C}-N(COCl)R$	1667–1672 1661	59 21

and it has been shown by the observed chemical shifts that the hydrogen is attached to the nitrogen adjacent to the alkyl group ([65]).

$$RSO_2N=\overset{\underset{\mid}{Cl}}{C}-NHR'$$

XLIV

IV. CHEMICAL BEHAVIOR

A. Reactions with Oxygen–Hydrogen Bonds

The reaction of chloroformamidines ([6,8]) and chloroformamidine hydrochlorides ([15,16]) with water yields the corresponding ureas.

$$RR^1\overset{\oplus}{N}{=}\overset{\underset{\displaystyle Cl}{|}}{C}{-}NR^2R^3]Cl^{\ominus} + H_2O \longrightarrow RR^1N{-}\overset{\underset{\displaystyle O}{\|}}{C}{-}NR^2R^3 + 2\,HCl$$

Symmetrically disubstituted chloroformamidine hydrochlorides eliminate hydrogen chloride to form the corresponding carbodiimides, which add water to yield the corresponding ureas. The addition of water to the carbodiimides is catalyzed by the generated hydrochloric acid.

$$R\overset{\oplus}{N}H{=}\overset{\underset{\displaystyle Cl}{|}}{C}{-}NHR]Cl^{\ominus} + H_2O \longrightarrow RN{=}C{=}NR + 2\,H_3O^{\oplus} + 2\,Cl^{\ominus}$$

$$\downarrow {\scriptstyle H_2O}$$

$$RNH{-}\overset{\underset{\displaystyle O}{\|}}{C}{-}NHR$$

If the generated acid is neutralized by alkali, and the carbodiimide is extracted into an organic solvent, carbodiimides can be obtained in good yield ([15,16]). This method provides a smooth transformation of thioureas into carbodiimides. The intermediate chloroformamidine hydrochlorides do not have to be isolated, i.e., their solution in an organic solvent is added to a two-phase aqueous base/organic solvent system, and the generated carbodiimide is isolated by fractional distillation or crystallization from the organic phase.

The reaction of chloroformamidinium chlorides with alkoxide ion yields urea acetals XLV; however, free alcohol has to be avoided, because solvolysis of the urea acetals to the ortho esters XLVI occurs ([15]).

$$RR^1\overset{\oplus}{N}{=}\overset{\underset{\displaystyle Cl}{|}}{C}{-}NR^2R^3]Cl^{\ominus} + 2\,NaOR^4 \longrightarrow$$

$$RR^1N{-}\overset{\underset{\displaystyle OR^4}{|}}{\overset{\overset{\displaystyle OR^4}{|}}{C}}{-}NR^2R^3 + R^4OH \longrightarrow RR^1NC(OR^4)_3$$

$$\text{XLV} \qquad\qquad\qquad\qquad\qquad\qquad \text{XLVI}$$

In contrast, pseudourea derivatives are obtained in the reaction of chloroformamidine hydrochlorides with alkoxide ion ([38,71]).

Similarly, arenesulfonylalkylchloroformamidines undergo reaction with alkoxide ion to yield the corresponding pseudourea derivatives (XLVII) ([33,65]).

$$RSO_2N{=}\overset{\underset{\displaystyle Cl}{|}}{C}{-}NHR^1 + NaOR^2 \longrightarrow RSO_2N{=}\overset{\underset{\displaystyle OR^2}{|}}{C}{-}NHR^1$$

$$\text{XLVII}$$

1-p-*Chlorophenylsulfonyl-2-ethyl-3*-n-*propylpseudourea* ([65]). To a solution of 8.8 g (0.03 mole) N-*p*-chlorophenylsulfonyl-N′-*n*-propylchloroformamidine in 90 ml of dry benzene a solution of 0.7 g sodium in 15 ml of absolute ethanol is added dropwise with stirring and cooling. After stirring for 30 min and removal of the precipitated sodium chloride by filtration, the solvent is removed under vacuum and the residue is extracted with diethyl ether. Evaporation of the diethyl ether and vacuum distillation of the residue yields 7.7 g (84.5%) of 1-*p*-chlorophenylsulfonyl-2-ethyl-3-*n*-propylpseudourea, b.p. 153–155°C/0.02 mm; m.p. 76–77°C.

The reaction of the chloroformamidinium chloride XLVIII with three moles of alcohol at 0°C produces the oligopseudo urea XLIX, which upon subsequent reaction with aqueous ammonium hydroxide at room temperature yields the triazine derivative L ([6]).

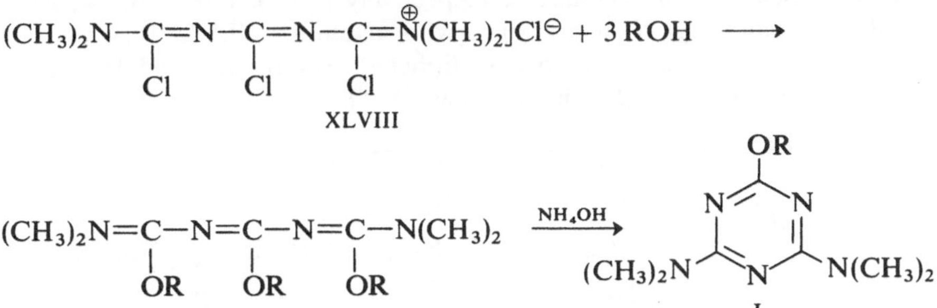

$$(CH_3)_2N-\underset{\underset{Cl}{|}}{C}=N-\underset{\underset{Cl}{|}}{C}=N-\underset{\underset{Cl}{|}}{C}=\overset{\oplus}{N}(CH_3)_2]Cl^{\ominus} + 3\,ROH \longrightarrow$$

$$\underset{\text{XLVIII}}{}$$

$$(CH_3)_2N=\underset{\underset{OR}{|}}{C}-N=\underset{\underset{OR}{|}}{C}-N=\underset{\underset{OR}{|}}{C}-N(CH_3)_2 \xrightarrow{NH_4OH}$$

$$\underset{\text{XLIX}}{}$$

In the reaction of XLVIII with excess *n*-butanol at 80°C, a mixture of *n*-butylchloride, di-*n*-butylcarbonate, N,N-dimethylurea and dimethylamine hydrochloride is obtained ([6]).

In contrast, arenesulfonyldialkylchloroformamidines can be recrystallized from ethanol ([47]). However, the geminate dichlorides LI (nonpolar form of a chloroformamidine hydrochloride), on reaction with methanol, yield the acetals LII ([65]).

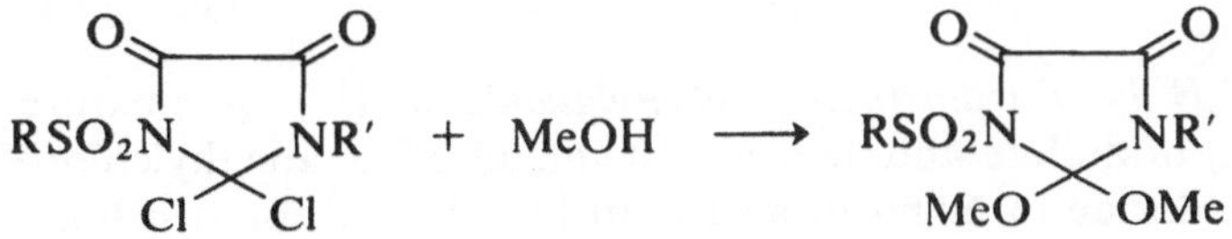

Similarly, reaction of chloroformamidines with phenols, in the presence of triethylamine as hydrogen chloride scavenger, affords pseudourea derivatives ([9,12]).

B. Reactions with Sulfur–Hydrogen Bonds

The reaction of chloroformamidine hydrochlorides with sodium hydrogen sulfide produces thiourea derivatives ([6,15,16,69]).

$$RR^1\overset{\oplus}{N}\!\!=\!\!\overset{\cdots}{\underset{Cl}{C}}\!\!-\!\!\overset{\cdots}{N}R^2R^3]Cl^{\ominus} + NaSH \longrightarrow RR^1N\!-\!\underset{\overset{\|}{S}}{C}\!-\!NR^2R^3$$

This reaction is perhaps more general, and thiolate ion could afford the corresponding pseudothioureas. For example, pseudothioureas are obtained from chloroformamidines and thiophenols in the presence of triethylamine ([9,12]).

Acylchloroformamidines, on treatment with thioamides and thioureas, afford nitriles and carbodiimides, respectively ([26,28]). For example, from 1-chloro-N-isopropyl-N′-acetylisopropylformamidine (LIII, $R = i\text{-}C_3H_7$) and diphenylthiourea a mixture of diphenylcarbodiimide and the acetyl thiourea (LIV, $R = i\text{-}C_3H_7$) is obtained ([26,28]).

$$RN\!=\!\underset{Cl}{C}\!-\!N(COCH_3)R + C_6H_5NH\!-\!\underset{\overset{\|}{S}}{C}\!-\!NHC_6H_5 \longrightarrow$$

LIII

$$C_6H_5N\!=\!C\!=\!NC_6H_5 + RNH\!-\!\underset{\overset{\|}{S}}{C}\!-\!N(COCH_3)R$$

LIV

C. Reactions with Nitrogen–Hydrogen Bonds

The reaction of chloroformamidines with primary and secondary amines provides a useful method of synthesis of guanidines LV ([5,6,9,12,15,16,35,43]).

$$RN\!=\!\underset{Cl}{C}\!-\!NR^1R^2 + R^3R^4NH \longrightarrow RN\!=\!\underset{NR^3R^4}{C}\!-\!NR^1R^2$$

LV

N,N,N′,N′-Tetramethyl-N″-phenylguanidine ([16]). A mixture of 9.3 g (0.1 mole) of aniline and 18.0 g (0.105 mole) of tetramethylchloroformamidinium chloride in 60 ml of acetronitrile is heated for two hours at 40°C and an additional hour at 60°C. After removal of the solvent under reduced pressure 50 ml of water and 20–30 ml of 30% sodium hydroxide is added. The precipitated oil is extracted with diethyl ether and evaporation of the solvent and vacuum distillation of the residue yields 14 g (73%) of N,N,N′,N′-tetramethyl-N″-phenylguanidine, b.p. 87–89°C/0.3 mm.

In the reaction of chloroformamidine-N-carbonyl chlorides LVI with arylamines a mixture of carbodiimides and isocyanates is obtained ([59]).

$$RN=\underset{\underset{Cl}{|}}{C}-N(COCl)R + RNH_2 \longrightarrow RN=C=NR + RNCO + 2\,HCl$$

LVI

Addition of LVI to excess methanolic ammonia gives the corresponding guanidine, most likely *via* the carbodiimide ([59]).

The reaction of the chloroformamidinium chloride LVII with ammonia produces, depending upon the reaction conditions, the triazine derivatives LVIII and LIX ([6]).

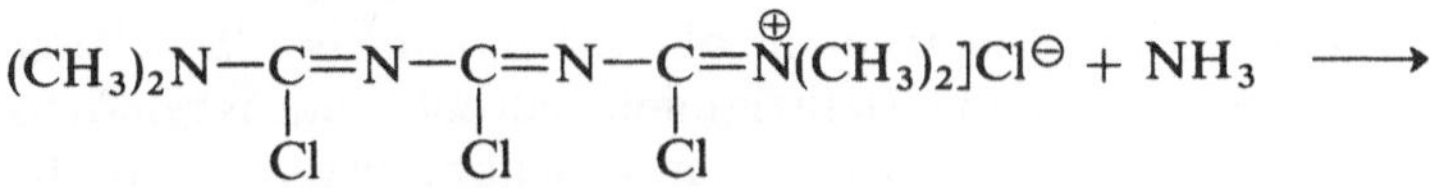

$$(CH_3)_2N-\underset{\underset{Cl}{|}}{C}=N-\underset{\underset{Cl}{|}}{C}=N-\underset{\underset{Cl}{|}}{C}=\overset{\oplus}{N}(CH_3)_2]Cl^{\ominus} + NH_3 \longrightarrow$$

LVII

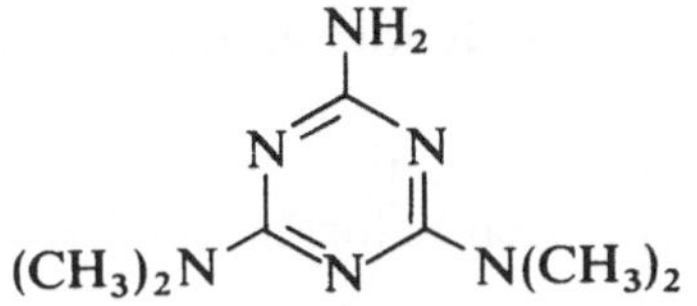

LVIII and LIX

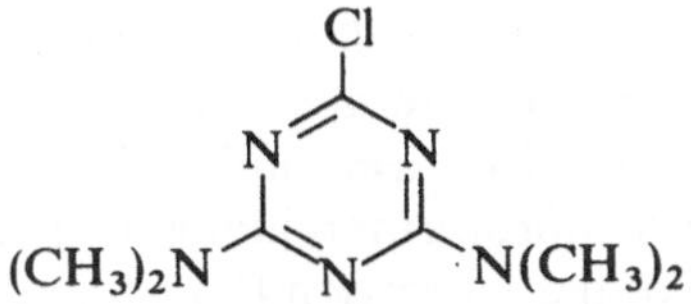

In the reaction of substituted chloroformamidine hydrochlorides with phenylhydrazine or with benzoic acid hydrazide the corresponding guanidines are obtained. Thus, from tetramethylchloroformamidinium chloride (LX) and phenylhydrazine, a guanidine LXI is obtained in good yield ([16]).

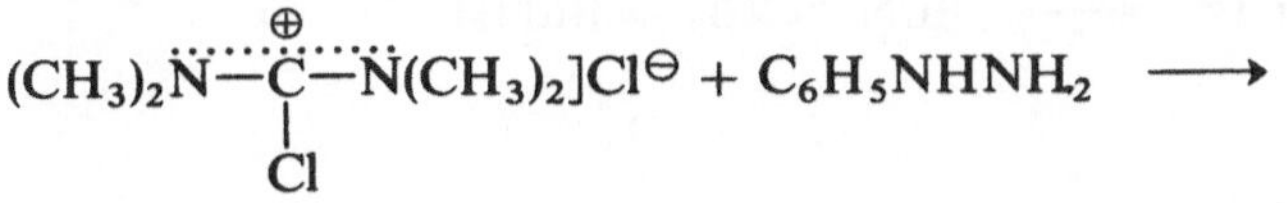

$$(CH_3)_2\overset{\cdots}{N}\overset{\oplus}{-}\underset{\underset{Cl}{|}}{C}\overset{\cdots}{-}\overset{\cdots}{N}(CH_3)_2]Cl^{\ominus} + C_6H_5NHNH_2 \longrightarrow$$

LX

$$(CH_3)_2N-\underset{\underset{N-NHC_6H_5}{\|}}{C}-N(CH_3)_2$$

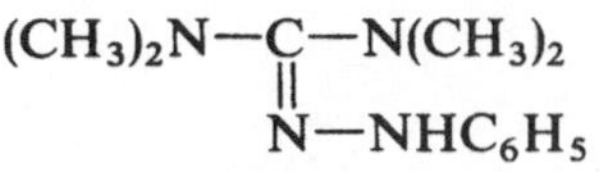

LXI

D. Addition and Elimination Reactions

The reactions of symmetrically disubstituted chloroformamidine hydrochlorides with a great variety of nucleophiles often proceed *via* initial elimination of hydrogen chloride to afford carbodiimide, the latter being attacked by the nucleophile.

The elimination of hydrogen chloride from diarylchloroformamidine hydrochlorides to produce diarylcarbodiimides proceeds already on heating in refluxing chlorobenzene ([66]). However, in the case of the aliphatic derivatives a stronger base, such as triethylamine, is being used as hydrogen chloride scavenger.

$$R\overset{..}{N}H-\overset{\oplus}{\underset{\underset{Cl}{|}}{C}}-\overset{..}{N}HR]Cl^{\ominus} \quad \xrightarrow{-2\,HCl} \quad RN=C=NR$$

Dialkylcarbodiimides–General Procedure ([62]). To a suspension of 0.1 mole of the N,N-dialkylchloroformamidine hydrochloride in 200 ml of benzene dropwise and with stirring 0.2 mole of triethylamine is added. After stirring for one hour the precipitated triethylamine hydrochloride is removed by filtration, and evaporation of the benzene and vacuum distillation of the residue affords the corresponding dialkylcarbodiimide in high yield. While secondary and tertiary dialkylcarbodiimides are stable at room temperature, primary dialkylcarbodiimides can be kept only for several days, and storage at low temperatures is recommended.

Likewise, arenesulfonylalkylchloroformamidines ([61,65,66]) and phosphorylchloroformamidines ([13]) eliminate hydrogen chloride quite readily to give the corresponding carbodiimides.

At more elevated temperature, and in the absence of a diluent, symmetrically disubstituted chloroformamidine hydrochlorides are in equilibrium with the carbonimidoyl dichloride, as evidenced by the isolation of tri-*n*-butylbiuret hydrochloride (LXII) in the pyrolysis of 1-chloro-N-*n*-butyl-N'-*n*-butylformamidine hydrochloride (LXIII) ([51]).

$$Bu\overset{..}{N}H-\overset{\oplus}{\underset{\underset{Cl}{|}}{C}}-\overset{..}{N}HBu]Cl^{\ominus} \quad \longleftrightarrow \quad BuN=CCl_2 + BuNH_2 \quad \longrightarrow$$

LXIII

$$Bu\overset{..}{N}H-\overset{\oplus}{\underset{\underset{NHBu}{|}}{C}}-\overset{..}{N}HBu]Cl^{\ominus}$$

LXII

Likewise, heating of an equimolar mixture of N,N'-dicyclohexylchloroformamidine hydrochloride (LXIV) and 1,3-dicyclohexylurea (LXV) produces N,N',N''-tricyclohexylguanidine hydrochloride (LXVI) in high yield ([22]).

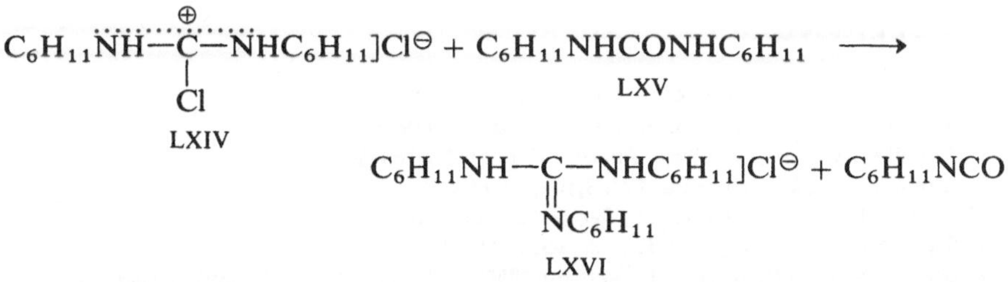

$$\text{C}_6\text{H}_{11}\overset{\oplus}{\ddot{\text{N}}\text{H}}\text{—}\overset{\cdots}{\ddot{\text{C}}}\text{—}\ddot{\text{N}}\text{HC}_6\text{H}_{11}]\text{Cl}^{\ominus} + \text{C}_6\text{H}_{11}\text{NHCONHC}_6\text{H}_{11} \longrightarrow$$
$$\underset{\text{Cl}}{|}$$

LXIV   LXV

$$\text{C}_6\text{H}_{11}\text{NH}\text{—}\overset{||}{\underset{\text{N}\text{C}_6\text{H}_{11}}{\text{C}}}\text{—NHC}_6\text{H}_{11}]\text{Cl}^{\ominus} + \text{C}_6\text{H}_{11}\text{NCO}$$

LXVI

Equilibrization of the dissociation products gives rise to the formation of the finally isolated LXVI and cyclohexyl isocyanate.

Heating of the chloroformamidinium chloride LXVII at 180–190°C produces the triazine derivative LXVIII in 80% yield ([6]).

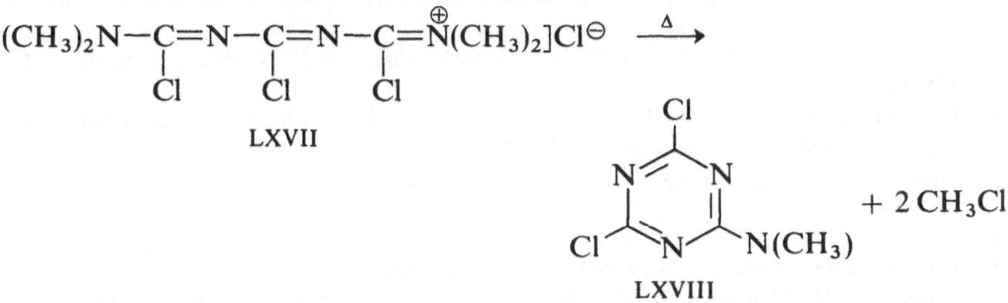

$$(\text{CH}_3)_2\text{N}\text{—}\underset{\text{Cl}}{\overset{|}{\text{C}}}\text{=N}\text{—}\underset{\text{Cl}}{\overset{|}{\text{C}}}\text{=N}\text{—}\underset{\text{Cl}}{\overset{|}{\text{C}}}\text{=}\overset{\oplus}{\ddot{\text{N}}}(\text{CH}_3)_2]\text{Cl}^{\ominus} \overset{\Delta}{\longrightarrow}$$

LXVII

$+ \, 2\,\text{CH}_3\text{Cl}$

LXVIII

The reaction of chloroformamidines with phenylmagnesium bromide results in the replacement of the chloro group by a phenyl group ([10,11]).

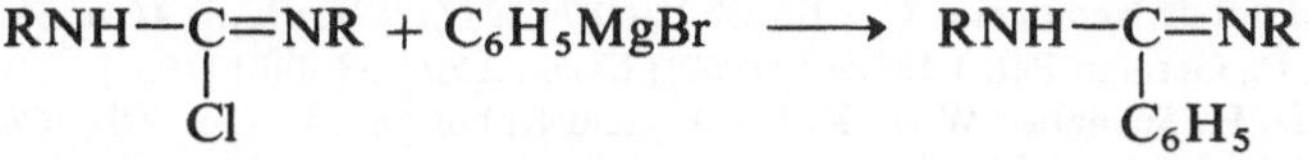

$$\text{RNH—}\underset{\text{Cl}}{\overset{|}{\text{C}}}\text{=NR} + \text{C}_6\text{H}_5\text{MgBr} \longrightarrow \text{RNH—}\underset{\text{C}_6\text{H}_5}{\overset{|}{\text{C}}}\text{=NR}$$

Weingarten and White ([68]) obtained tetrakis(dimethylamino)methane (LXIX), the first tetraminomethane derivative, from the reaction of tetramethylchloroformamidinium chloride with lithium dimethylamide.

$$(\text{CH}_3)_2\overset{\oplus}{\ddot{\text{N}}}\text{—}\overset{\cdots}{\ddot{\text{C}}}\text{—}\ddot{\text{N}}(\text{CH}_3)_2]\text{Cl}^{\ominus} + 2\,\text{LiN}(\text{CH}_3)_2 \longrightarrow$$
$$\underset{\text{Cl}}{|}$$

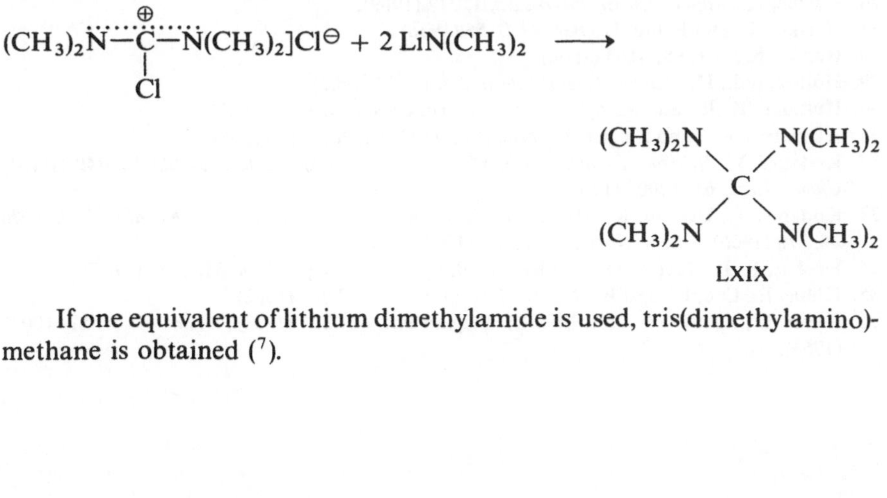

LXIX

If one equivalent of lithium dimethylamide is used, tris(dimethylamino)-methane is obtained ([7]).

V. REFERENCES

1. Allenstein, E., *Z. Anorg. Allgem. Chem.* **322**, 265 (1963).
2. Allenstein, E., *Z. Anorg. Allgem. Chem.* **322**, 276 (1963).
3. Aumüller, W., *Angew. Chem. Intern. Ed. Engl.* **2**, 616 (1963).
4. Aumüller, W., United States Pat. 3,169,138 (1965).
5. Bredereck, K., and Richter, R., *Ber.* **99**, 2454 (1966).
6. Bredereck, K., and Richter, R., *Ber.* **99**, 2461 (1966).
7. Bredereck, K., Effenberger, F., and Brendle, T., German Pat. 1,217,391; *Chem. Abstr.* **65**, 5366 (1966).
8. Derkach, G. I., Zhuravleva, L. P., and Kirsanov, A. V., *Zh. Obshch. Khim.* **32**, 879 (1962); *Chem. Abstr.* **58**, 2388 (1963).
9. Derkach, G. I., Narbut, A. V., and Kirsanov, A. V., Union of Soviet Socialist Republics Pat. 166,340; *Chem. Abstr.* **62**, 10461 (1965).
10. Derkach, G. I., and Narbut, A. V., *Zh. Obshch. Khim.* **35**, 1006 (1965); *Chem. Abstr.* **63**, 9857 (1965).
11. Derkach, G. I., and Narbut, A. V., *Zh. Obshch. Khim.* **36**, 322 (1966); *Chem. Abstr.* **64**, 15738 (1966).
12. Derkach, G. I., Narbut, A. V., and Kirsanov, A. V., *Probl. Organ. Senteza, Akad. Nauk SSSR, Otd. Obshch. i Tekhn. Khim.* 278 (1965).
13. Derkach, G. I., and Liptuga, N. I., *Zh. Obshch. Khim.* **36**, 461 (1966); *Chem. Abstr.* **65**, 634 (1966).
14. Drechsel, E., *J. Prakt. Chem.* [2] **11**, 315 (1875).
15. Eilingsfeld, H., Seefelder, M., and Weidinger, H., *Angew. Chem.* **72**, 836 (1960).
16. Eilingsfeld, H., Neubauer, G., Seefelder, M., and Weidinger, H., *Ber.* **97**, 1232 (1964).
17. Emeleus, H. J., and Hurst, G. L., *J. Chem. Soc.* 396 (1964).
18. Farbenfabriken Bayer, A. G., British Pat. 888,646 (1962); *Chem. Abstr.* **57**, 13696 (1962).
19. Farbenfabriken Bayer, A. G., British Pat. 959,977 (1964); *Chem. Abstr.* **61**, 6924 (1964).
20. E. I. DuPont de Nemours & Co., British Pat. 874,924 (1959); *Chem. Abstr.* **56**, 9978 (1962).
21. Fischer, P., German Pat. 1,131,661 (1962); *Chem. Abstr.* **58**, 1401 (1963).
22. Gavin, D. F., Schnabel, W. J., Kober, E., and Robinson, M. A., *J. Org. Chem.* **32**, 2511 (1967).
23. Gysin, H., Knuesli, E., and Rumpf, J., United States Pat. 3,053,843 (1962); *Chem. Abstr.* **58**, 6846 (1963).
24. Hantzsch, A., and Vogt, A., *Ann.* **314**, 366 (1901).
25. Hartke, K., and Bartulin, J., *Angew. Chem. Intern. Ed.* **1**, 211 (1962).
26. Hartke, K., *Angew. Chem. Intern. Ed.* **1**, 212 (1962).
27. Hartke, K., and Palou, E., *Ber.* **99**, 3155 (1966).
28. Hartke, K., *Ber.* **99**, 3163 (1966).
29. Holtschmidt, H., *Angew. Chem. Intern. Ed.* **1**, 632 (1962).
30. Huffman, K. R., and Schaefer, F. C., *J. Org. Chem.* **29**, 1816 (1963).
31. Kleineberg, G., and Ziegler, E., *Monatshefte Chem.* **96**, 1352 (1965).
32. Kodama, Y., Sekiba, T., and Kato, M., *Yuki Gosei Kagaku Kyokai Shi* **22**, 749 (1964); *Chem. Abstr.* **61**, 12002 (1964).
33. Kodama, T., Uehara, K., Hisada, K., and Shinohara, S., *Yuki Gosei Kagaku Kyokai Shi* **24**, 778 (1966); *Chem. Abstr.* **65**, 16886 (1966).
34. Koshar, R. J., Husted, D., and Meiklejohn, R. A., *J. Org. Chem.* **31**, 4232 (1966).
35. Kühle, E., Eue, L., and Bayer, O., German Pat. 1,137,000 (1962).
36. Kühle, E., Klauke, E., and Eue, L., Belgian Pat. 621,295 (1963); *Chem. Abstr.* **59**, 10071 (1963).

37. Kühle, E., and Wegler, R., French Addn. 81,999 (1963); Add. to French Pat. 1,256,873; *Chem. Abstr.* **60**, 11955 (1964).
38. Kühle, E., and Wegler, R., German Pat. 1,219,020 (1966); *Chem. Abstr.* **65**, 13605 (1966).
39. Kuhn, M., and Mecke, R., *Ber.* **94**, 3016 (1961).
40. Latscha, H. P., *Z. Anorg. Allgem. Chem.* **346**, 116 (1966).
41. Lecher, H. Z., and Kosloski, C. L., United States Pat. 2,727,922 (1955); *Chem. Abstr.* **50**, 15578 (1956).
42. Lecher, H. Z., Hardy, E. M., and Kosloski, C. L., United States Pat. 2,845,458 (1958); *Chem. Abstr.* **53**, 231 (1959).
43. Lecher, H. Z., Hardy, E. M., and Kosloski, C. L., United States Pat. 2,845,459 (1958); *Chem. Abstr.* **53**, 231 (1959).
44. Lengfeld, F., and Stieglitz, J., *Am. Chem. J.* **17**, 108 (1895).
45. Makarov, S. P., Shpanskii, V. A., Ginsburg, V. A., Shchekotikhin, A. K., Filatov, A. S., Martynova, L. C., Parlovskaya, I. V., Golovaneva, A. F., and Yakubovich, A. Y., *Dokl. Akad. Nauk SSSR* **142**, 596 (1962); *Chem. Abstr.* **57**, 4528 (1962).
46. Mulder, E., and Smit, J. A. R., *Ber.* **7**, 1634 (1874).
47. Neidlein, R., and Haussman, W., *Tetrahedron Letters*, 2217 (1966).
48. Newhall, W. F., Poos, G. I., Rosenau, J. D., and Suh, J. T., *J. Org. Chem.* **29**, 1809 (1964).
49. Pink, L. A., and Hetherington, H. C., *Ind. Eng. Chem.* **18**, 629 (1926).
50. Sauermilch, W., *Explosivstoffe* **9**, 71 (1961); *Chem. Abstr.* **55**, 21589 (1961).
51. Sayigh, A. A. R., Tilley, J. N., and Ulrich, H., *J. Org. Chem.* **29**, 3344 (1964).
52. Seefelder, M., German Pat. 1,119,258 (1961); *Chem. Abstr.* **56**, 11450 (1961).
53. Shipton, G. O., British Pat. 945,855 (1964); *Chem. Abstr.* **60**, 10700 (1964).
54. Stachel, H. D., *Angew. Chem.* **71**, 246 (1958).
55. Steindorff, A., *Ber.* **37**, 964 (1904).
56. Stieglitz, J., and McKee, R. H., *Ber.* **33**, 1517 (1900).
57. Stoffel, P. J., *J. Org. Chem.* **29**, 2794 (1964).
58. Tullock, C. W., Coffman, D. D., and Mutterties, E. L., *J. Am. Chem. Soc.* **86**, 357 (1964).
59. Ulrich, H., and Sayigh, A. A. R., *J. Org. Chem.* **28**, 1427 (1963).
60. Ulrich, H., and Sayigh, A. A. R., *Angew. Chem. Intern. Ed.* **3**, 585 (1964).
61. Ulrich, H., and Sayigh, A. A. R., *Angew. Chem. Intern. Ed.* **3**, 639 (1964).
62. Ulrich, H., Tilley, J. N., and Sayigh, A. A. R., *J. Org. Chem.* **29**, 2401 (1964).
63. Ulrich, H., and Sayigh, A. A. R., *J. Org. Chem.* **30**, 2779 (1965).
64. Ulrich, H., and Sayigh, A. A. R., *J. Org. Chem.* **30**, 2781 (1965).
65. Ulrich, H., Tucker, B., and Sayigh, A. A. R., *Tetrahedron* **22**, 1565 (1966).
66. Ulrich, H., and Sayigh, A. A. R., *Angew. Chem. Intern. Ed.* **5**, 704 (1966).
67. Ulrich, H., *Cycloaddition Reactions of Heterocumulenes*, Academic Press, New York (1967), pp. 259–260.
68. Weingarten, H., and White, W. A., *J. Am. Chem. Soc.* **88**, 2885 (1966).
69. Weidinger, H., and Eilingsfeld, H., German Pat. 1,119,843 (1961).
70. Will, W., *Ber.* **14**, 1486 (1881).
71. Winthrop, S. O., and Gavin, G., *J. Org. Chem.* **24**, 1936 (1959).
72. Young, J. A., Durell, W. S., and Dresdner, R. D., *J. Am. Chem. Soc.* **81**, 1587 (1959).
73. Young, J. A., Tsoukalas, S. N., and Dresdner, R. D., *J. Am. Chem. Soc.* **82**, 396 (1960).

Chapter 5

1-HALOFORMIMIDATES AND
1-HALOTHIOFORMIMIDATES

I. INTRODUCTION

The compounds under discussion are derivatives of 1-haloformimidic acid (I) and 1-halothioformimidic acid (II), respectively.

$$HN=\underset{\underset{\text{I}}{\overset{|}{X}}}{C}-OH \qquad\qquad HN=\underset{\underset{\text{II}}{\overset{|}{X}}}{C}-SH$$

The free acids are not known although their formation from isocyanic or isothiocyanic acid and one mole of a hydrogen halide can be visualized. While the esters of isocyanic acid, the isocyanates III, add one mole of hydrogen halide to afford the corresponding carbamic acid chlorides (IV), methyl isothiocyanate (V) adds hydrogen bromide to form a mixture of two isomers, to which 1-bromothioformidic acid structures VI and VII have been assigned ([1]).

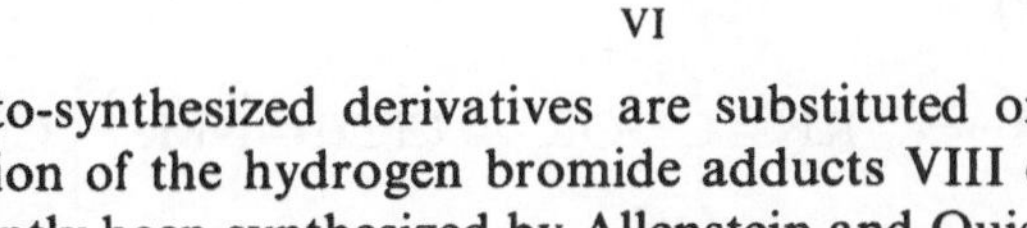

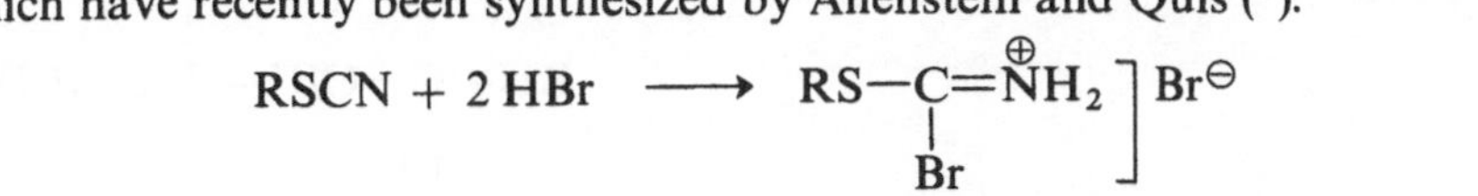

$$RN=C=O + HX \longrightarrow RNH-\underset{\underset{\text{IV}}{\overset{|}{X}}}{C}=O$$

$$CH_3N=C=S + HBr \longrightarrow CH_3\overset{\oplus}{N}H=\underset{\underset{\text{VI}}{\overset{|}{Br}}}{C}-S^{\ominus} + CH_3N=\underset{\underset{\text{VII}}{\overset{|}{Br}}}{C}-SH$$

The hitherto-synthesized derivatives are substituted on the nitrogen, with the exception of the hydrogen bromide adducts VIII of thiocyanates which have recently been synthesized by Allenstein and Quis ([1]).

$$RSCN + 2\,HBr \longrightarrow \left. RS-\underset{\underset{\text{VIII}}{\overset{|}{Br}}}{C}=\overset{\oplus}{N}H_2 \right] Br^{\ominus}$$

Another class of thioformimidic acid derivatives, the S-chlorides IX, is obtained by controlled addition of chlorine to isothiocyanates ([10,29,30]) or thiocyanogen chloride ([2,6]). Ottmann and Hooks ([29,30]) used the name S-chloroisothiocarbamoyl chlorides for compounds IX, while *Chemical Abstracts* relates their name to formimidoyl chloride, i.e., 1-(chlorothio)-formimidoyl chloride.

$$RN{=}C{=}S + Cl_2 \longrightarrow RN{=}\underset{\underset{\displaystyle Cl}{|}}{C}{-}SCl$$

$$IX$$

In order to preserve uniformity of nomenclature, *Chemical Abstracts* terminology is being used for compounds IX.

1-Haloformimidates were first synthesized by Nef ([23–25]), Hantzsch and Mai ([8a]), Lengfeld and Stieglitz ([20]), and Smith ([35a]) in the last decade of the nineteenth century.

The displacement of one chloro group in carbonimidoyl dichlorides by alkoxide, phenoxide, or thiolate ion is the best general method of synthesis of 1-haloformimidates or 1-halothioformimidates. However, several other methods of synthesis are available and since nucleophilic displacement of the halo group is easily accomplished, further utilization of these reactive intermediates in organic chemistry is anticipated.

II. SYNTHESIS

A. From Carbamates and Thiocarbamates

By analogy with the general synthesis of imidoyl chlorides it can be expected that carbamates and thiocarbamates undergo reaction with a variety of acid halides to afford 1-haloformimidates and 1-halothio-formimidates, respectively. For example, carbamates have been reacted with carbonyl chloride ([36]), pyrocatecholphosphorus trichloride ([8]), and phosphorus pentachloride ([7]), and isocyanates were obtained. In view of the catalytic effect of N,N-dimethylformamide in the phosgenation of carbamates to isocyanates, the intermediacy of 1-chloroformimidates X is anticipated ([36]).

$$RNHCOOR' \longleftrightarrow \left[RN{=}\underset{\underset{\displaystyle OH}{|}}{C}{-}OR' + ClCH{=}\overset{\oplus}{N}Me_2 \right] Cl^{\ominus} \longrightarrow$$

$$RN{=}\underset{\underset{\displaystyle Cl}{|}}{C}{-}OR' + Me_2NCHO$$

$$X$$

The thermolysis of 1-chloroformimidates to isocyanates and alkyl halides occurs quite readily ([22]), thus explaining the formation of the isolated isocyanates. Of course, dithiocarbamates react more readily, and Eilingsfeld and Möbius ([4,5]) have synthesized a large number of 1-chlorothio-formimidinium chlorides by treating N,N-disubstituted dithiocarbamates with phosgene or phosphorus pentachloride.

1-Chlorothioformimidinium chlorides (general procedure) ([4]). To the solution of 1 mole of the N,N-disubstituted dithiocarbamate in 400–700 ml of toluene at 20–30°C approximately 1.5–2.0 moles of phosgene is added. After standing for several hours at room temperature, dry diethyl ether is added and the obtained 1-chlorothioformimidinium chlorides are collected by filtration. The obtained yields range from 72–100%.

The compounds thus obtained are very hygroscopic and moisture has to be excluded during the work-up.

The corresponding cyclic dithiocarbamates (2-mercaptothiazoles) are phosgenated in a similar manner to afford 2-chlorothiazolidinium salts ([4,5,37]). For example, from 2-thioxo-3-methylthiazolidine (XI) and phosgene, 2-chloro-3-methylthiazolidinium chloride (XII) is obtained ([4]).

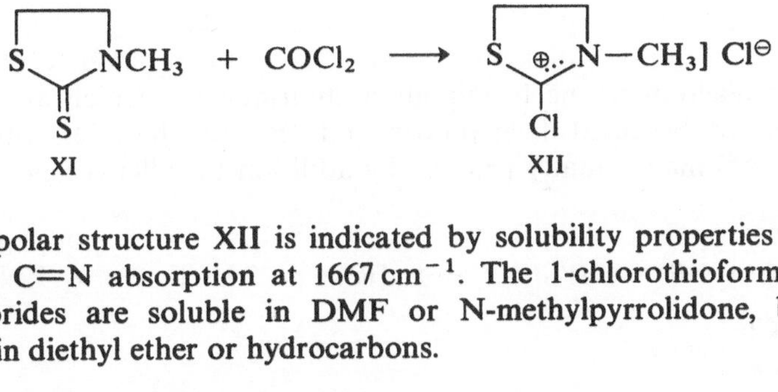

The polar structure XII is indicated by solubility properties as well as by the $C{=}N$ absorption at $1667\ cm^{-1}$. The 1-chlorothioformimidate hydrochlorides are soluble in DMF or N-methylpyrrolidone, but are insoluble in diethyl ether or hydrocarbons.

B. From Carbonimidoyl Dihalides

The reaction of carbonimidoyl dichlorides with alkoxide or phenoxide ion is perhaps the most general synthetic procedure for 1-chloroformimidates XIII. While generally the alkoxides or phenoxides are used ([8a,20,35,35a]), Kühle ([5]) has shown that aqueous sodium hydroxide and alcohols work equally well.

$$RN{=}CCl_2 + R'OH + NaOH \longrightarrow RN{=}\underset{\underset{Cl}{|}}{C}{-}OR' + NaCl$$

XIII

Similarly, reaction of thiolate ion with carbonimidoyl dichlorides affords the 1-chlorothioformimidates XIV ([14,18]).

$$RN{=}CCl_2 + NaSR' \longrightarrow RN{=}\underset{\underset{XIV}{Cl}}{\overset{|}{C}}{-}SR' + NaCl$$

C. From Isocyanides

The addition of ethyl hypochlorite to isocyanides has been used by Nef ([25]) to synthesize 1-haloformimidates.

$$RNC + R'OCl \longrightarrow RN{=}\underset{Cl}{\overset{|}{C}}{-}OR'$$

Similarly, addition of reactive halides, such as acyl halides, to isocyanides is feasible, and Nef has synthesized a variety of formimidoyl chlorides by this method ([23,24]). Recently, Havlik and Wald ([9]) have demonstrated that sulfenyl chlorides could also be added to isocyanides to afford 1-chlorothioformimidates.

$$RNC + R'SCl \longrightarrow RN{=}\underset{Cl}{\overset{|}{C}}{-}SR'$$

The reaction of the formimidoyl chlorides XV, which are generated from monosubstituted formamides and thionyl chloride, with sulfenyl chloride ([18]) may actually proceed by addition of sulfenyl chloride to isocyanide.

$$\overset{\oplus}{R}NH{=}\underset{\underset{XV}{Cl}}{\overset{|}{C}}{-}H]Cl^{\oplus} + R'SCl \longrightarrow [RNC + 2\,HCl] \longrightarrow RN{=}\underset{Cl}{\overset{|}{C}}{-}SR'$$

This reaction is a general synthetic method for 1-chlorothioformimidates, and good yields are reported ([18]).

If formamide is used instead of the monosubstituted derivatives, thiocyanates XVI are obtained in good yield ([18]).

$$H_2N{=}\underset{Cl}{\overset{|}{C}}{-}H]Cl^{\ominus} + RSCl \longrightarrow [HN{=}\underset{Cl}{\overset{|}{C}}{-}SR] + HCl$$

$$\downarrow{-}HCl$$

$$\underset{XVI}{RSCN}$$

The reaction of monosubstituted formamides, such as 4-chloro-formanilide (XVII), with sulfur chloride gives a mixture of 4-chlorophenyl-carbonimidoyl dichloride (XVIII) and 4-chlorophenyl isothiocyanate (XIX) ([18]).

$$4\text{-}ClC_6H_4NHCHO + SOCl_2 + S_2Cl_2 \longrightarrow$$
$$\phantom{4\text{-}ClC_6H_4}\text{XVII}$$

$$4\text{-}ClC_6H_4N\text{=}CCl_2 + 4\,ClC_6H_4NCS$$
$$\phantom{4\text{-}ClC_6H_4N}\text{XVIII} \text{XIX}$$

D. From Thiocyanates and Isothiocyanates

Recently Allenstein and Quis ([1]) investigated the addition of hydrogen halides to thiocyanates. While with hydrogen chloride only labile adducts have been observed, hydrogen bromide adds to methyl- and phenylthio-cyanate to afford 2:1 adducts, for which structure XX has been assigned on the basis of infrared spectroscopy ([1]).

$$RSCN + 2\,HBr \longrightarrow RS\text{—}\overset{\underset{|}{Br}}{C}\text{=}\overset{\oplus}{N}H_2]Br^{\ominus}$$
$$\text{XX}$$

The corresponding 1-chloroderivative may be an intermediate in the transformation of formimidoyl chloride to thiocyanates by means of sulfenyl chloride ([18]) (see page 142).

The above 1-bromothioformimidates XX are the only known derivatives of thioformimidic acid, which are not substituted on the nitrogen.

The addition of hydrogen bromide to methyl isothiocyanate affords, depending upon reaction conditions, two isomers, XXI and XXII ([1]). The latter is the predominant product and it is difficult to obtain pure XXI. Actually, Allenstein and Quis ([1]) could not repeat the formation of this compound.

$$CH_3NCS + HBr \longrightarrow CH_3N\text{=}\overset{\underset{|}{Br}}{C}\text{—}SH + CH_3\overset{\oplus}{N}H\text{=}\overset{\underset{|}{Br}}{C}\text{—}S^{\ominus}$$
$$\text{XXI} \text{XXII}$$

From isothiocyanates and one equivalent of chlorine, relatively stable 1:1 adducts can be obtained ([10,29,30]). While Dyson and Harrington ([3]) in 1942 postulated the intermediacy of such 1:1 adducts in the conversion of isothiocyanates into carbonimidoyl dichlorides by chlorine (see Chapter 2), Ottmann and Hooks ([29,30]) have isolated 1:1 adducts from alkyl and aryl isothiocyanates and chlorine, and structure XXIII has been proposed for these adducts.

$$RN{=}C{=}S + Cl_2 \longrightarrow RN{=}\underset{\underset{Cl}{|}}{C}{-}SCl$$

XXIII

Likewise, aroyl isothiocyanates add chlorine across the C=S double bond to form the corresponding 1-(chlorothio)formimidoyl chlorides XXIV in low yield ([10]).

$$RCON{=}C{=}S + Cl_2 \longrightarrow RCON{=}\underset{\underset{Cl}{|}}{C}{-}SCl$$

XXIV

E. Miscellaneous Methods

In 1966 Neidlein, Haussmann, and Heukelbach ([28]) investigated the reaction of thioacetals, XXV with bromine and chlorine, and the authors obtained good yields of the corresponding N-arenesulfonyl-1-halothioformimidates (XXVI).

$$\underset{\text{XXV}}{RSO_2N{=}C(SR')_2} + X_2 \longrightarrow \underset{\underset{X}{|}}{RSO_2N{=}C}{-}SR' + R'SCl$$

XXVI

$$(R = C_6H_5, R' = CH_3, X = Br, Cl)$$

The halogen has to be applied in stoichiometric amounts, and therefore, instead of chlorine, sulfuryl chloride has been used, which is easier to handle in the laboratory ([26,28]). If XXVI is treated with excess chlorine, the corresponding sulfonylcarbonimidoyl dichloride is obtained (see Chapter 2).

The addition of 1-(chlorothio)formimidoyl chloride to olefins ([31]) (see Section IVC) constitutes another method of synthesis of 1-chlorothioformimidates.

The hitherto reported 1-haloformimidates, 1-halothioformimidates, and 1-(chlorothio)formimidoyl chlorides are listed in Tables I–III.

TABLE I
1-Chloroformimidates

$$RN{=}\overset{\displaystyle |}{\underset{\displaystyle Cl}{C}}{-}OR'$$

R	R'	Method of preparation	B.p., °C/mm (M.p., °C)	Yield, %	Reference
C_6H_5	CH_3	B	104/15	—	35a
C_6H_5	C_2H_5	B	105/12	—	20, 35a
$4\text{-}ClC_6H_4$	CH_3	B	128–134/16	85	15
$2,3\text{-}Cl_2C_6H_3$	CH_3	B	134–140/9	89	15
$2,4\text{-}Cl_2C_6H_3$	CH_3	B	128–131/9	85.5	15
$2,5\text{-}Cl_2C_6H_3$	CH_3	B	129–135/9	88	15
$2,4,5\text{-}Cl_3C_6H_2$	CH_3	B	151–155/9	86.5	15
$2,4,5\text{-}Cl_3C_6H_2$	C_2H_5	B	Oil	94	15
$2,4,5\text{-}Cl_3C_6H_2$	$CH_2{=}CHCH_2$	B	Oil	76	15
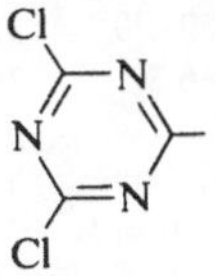 triazinyl (Cl,Cl)	CH_3	B	(95–97)	—	13
C_6H_5	C_6H_5	B	180/15 (43)	—	8a
C_6H_5	$4\text{-}BrC_6H_4$	B	227 (45)	—	8a

TABLE II
1-Bromothioformimidates

$$RN{=}\overset{\displaystyle |}{\underset{\displaystyle Br}{C}}{-}SR'$$

R	R'	Method of preparation	M.p., °C.	Yield, %	Reference
H	CH_3	D	—*†	—	1
H	C_6H_5	D	—*†	—	1
CH_3	H	D	—†	—	1

* Hydrobromide.
† Hydrolytically unstable, only infrared and bromine analyses were reported.

TABLE III
1-Chlorothioformimidates and 1-(Chlorothio)formimidoyl Chlorides*

$$RN=\underset{\underset{Cl}{|}}{C}-SR'$$

R	R'	Method of preparation	B.p., °C/mm (M.p., °C)	Yield, %	Reference
Cl	Cl	D	40–41/9	70–80	2, 6
CH_3	Cl	D	Oil	68	30
C_2H_5	Cl	D	Oil	93	30
$n\text{-}C_4H_9$	Cl	D	Oil	92	30
$n\text{-}C_7H_{15}$	Cl	D	Oil	94	30
C_6H_{11}	Cl	D	Oil	98	30
C_6H_5	Cl	D	Oil	91–95	30
$4\text{-}ClC_6H_4$	Cl	D	(41–42)	69–90	30
$4\text{-}CH_3OC_6H_4$	Cl	D	Oil	46.5	30
$4\text{-}O_2NC_6H_4$	Cl	D	(49)	70	30
$4\text{-}ClS-\underset{\underset{Cl}{\mid}}{C}=NC_6H_4$	Cl	D	(82–83)	67–97	30
$2,5\text{-}Cl_2C_6H_3$	Cl	D	(31)	71	30
$2,5\text{-}CH_3(ClS-\underset{\underset{Cl}{\mid}}{C}=N)C_6H_3$	Cl	D	(75–75.5)	65–81	30
$2,4,5\text{-}CH_3(Cl)(ClS-\underset{\underset{Cl}{\mid}}{C}=N)C_6H_2$	Cl	D	(105–106)	55–83	30
C_6H_5CO	Cl	D	(91–92)	12	10
$4\text{-}ClC_6H_4CO$	Cl	D	(112–114)	10	10
CH_3	CCl_3	C	98–101/30	46	18
CH_3	$CFCl_2$	C	64–65/14	52	18
CH_3	C_6Cl_5	C	(116)	100	18
CH_3	2-Benzothiazolyl	C	(192-196)	53	18
C_2H_5	C_6H_5	C	151–158/30	21	18
C_3H_7	CCl_3	C	127–128/15	55	18
$n\text{-}C_4H_9$	$-CH(CH_2)_4C(Cl)H-$	D	Oil	78	31
C_6H_5	CCl_3	C	183–187/30	55	18
C_6H_5	$CFCl_2$	C	148–150/18	79.5	18
C_6H_5	C_6H_5	C	(56–58)	71	18
C_6H_5	C_6Cl_5	C	(124)	100	18
C_6H_5	CH_2CH_2Cl	D	Oil	83	31
C_6H_5	$CH_2CH(CH_2Cl)Cl$	D	Oil	83	31
C_6H_5	$CH_2CH(C_3H_7)Cl$	D	Oil	91	31
C_6H_5	$CH_2CH(C_6H_5)Cl$	D	(49–50)	65	31
C_6H_5	$-CH(CH_2)_4C(Cl)H-$	D	(41–42)	92	31

* A number of 1-chloroformimidinium chlorides are listed in Reference 4; however, no physical constants are reported.

Table III—*continued*

R	R'	Method of preparation	B.p., °C/mm (M.p., °C)	Yield, %	Reference
4-$CH_3C_6H_4$	2,4-$(O_2N)C_6H_3$	C	(114–115)	—	9
2-ClC_6H_4	C_6Cl_5	C	(92)	96	18
4-ClC_6H_4	C_6Cl_5	C	(134–138)	100	18
4-ClC_6H_4	$-CH(CH_2)_4C(Cl)H-$	D	(43–44)	97	31
2,5-$Cl_2C_6H_3$	$-CH(CH_2)_4C(Cl)H-$	D	(58–59)	97	31
4-$ClCOC_6H_4$	C_6H_5	C	186–193/0.2	78	18
3-$O_2NC_6H_4$	C_6H_5	C	192–194/0.2	70	18
4-$O_2NC_6H_4$	C_6Cl_5	C	(157–160)	95	18
$Cl_5C_6-\underset{\underset{Cl}{\vert}}{C}=NC_6H_4$	C_6Cl_5	C	(270–271)	72	18
2,4-$Cl_2C_6H_3$	4-$O_2NC_6H_5$	C	(122–124)	100	18
2,4-$Cl_2C_6H_3$	C_6Cl_5	C	(102)	92	18
α-C_7H_{10}	C_6Cl_5	C	(266)	20	18
2-Benzothiazolyl	C_6Cl_5	C	(212)	33	18
CH_3SO_2	CH_3	E	(52–53)	74	28
$C_6H_5SO_2$	CH_3	E	(65–66)	93	28
4-$CH_3C_6H_4SO_2$	CH_3	E	(89–90)	89	28
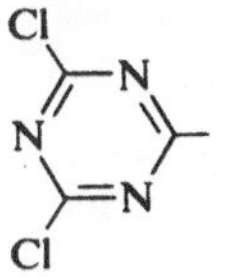	C_2H_5	B	144–147/3	95	14
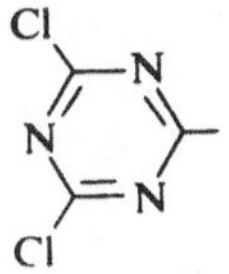	n-C_4H_9	B	158–159/4	95	14

III. PHYSICOCHEMICAL PROPERTIES

The 1-chloroformimidates and the 1-chlorothioformimidates are in general liquids or low-melting solids, which can be purified by vacuum distillation. In contrast, 1-bromothioformimidates, obtained as hydrobromides from thiocyanates and hydrogen bromide, or from isothiocyanates and hydrogen bromide, are hygroscopic salt-like compounds which are difficult to handle.

The 1:1 adducts of chlorine and isothiocyanates are yellow liquids or low-melting solids which can be handled conveniently at room temperature. However, upon heating, dissociation readily occurs, with partial regeneration of the starting components.

The haloformimidates and thioformimidates show their characteristic C=N stretching vibration in the infrared at 1527–1662 cm^{-1}. The 1-(chlorothio)formimidoyl chlorides have in addition to the C=N absorption at 1650 cm^{-1}, a second absorption of unknown origin at 1700 cm^{-1} ([30]). The C=N stretching of several groups of thioformimidates is listed in Table IV.

TABLE IV
Asymmetrical Stretching Vibration of the
C=N Group in 1-Haloformimidates and
1-Halothioformimidates

Class of compounds	$\nu_{C=N}$ (cm^{-1})	Reference
H$_2$N$^{\oplus}$=C(—Br)—SR]Br$^{\ominus}$	1587–1616	1
D$_2$N$^{\oplus}$=C(—Br)—SCH$_3$]Br$^{\ominus}$	1545	1
RN=C(—Br)—SH	1662	1
RNH$^{\oplus}$=C(—Br)—S$^{\ominus}$	1642	1
RN=C(—Cl)—SR′	1650	30
RSO$_2$N=C(—Br)—SCH$_3$	1538	28
RSO$_2$N=C(—Cl)—SCH$_3$	1527–1563	28
ClN=C(—Cl)—SCl	1600	2, 6
RN=C(—Cl)—SCl	1650	30

IV. CHEMICAL BEHAVIOR

A. Reactions with Oxygen–Hydrogen Bonds

The reaction of haloformimidates with water affords the corresponding carbamate (XXVII), and the reaction is catalyzed by the generated acid ([21]).

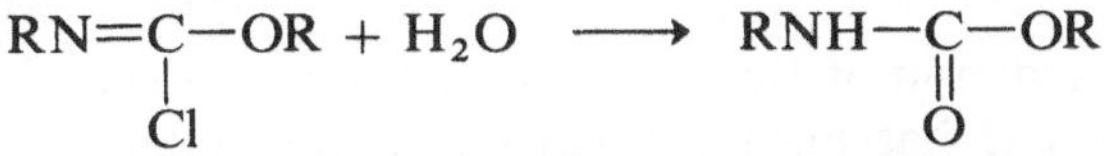

$$RN{=}\underset{\underset{Cl}{|}}{C}{-}OR + H_2O \longrightarrow RNH{-}\underset{\underset{O}{\|}}{C}{-}OR$$

XXVII

The hydrolysis of 1-chlorothioformimidinium chlorides with excess sodium carbonate affords N,N-disubstituted thiocarbamic acid S-esters XXVIII ([4]).

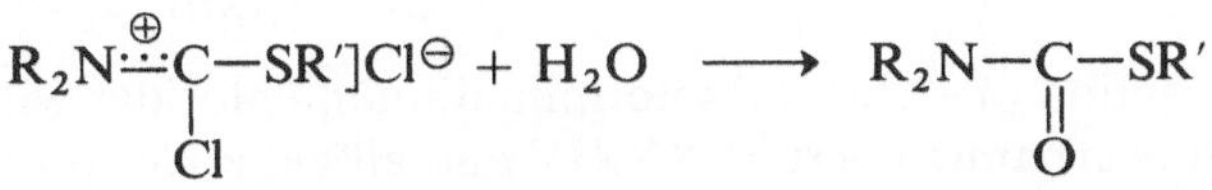

$$R_2N{\overset{\oplus}{=\!\!\cdot}}\underset{\underset{Cl}{|}}{C}{-}SR']Cl^{\ominus} + H_2O \longrightarrow R_2N{-}\underset{\underset{O}{\|}}{C}{-}SR'$$

XXVIII

Thiocarbamic acid S-esters (general procedure) ([4]). 1-Chlorothioformimidinium chlorides are added to excess 20% aqueous sodium carbonate. After stirring for ten min the thiocarbamic acid ester is extracted with methylene chloride and after evaporation of the solvent either crystallized or purified by vacuum distillation. The reported yields range from 65–100%.

Likewise, 1-chlorothioformimidates XXIX, obtained by the addition of 1-(chlorothio)formimidoyl chlorides to olefins ([31]), are hydrolyzed by water to form the corresponding thiocarbamates (XXX). In some cases (R = electron-attracting group) ring closure to the thiazolidinone XXXI occurs ([31]).

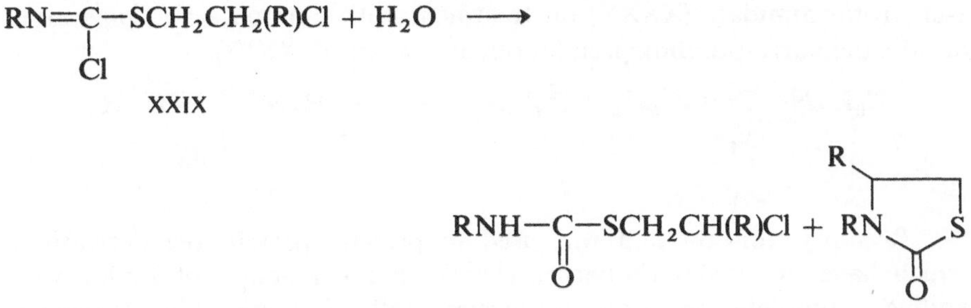

$$RN{=}\underset{\underset{Cl}{|}}{C}{-}SCH_2CH_2(R)Cl + H_2O \longrightarrow$$

XXIX

$$RNH{-}\underset{\underset{O}{\|}}{C}{-}SCH_2CH(R)Cl + RN{<}...{>}S$$

XXX   XXXI

In the reaction of N-phenyl-1-(chlorothio)formimidoyl chloride with water 4-phenyl-5-phenylimino-1,2,4-dithiazolidin-3-one (XXXII) has been obtained in 67% yield ([30]).

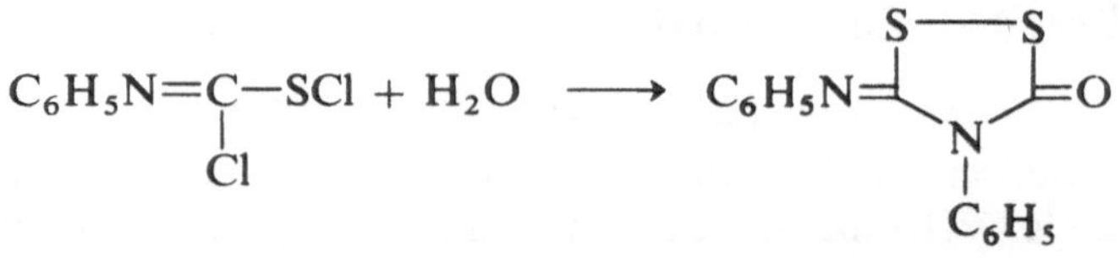

$$C_6H_5N{=}C{-}SCl + H_2O \longrightarrow$$

XXXII

The facile reaction of 1-chloroformimidates with alkoxide ion to yield the acetals XXXIII has been discussed in more detail in Chapter 2 (see page 40).

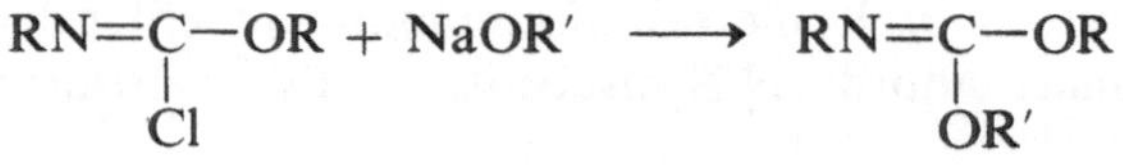

$$RN{=}C{-}OR + NaOR' \longrightarrow RN{=}C{-}OR$$

XXXIII

In the reaction of 1-chlorothioformimidinium chlorides with phenols a mixture of thiocarbamic O-esters XXXIV and alkyl halides is obtained ([4]).

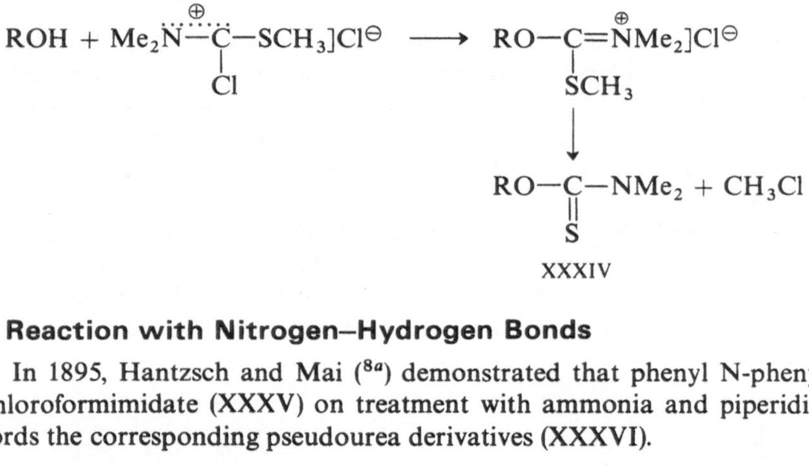

$$ROH + Me_2\overset{\oplus}{N}{-}C{-}SCH_3]Cl^{\ominus} \longrightarrow RO{-}C{=}\overset{\oplus}{N}Me_2]Cl^{\ominus}$$

$$RO{-}C{-}NMe_2 + CH_3Cl$$

XXXIV

B. Reaction with Nitrogen–Hydrogen Bonds

In 1895, Hantzsch and Mai ([8a]) demonstrated that phenyl N-phenyl-1-chloroformimidate (XXXV) on treatment with ammonia and piperidine affords the corresponding pseudourea derivatives (XXXVI).

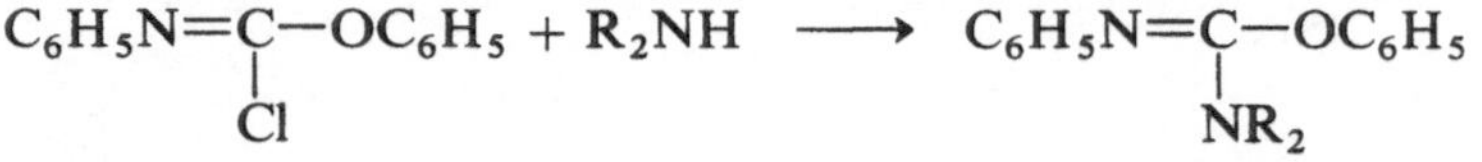

$$C_6H_5N{=}C{-}OC_6H_5 + R_2NH \longrightarrow C_6H_5N{=}C{-}OC_6H_5$$

XXXV   XXXVI

Recently, this method was used to prepare pseudourea derivatives, which have herbicidal properties ([16,17]). In the reaction of XXXV with aniline, complete substitution occurs with formation of triphenyl-guanidine ([35a]).

An analogous reaction occurs when N-arenesulfonyl-1-chlorothio-imidates XXXVII are treated with piperidine, and the corresponding pseudothioureas (XXXVIII) are obtained ([28]).

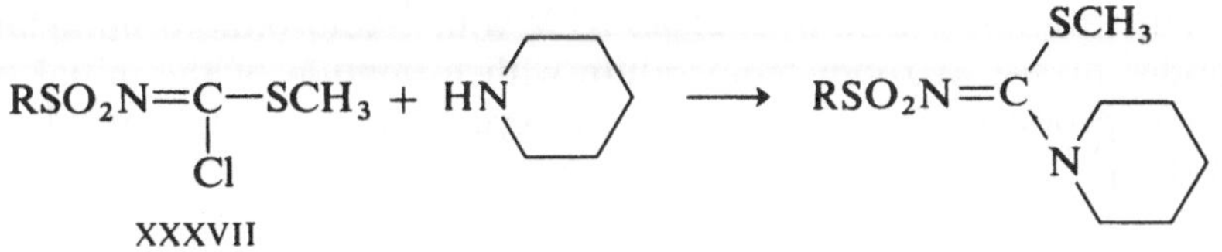

$$RSO_2N{=}\underset{\underset{Cl}{|}}{C}{-}SCH_3 + HN\bigcirc \longrightarrow RSO_2N{=}C\underset{\diagdown N \bigcirc}{\overset{\diagup SCH_3}{}}$$

XXXVII

XXXVIII

1-Phenyl-2-methyl-3-dimethylpseudourea ([16]). To 20 g of methylphenyl-1-chloroformimidate, dissolved in 150 ml of benzene, dropwise and with stirring 11 g of dimethylamine is added at room temperature. The reaction mixture is extracted with water (to remove dimethylamine hydrochloride) and the benzene layer is dried with sodium sulfate. Evaporation of the benzene and vacuum distillation of the residue yields 20 g (94 %) of 1-phenyl-2-methyl-3-dimethylpseudourea, b.p. 126–128°C/14 mm.

The reaction of 1-chlorothioformimidinium chlorides with primary amines or sulfonamides in the presence of a base affords the isothiourea derivatives XXXIX ([4]).

$$R_2\overset{\oplus}{\underset{\underset{Cl}{|}}{N}}{\cdots}\overset{}{C}{-}SR']Cl^{\ominus} + R''NH_2 \longrightarrow R_2N{-}\underset{\underset{SR'}{|}}{C}{=}NR''$$

XXXIX

In the reaction of XL with O-benzylhydroxylamine, the corresponding pseudothiourea derivative (XLI) has been obtained ([28]).

$$RSO_2N{=}\underset{\underset{Cl}{|}}{C}{-}SCH_3 + H_2NOCH_2C_6H_5 \longrightarrow$$

XL

$$RSO_2N{=}C\underset{\diagdown NHOCH_2C_6H_5}{\overset{\diagup SCH_3}{}}$$

XLI

Oximes, on treatment with 1-chloroformimidates, afford carbamates, i.e., the reaction amounts to a dehydration of the oxime to the corresponding nitrile. Thus, benzaldoxime (XLII) and ethyl N-phenyl-1-chloroformimidate (XLIII) combine to form benzonitrile and ethyl phenylcarbamate (XLIV) ([22]).

$$C_6H_5CH{=}NOH + C_6H_5N{=}\underset{\underset{Cl}{|}}{C}{-}OEt \longrightarrow$$

XLII   XLIII

$$C_6H_5CN + C_6H_5NHCOOEt$$

XLIV

The reaction of acid hydrazides **XLV** with 1-chlorothioformimidinium chlorides occurs at room temperature with formation of the initial polar reaction products **XLVI**, which readily cyclize to form 1,3,4-oxadiazoles **XLVII** ([4]).

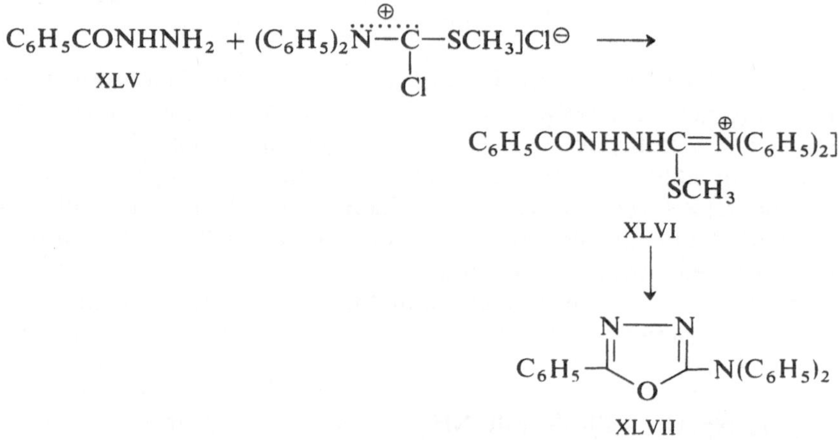

$$C_6H_5CONHNH_2 + (C_6H_5)_2\overset{\oplus}{\ddot{N}}\text{---}\overset{..}{\underset{\underset{Cl}{|}}{C}}\text{---}SCH_3]Cl^{\ominus} \longrightarrow$$

$$\underset{XLV}{}$$

$$C_6H_5CONHNHC\overset{\oplus}{=}\underset{\underset{SCH_3}{|}}{\ddot{N}}(C_6H_5)_2]Cl^{\ominus}$$

$$\underset{XLVI}{}$$

$$\underset{XLVII}{C_6H_5\text{---}\!\!\!\!\overset{N\text{---}N}{\underset{O}{\diagup\!\!\diagdown}}\!\!\!\!\text{---}N(C_6H_5)_2}$$

The choice of the basicity of the solvent used in this reaction (pyridine, N-methylpyrrolidone) determines its course. While in pyridine the amino-1,3,4-oxadiazoles are formed, 2-mercapto derivatives are obtained in N-methylpyrrolidone. Similarly, amidoximes **XLVIII**, on reaction with 1-chlorothioformimidinium chlorides, yield either 2-amino-1,2,4-oxadiazoles **XLIX** or the 2-mercapto derivatives **L** ([4]).

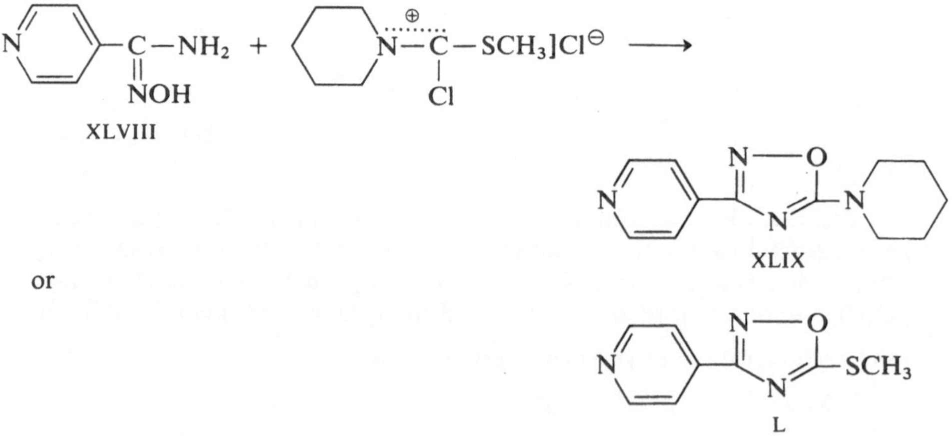

XLVIII

XLIX

or

L

Similar cyclizations occur if thiosemicarbazides or amidehydrazides are used as the substrate, and from *o*-phenylenediamine, *o*-aminophenol, and *o*-aminothiophenol the corresponding five-membered ring hetero-cycles, having a 2-amino group, are obtained. In certain special cases the reaction again can be directed toward formation of the 2-mercapto derivatives ([4]).

The reaction of sulfonyl-1-chlorothioimidates LI with sodium azide gives rise to the formation of the corresponding azides (LII) ([27]).

$$RSO_2N{=}\underset{\underset{LI}{\overset{|}{Cl}}}{C}{-}SCH_3 + NaN_3 \longrightarrow RSO_2N{=}\underset{\underset{LII}{\overset{|}{N_3}}}{C}{-}SCH_3$$

Treatment of the azides LII with triphenylphosphorus results in the formation of N-sulfonylphosphinimines LIII and methyl thiocyanate ([27]).

$$RSO_2N{=}\underset{\underset{LII}{\overset{|}{N_3}}}{C}{-}SCH_3 + (C_6H_5)_3P \longrightarrow RSO_2N{=}P(C_6H_5)_3 + CH_3SCN$$
$$LIII$$

C. Reactions with Carbon–Hydrogen Bonds

The reaction of 1-chlorothioformimidinium chlorides with nucleophilic carbon atoms results in the formation of a carbon–carbon bond. For example, reaction of 1-chlorothioformimidinium chlorides with malonitrile yields the ketene S,N-acetals LIV ([4]).

$$(NC)_2CH_2 + R_2\overset{\oplus}{N}{\cdots}\underset{\overset{|}{Cl}}{C}{-}SR']Cl^{\ominus} \longrightarrow (NC)_2C{=}C(NR_2)SR'$$
$$LIV$$

D. Addition and Elimination Reactions

Recently, Ottmann and Hooks ([31]) added 1-(chlorothio)formimidoyl chlorides to a variety of olefins. The reaction occurs already at ambient temperature, and olefins, such as ethylene, 1,2-cyclohexene and styrene, have been used as substrates. The 1:1 adducts LV are thus obtained in high yield ([31]).

$$RN{=}\underset{\overset{|}{Cl}}{C}{-}SCl + CH_2{=}CHR \longrightarrow RN{=}\underset{\overset{|}{Cl}}{C}{-}SCH_2CH(R)Cl$$
$$LV$$

Similarly, addition of 1-(chlorothio)formimidoyl chlorides to iso-cyanates has been observed, and the 1:1 adducts LVI are obtained ([32,34]).

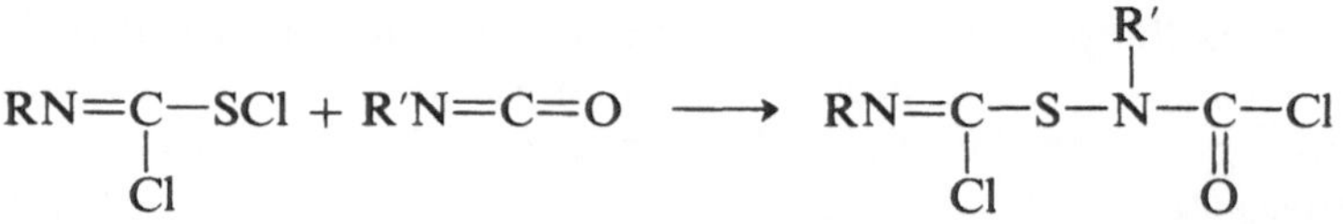

$$RN{=}\overset{\underset{Cl}{|}}{C}{-}SCl + R'N{=}C{=}O \longrightarrow RN{=}\overset{\underset{Cl}{|}}{C}{-}S{-}\overset{\overset{R'}{|}}{N}{-}\overset{\overset{}{\underset{O}{\parallel}}}{C}{-}Cl$$

LVI

Subsequent reaction of the adducts LVI with water yields 1,2,4-thia-diazolidine-3,5-diones LVII ([32]), and with primary amines, 5-imino-1,2,4-thiazolidin-3-ones LVIII are obtained ([33]).

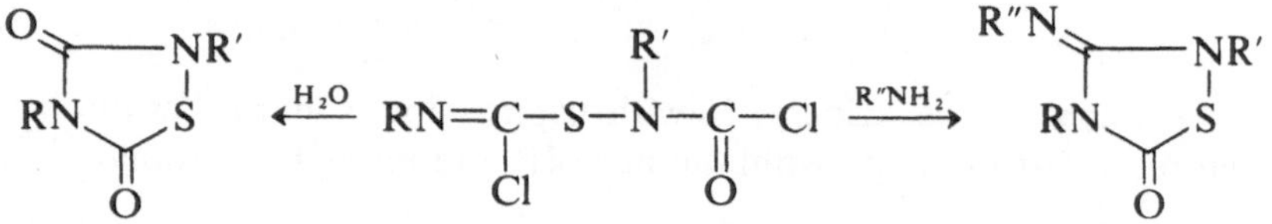

LVII   LVIII

The high reactivity of 1-(chlorothio)formimidoyl chlorides toward addition to double bond systems is further verified by their facile reaction with arylalkylketones ([34a]). Thus, the previously unreported 2-imino-1,3-oxathioles LIX are obtained on addition of 1-(chlorothio)formimidoyl chlorides to arylalkylketones in diethyl ether at room temperature, and the reported yields range from 34–64 % ([34a]).

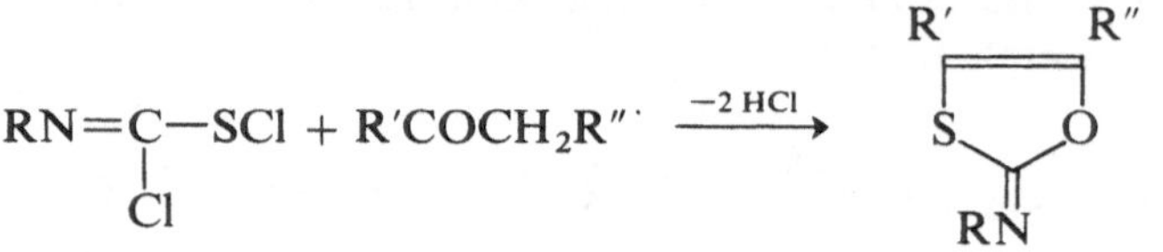

LIX

Elimination of alkylhalide from 1-haloformimidates and 1-halothio-formimidates yields isocyanates and isothiocyanates, respectively ([12,22]).

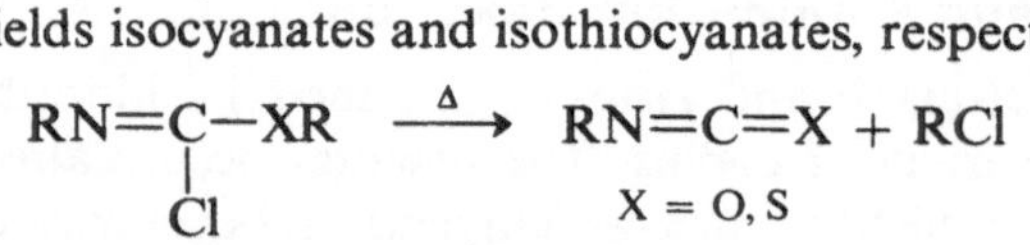

$$RN{=}\overset{\underset{Cl}{|}}{C}{-}XR \overset{\Delta}{\longrightarrow} RN{=}C{=}X + RCl$$
$$X = O, S$$

V. REFERENCES

1. Allenstein, E., and Quis, P., Ber. **97**, 3162 (1964).
2. Bacon, R. G. R., Irwin, R. S., Pollock, J. McC., and Pullin, A. D. E., J. Chem. Soc. 764 (1958).
3. Dyson, G. M., and Harrington, T., J. Chem. Soc. 374 (1942).
4. Eilingsfeld, H., and Möbius, L., Ber. **98**, 1293 (1965).
5. Eilingsfeld, H., and Möbius, L., Belgian Pat. 660,941 (1965); Chem. Abstr. **64**, 3364 (1966).
6. Feher, F., and Weber, H., Ber. **91**, 2523 (1958).
7. Folin, O., Amer. Chem. J. **19**, 323 (1897).

8. Gross, H., and Gloede, J., *Ber.* **96**, 1387 (1963).

8a. Hantzsch, A., and Mai, L., *Ber.* **28**, 977 (1895).

9. Havlik, A. J., and Wald, M. M., *J. Am. Chem. Soc.* **77**, 5171 (1951).

10. Ivanova, Zh. M., Kirsanova, N. A., and Derkach, G. I., *Zh. Organ. Khim.* **1**, 2186 (1965); *Chem. Abstr.* **64**, 11123 (1966).

11. Kodama, Y., *Yuki Gosei Kagaku Kyokai Shi* **23**, 57 (1965); *Chem. Abstr.* **62**, 9132 (1965).

12. Kodama, Y., and Sekiba, T., Japan Pat. 11,389 (1965); *Chem. Abstr.* **63**, 13293 (1965).

13. Kodama, Y., and Sekiba, T., Japan Pat. 430 (1966); *Chem. Abstr.* **64**, 11230 (1966).

14. Kodama, Y., and Sekiba, T., Japan Pat. 431 (1966); *Chem. Abstr.* **64**, 11231 (1966).

15. Kühle, E., German Pat. 1,126,380 (1962).

16. Kühle, E., and Eue, L., German Pat. 1,138,039 (1962).

17. Kühle, E., Eue, L., and Bayer, O., German Pat. 1,137,000 (1962).

18. Kühle, E., *Angew. Chem. Intern. Ed.* **1**, 647 (1962).

19. Kühle, E., Belgian Pat. 660,171 (1965); *Chem. Abstr.* **64**, 603 (1966).

20. Lengfeld, F., and Stieglitz, J., *Am. Chem. J.* **16**, 70 (1894).

21. Mukaiyama, T., Fujisawa, T., and Hyugaji, T., *Bull. Chem. Soc. Japan* **35**, 687 (1962).

22. Mukaiyama, T., Fujisawa, T., and Mitsunobo, O., *Bull. Chem. Soc. Japan* **35**, 1104 (1962).

23. Nef, J. U., *Ann.* **270**, 295 (1892).

24. Nef, J. U., *Ann.* **280**, 298 (1894).

25. Nef, J. U., *Ann.* **287**, 301 (1895).

26. Neidlein, R., and Haussmann, W., *Angew. Chem. Intern. Ed.* **4**, 521 (1965).

27. Neidlein, R., and Haussmann, W., *Tetrahedron Letters* 5401 (1966).

28. Neidlein, R., Haussmann, W., and Heukelbach, E., *Ber.* **99**, 1252 (1966).

29. Ottmann, G., and Hooks, H., *Angew. Chem. Intern. Ed.* **4**, 432 (1965).

30. Ottmann, G., and Hooks, H., *J. Org. Chem.* **31**, 838 (1966).

31. Ottmann, G., and Hooks, H., *Angew. Chem. Intern. Ed.* **5**, 250 (1966).

32. Ottmann, G., and Hooks, H., *Angew. Chem. Intern. Ed.* **5**, 672 (1966).

33. Ottmann, G., and Hooks, H., United States Pat. 3,282,950 (1966); *Chem. Abstr.* **66**, 10939 (1967).

34. Ottmann, G., and Hooks, H., United States Pat. 3,301,894 (1967).

34a. Ottmann, G., and Hooks, H., *Angew. Chem.* **79**, 470 (1967).

35. Shipton, G. O., British Pat. 945,835; *Chem. Abstr.* **60**, 10700 (1964).

35a. Smith, W. R., *Am. Chem. J.* **16**, 392 (1894).

36. Ulrich, H., Tucker, B., and Sayigh, A. A. R., *Angew. Chem. Intern. Ed.* **6**, 636 (1967).

37. Weissauer, H., and Weiser, D., German Pat. 1,187,620 (1965); *Chem. Abstr.* **62**, 16254 (1965).

Chapter 6

HYDROXAMOYL HALIDES

I. INTRODUCTION

Substituted hydroxamoyl chlorides were synthesized for the first time by Werner and his students in 1894. In the same year, the parent hydroxamoyl chloride was obtained by Nef.

The nomenclature of this class of compounds is not at all clear. For example, the following names can be found in the literature: hydroxamic acid chlorides, hydroxamic chlorides, hydroximic chlorides (Beilstein lists them as "Hydroximsäure chloride"), and acyl and aroylchloride oximes. The latter names are used by *Chemical Abstracts*. Although the name benzoyl chloride oxime for benzhydroxamoyl chloride is formally correct, it does not reflect the close relationship between hydroxamoyl chlorides and hydroxamic acids, their hydrolysis products. In order to be consistent with the nomenclature used in the earlier chapters, I prefer the term hydroxamoyl to hydroxamic chloride. The interrelationship of the halides with the corresponding acids is shown below, and R is representative of the alkyl, aryl, acyl, aroyl, and carbalkoxy group.

$$R-\underset{\underset{OH}{|}}{C}=NOH \qquad\qquad R-\underset{\underset{Cl}{|}}{C}=NOH$$

Hydroxamic acid Hydroxamoyl chloride

$$R-\underset{\underset{OH}{|}}{C}=NOR' \qquad\qquad R-\underset{\underset{Cl}{|}}{C}=NOR'$$

Hydroxamic acid ester Hydroxamic ester chloride

The standard displacement reactions of the chloro group and the elimination to nitrile oxides have been studied by Werner and his co-workers. Weygand and Bauer in 1927 studied the reaction of hydroxamoyl chlorides with alkali salts of ethyl cyanoacetate; they obtained isoxazole derivatives. Likewise, Quilico and his students utilized hydroxamoyl chlorides to synthesize numerous isoxazole derivatives.

One of the most important reactions of hydroxamoyl chlorides involves dehydrochlorination to nitrile oxides, and the renewed interest in the 1,3-cycloaddition reactions has focused attention again on the nitrile oxide precursors.

In view of the ready availability of hydroxamoyl chlorides, and their numerous possibilities for transformations, they will most certainly become more important in the years to come.

II. SYNTHESIS

A. Halogenation of Aldoximes

The chlorination of aldoximes I is the classical method of synthesis of hydroxamoyl chlorides II. Werner and his co-workers in 1894 ([65–69]) used this method to synthesize a variety of aromatic hydroxamoyl chlorides.

$$\underset{\text{I}}{RCH{=}NOH} + Cl_2 \longrightarrow \underset{\text{II}}{RCCl{=}NOH} + HCl$$

This reaction can be conducted using chlorine as the chlorinating agent in a variety of solvents, such as chloroform ([4,66]), diethylether ([31]), or, preferably, in concentrated hydrochloric acid ([40,41,66,75]). In a recent patent the use of t-butylhypochlorite in methanol is described. Rheinboldt ([51]) used nitrosyl chloride as the chlorinating agent ([11]). If the aryl groups in arylaldoximes contain aliphatic side-chains, competition with side-chain chlorination is encountered, and addition of hydrogen chloride to the corresponding nitrile oxide is the method of choice ([19]). Chlorination of the ring has been observed, as for example, in the chlorination of salicylaldoxime and 2-methoxybenzaldoxime ([75]). Aliphatic aldoximes are similarly chlorinated in aqueous hydrochloric acid to afford the hydroxamoyl chlorides in good yield ([10,76,77]). However, further chlorination, or addition of chlorine to double bonds, has been encountered in some instances ([20]).

4-Phenylbenzhydroxamoyl Chloride ([75]). To 10 g 4-phenylbenzaldoxime suspended in 50 ml of 8N hydrochloric acid at 0°C an equivalent amount of chlorine is added. The resulting crude hydroxamoyl chloride (11 g, 94%) is removed by filtration, and recrystallization from cyclohexane yields pure 4-phenylbenzhydroxamoyl chloride, m.p. 129–130°C.

Trimethylacethydroxamoyl Chloride ([77]). To a suspension of 50.5 g (0.5 mole) of trimethylacetaldoxime in 150 ml of conc. hydrochloric acid and 350 g ice, 35.5 g (0.5 mole) of chlorine is added. After stirring for two hours the reaction mixture is extracted several times with diethyl ether (300 ml total). The dried ether extract is evaporated and the residue is dissolved in 100 ml of petroleum ether. Upon chilling with dry ice 63.0 g (93%) of trimethylacethydroxamoyl chloride, m.p. 33°C, is obtained.

The volatile compound should be handled under a hood because it strongly irritates the mucous membranes.

If the chlorination of aliphatic oximes is conducted in diethyl ether at $-60°C$, aliphatic nitroso compounds III are obtained, which are in equilibrium with their dimeric form IV. The nitroso compounds on standing gradually isomerize to the hydroxamoyl chlorides II ([6]).

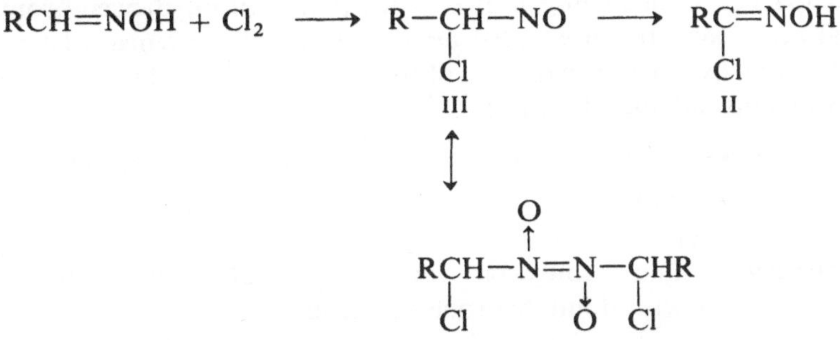

$$RCH{=}NOH + Cl_2 \longrightarrow \underset{\underset{III}{\overset{|}{Cl}}}{R{-}CH{-}NO} \longrightarrow \underset{\underset{II}{\overset{|}{Cl}}}{RC{=}NOH}$$

$$\underset{IV}{\underset{\overset{|}{Cl}\quad\overset{|}{O}\ \overset{|}{Cl}}{RCH{-}N{=}N{-}CHR}}$$

However, isonitrosoacetone reacts with chlorine in chloroform with formation of the oxime V rather than the hydroxamoyl chloride ([2]). The formation of the hydroxamoyl chloride has previously been postulated ([54]).

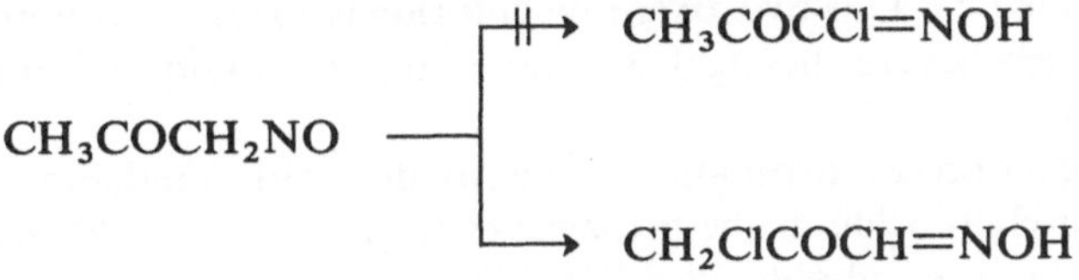

$$CH_3COCH_2NO \left\{ \begin{array}{l} \overset{|}{\nrightarrow} CH_3COCCl{=}NOH \\[2em] \longrightarrow CH_2ClCOCH{=}NOH \\ \qquad\qquad\qquad V \end{array} \right.$$

The esters of hydroxamic acids upon treatment with phosphorus pentachloride yield hydroxamic ester chlorides VI ([32]).

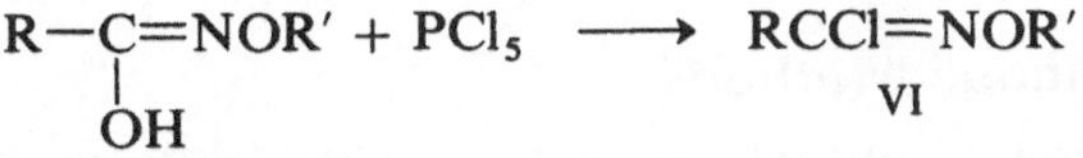

$$\underset{\overset{|}{OH}}{R{-}C{=}NOR'} + PCl_5 \longrightarrow \underset{VI}{RCCl{=}NOR'}$$

B. From Nitrile Oxides

The addition of hydrogen chloride to nitrile oxides VII produces hydroxamoyl chlorides in high yield ([19,21]). However, this method is only of limited interest because nitrile oxides are normally prepared from hydroxamoyl halides.

$$\underset{VII}{RCNO} + HCl \longrightarrow RCCl{=}NOH$$

Hydroxamoyl chlorides from nitrile oxides. General procedure ([19]). To a solution of 0.01 mole of the corresponding nitrile oxide in 10 ml of methylene chloride and 5 ml of anhydrous diethyl ether, excess hydrogen chloride is added with ice-cooling. Evaporation of the solvents under vacuum, and recrystallization from benzene/ligroine affords the hydroxamoyl chlorides, usually in high yield.

The reaction of cyanogen-di-N-oxide (VIII) with hydrogen bromide in glacial acetic acid has been used as a quantitative analytical method ([17]), and aqueous hydrochloric or hydrobromic acid can be added to afford the corresponding dihalide (IX) ([16,17,22]).

$$\underset{\text{VIII}}{\underset{\text{X = Cl, Br}}{ONC-CNO}} + 2\,HX \longrightarrow \underset{\text{IX}}{HON=\underset{X}{\overset{|}{C}}-\underset{X}{\overset{|}{C}}=NOH}$$

The parent hydroxamoyl chloride X was obtained by heating salts of fulminic acid XI with dilute hydrochloric acid ([35]).

$$\underset{\text{XI}}{NaON=C:} + HCl \longrightarrow \underset{\text{X}}{HON=CHCl}$$

The hydroxamoyl chloride X is stable at 0°C for some time, but upon warming rapid decomposition occurs, with formation of hydroxylamine hydrochloride ([35]). This hydroxamoyl chloride is a lachrimator and vesicant, and it causes severe headaches; therefore, it should be handled with caution ([35]).

The reaction of aromatic nitrile oxides with methanesulfonyl and benzylsulfonyl chloride in the presence of triethylamine yields the sulfonate esters of hydroxamoyl chlorides XII ([61]).

$$RCNO + R'SO_2Cl \longrightarrow \underset{\text{XII}}{R-CCl=NOSO_2R'}$$

C. Miscellaneous Methods

The reaction of phenylnitromethane with dry hydrogen chloride in diethylether affords benzhydroxamoyl chloride ([36,58]).

$$C_6H_5CH_2NO_2 + HCl \longrightarrow C_6H_5CCl=NOH$$

Instead of the hydrogen chloride other acid halides and phosphorus oxychloride can be used ([36]).

The direct synthesis of aroylhydroxamoyl chlorides from acetophenones using nitrosyl chloride ([50]), isoamyl nitrite ([7]), or isopropyl nitrite ([5]) has been reported. Isopropyl nitrite, especially in the presence of gaseous hydrogen chloride, can be used to prepare a variety of aroylhydroxamoyl chlorides XIII in good yield ([5]).

$$RC_6H_4COCH_3 + R'ONO \xrightarrow{HCl} RC_6H_4COCCl=NOH$$
                                                            XIII

Aroylhydroxamoyl chlorides. General procedure ([5]). To 0.025 moles of the acetophenone derivative in 100 ml of anhydrous diethyl ether, dry hydrogen chloride and a solution of 0.0375 moles of isopropyl nitrite in 40 ml of anhydrous diethyl ether is added simultaneously over a period of 30 min at 10–20°C. After addition of hydrogen chloride for an additional 2 hours, the solvent is removed by evaporation, and the residue is dissolved again in 100 ml of diethyl ether and an additional amount of 0.025 mole of isopropyl nitrite is added dropwise and simultaneously with hydrogen chloride at 20°C. Addition of hydrogen chloride was continued for approximately two hours, the solvent was evaporated, and the aroylhydroxamoyl chloride was purified by crystallization from benzene or a mixture of benzene and carbon tetrachloride.

This stepwise addition of isopropylnitrite was necessitated by the formation of by-products due to the presence of the generated isopropanol. Yields: 31–89%.

Reduction of trihalonitrosomethane with hydrogen sulfide yields halohydroxamoyl halides ([44,45]).

In perfluorohydroxamoyl fluorides, ready displacement of the mobile fluoro group has been observed ([10]). For example, trifluoroacethydroxamoyl fluoride (XIV) undergoes reaction with hydrogen chloride and hydrogen bromide at −78°C to yield trifluoroacethydroxamoyl halides XV ([10]).

$$\underset{\substack{| \\ F \\ XIV}}{CF_3C=NOH} + HX \longrightarrow \underset{\substack{| \\ X \\ XV}}{CF_3C=NOH} + HF$$

X = Cl, Br

The hitherto-obtained hydroxamoyl halides are listed in Tables I–III.

TABLE I
Aliphatic Hydroxamoyl Halides

$$R-C=NOH$$
$$\underset{X}{|}$$

R	X	M.p., °C (B.p./°C)	Yield, %	Reference
H	Br	37	—	38
H	I	65*	—	33
Cl	Cl	39–40 (47/18)	—	44
Br	Br	68–69 (73–75/14)	—	45
I	I	69	—	45
CN	Cl	55–56	—	24, 57
HON=CCl	Cl	204	—	57, 75
HON=CBr	Br	110*	—	24
HON=CBr	Cl	222–223	—	42, 43, 57
HON=CI	I	172	—	39
HON=CNH$_2$	Cl	109*	—	56
CF$_3$	Cl	98–102	55	10
CF$_3$	Br	52–55†	—	10
CH$_3$	Cl	−2	—	41, 76
(CH$_3$)$_3$C	Cl	33	93	76, 77
(C$_2$H$_5$)$_2$CH	Cl	17	—	76

* Melts with decomposition.
† B.p. at 72 mm.

TABLE II
Aromatic Hydroxamoyl Halides

$$R-\underset{\underset{X}{|}}{C}=NOH$$

R	X	M.p., °C	Method of synthesis	Yield, %	Reference
C_6H_5	Cl	45	A	—	65, 75
$4\text{-}CH_3C_6H_4$	Cl	68–71	A	—	51
$2,4,6\text{-}(CH_3)_3C_6H_2$	Cl	72	B	76	19
$2,3,5,6\text{-}(CH_3)_4C_6H$	Cl	124–126	B	73	19
$3\text{-}ClC_6H_4$	Cl	75–80	A	—	75
$4\text{-}ClC_6H_4$	Cl	82–86	A	—	75
$2,4\text{-}Cl_2C_6H_3$	Cl	100–105	A	—	75
$3,5\text{-}Cl_2C_6H_3$	Cl	172–174	A	—	75
$2\text{-}(CH_3O)\text{-}5\text{-}ClC_6H_3$	Cl	115–125	A	—	75
$4\text{-}(CH_3O)\text{-}3\text{-}ClC_6H_3$	Cl	106.5	A	36	28
$4\text{-}CH_3OC_6H_4$	Cl	88–89	A	80	28
$2,4,6\text{-}(CH_3O)_3C_6H_2$	Cl	108–111	B	85	19
$2\text{-}O_2NC_6H_4$	Cl	90–93	A	—	65, 75
$3\text{-}O_2NC_6H_4$	Cl	96	A	—	65, 75
$4\text{-}C_6H_5C_6H_4$	Cl	129–130	A	94	75
$4\text{-}(HON{=}ClC)C_6H_4$	Cl	188	A	—	65, 75
$4\text{-}(HON{=}ClC)\text{-}2,3,5,6\text{-}(CH_3)_4C_6$	Cl	158–160	B	83	19
$2\text{-}C_5H_4N^*$	Cl	144	A	70	39
$3\text{-}C_5H_4N^*$	Cl	165	A	70	39
$4\text{-}C_5H_4N^*$	Cl	191	A	52	39

* C_5H_4N = Pyridyl.

TABLE III
Aroylhydroxamoyl Chlorides

$$R-\underset{\underset{O}{\|}}{C}-\underset{\underset{X}{|}}{C}=NOH$$

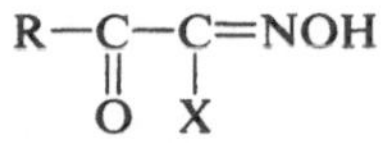

R	X	M.p., °C	Yield, %	Reference
H_2N	Cl	166*	—	55
H_2N	I	154–155	—	57
HO	Cl	129*	—	57
C_2H_5O	Br	93	—	30
C_2H_5O	I	95–96	—	57
C_6H_5	Cl	132	67	5, 29
$4\text{-}ClC_6H_4$	Cl	118–119	31	5, 29
$4\text{-}BrC_6H_4$	Cl	116	63	5
$3\text{-}CH_3OC_6H_4$	Cl	110.5	83	5
$3\text{-}O_2NC_6H_4$	Cl	149.5	89	5
$4\text{-}O_2NC_6H_4$	Cl	137.5	48	5
$3\text{-}AcOC_6H_4$	Cl	111	30	5
$3\text{-}PhCOOC_6H_4$	Cl	132–133	81	5

* Melts with decomposition.

III. PHYSICOCHEMICAL PROPERTIES

Hydroxamoyl halides are colorless-to-yellow solid compounds, which sometimes melt with decomposition because of elimination of hydrogen halide. Exceptions are some of the lower aliphatic derivatives, which are distillable liquids.

The infrared spectra of several hydroxamoyl chlorides were investigated by Navech and co-workers ([34]), and it was demonstrated by ultraviolet spectroscopy that the hydrolysis of hydroxamoyl chlorides to nitrile oxides, as well as the reverse reaction, obeys second-order kinetics ([53,54]).

IV. CHEMICAL BEHAVIOR

A. Reaction with Oxygen–Hydrogen Bonds

The reaction of hydroxamoyl chlorides with water affords hydroxamic acids XVI ([3,65,69]).

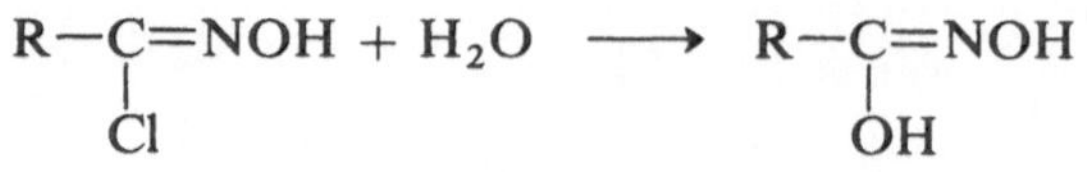

$$R-\underset{\underset{Cl}{|}}{C}=NOH + H_2O \longrightarrow R-\underset{\underset{OH}{|}}{C}=NOH$$

XVI

Nucleophilic attack of hydroxamoyl chlorides by alkoxide ion similarly affords the corresponding ethers (XVII) ([65,67]).

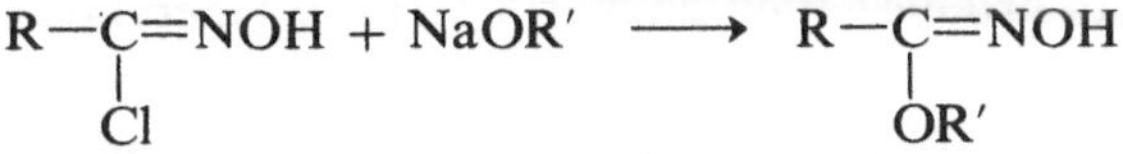

$$\begin{matrix} R-C=NOH \\ | \\ Cl \end{matrix} + NaOR' \longrightarrow \begin{matrix} R-C=NOH \\ | \\ OR' \end{matrix}$$

XVII

If hydroxamoyl chlorides are treated with the silver salt of benzoic acid the N-benzoate XVIII is obtained *via* rearrangement of the initially formed O-benzoate XIX ([59,68]).

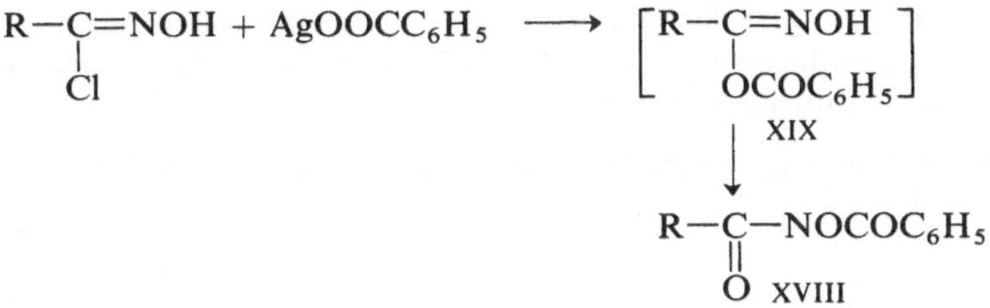

$$\begin{matrix} R-C=NOH \\ | \\ Cl \end{matrix} + AgOOCC_6H_5 \longrightarrow \left[\begin{matrix} R-C=NOH \\ | \\ OCOC_6H_5 \end{matrix}\right]$$

XIX

$$\begin{matrix} R-C-NOCOC_6H_5 \\ \| \\ O \end{matrix}$$ XVIII

B. Reactions with Sulfur–Hydrogen Bonds

The reaction of aliphatic hydroxamoyl chlorides with thiophenols in the presence of triethylamine was used by Benn to synthesize the thiohydroxamic ethers XX ([4]).

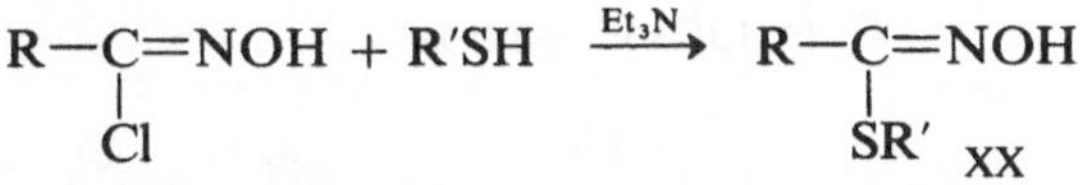

$$\begin{matrix} R-C=NOH \\ | \\ Cl \end{matrix} + R'SH \xrightarrow{Et_3N} \begin{matrix} R-C=NOH \\ | \\ SR' \end{matrix}$$ XX

Similarly, displacement of the chloro group in XXI by the sodium salt of 1-thio-β-D-glucopyranosyl to produce XXII has been observed ([11]).

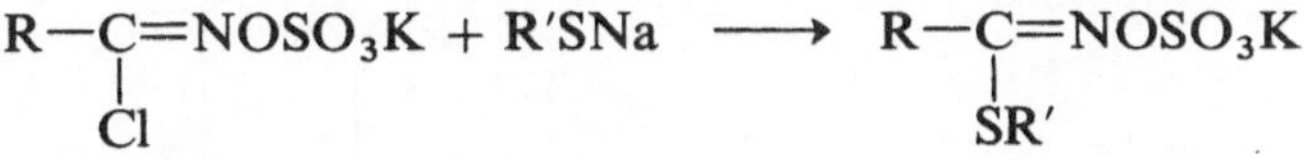

$$\begin{matrix} R-C=NOSO_3K \\ | \\ Cl \end{matrix} + R'SNa \longrightarrow \begin{matrix} R-C=NOSO_3K \\ | \\ SR' \end{matrix}$$

XXI   XXII

However, reaction of benzhydroxamoyl chloride with xanthates yields phenyl isothiocyanate and carbonyl sulfide ([14]).

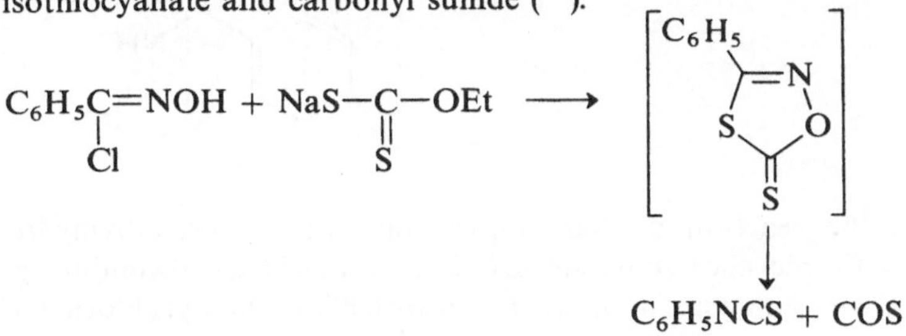

$$\begin{matrix} C_6H_5C=NOH \\ | \\ Cl \end{matrix} + \begin{matrix} NaS-C-OEt \\ \| \\ S \end{matrix} \longrightarrow \left[\begin{matrix} C_6H_5 \\ \diagup = N \\ S \diagdown \diagup O \\ \| \\ S \end{matrix}\right]$$

$$C_6H_5NCS + COS$$

C. Reaction with Nitrogen–Hydrogen Bonds

The reaction of hydroxamoyl chlorides with ammonia, or alkylamines, or arylamines to produce amid-oximes XXIII is a general reaction ([3,9,10,30, 35,65,66,69]).

$$R-\underset{\underset{Cl}{|}}{C}=NOH + R'NH_2 \longrightarrow R-\underset{\underset{NHR'}{|}}{C}=NOH$$

$$\text{XXIII}$$

Aminocarboxylic acid esters undergo reaction with hydroxamoyl chloride in a similar fashion to yield the substitution products XXIV; however, saponification of the ester group occurs, unless carboxyethyl-hydroxamoyl chloride is being used ([9]).

$$R-\underset{\underset{Cl}{|}}{C}=NOH + H_2NCH_2COOR' \longrightarrow R-\underset{\underset{NHCH_2COOH}{|}}{C}=NOH$$

$$\text{XXIV}$$

From methyl anthranilate and carboxyethylhydroxamoyl chloride the corresponding amide-oxime (XXV) is obtained, which upon saponification with a base or an acid yields the quinazoline derivative XXVI ([9]).

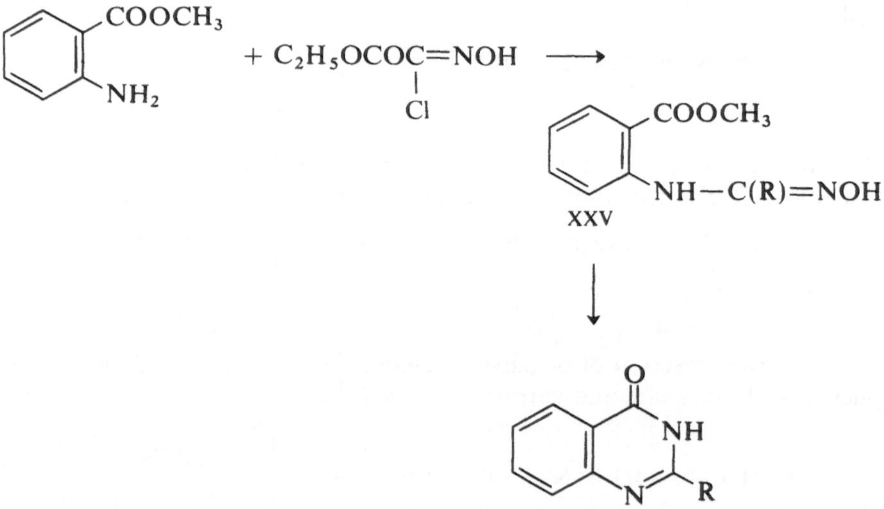

In the reaction of 2-aminopyridine with carboxyethylhydroxamoyl chloride the bicyclic compound XXVII is obtained ([9]), and amidines generally form 1,2,4-oxadiazoles upon treatment with hydroxamoyl chlorides ([33]).

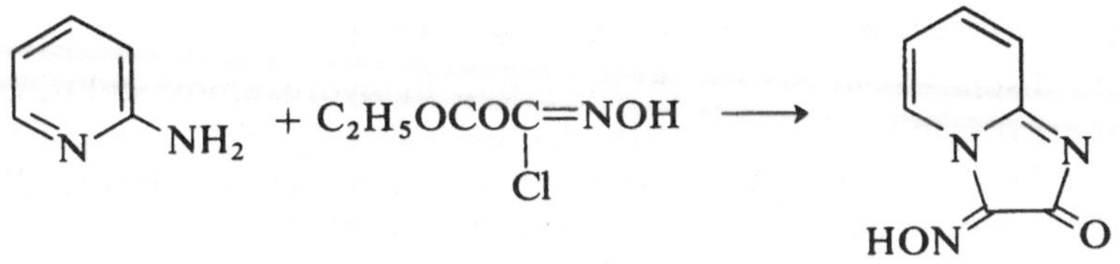

XXVII

When hydroxylamine is allowed to react with hydroxamoyl chlorides, attack on nitrogen occurs with formation of the hydroxylamino-oximes XXVIII ([30]).

$$R-C=NOH + H_2NOH \longrightarrow R-C=NOH$$

XXVIII

The reaction of hydroxamoyl chlorides with hydrazine has also been investigated and hydrazide-oximes XXIX ([68,69]) are obtained. However, the hydrazine derivative XXX yields N-substituted 2-amino-5-methyl-mercapto-1,3,4-thiadiazoles XXXI upon reaction with hydroxamoyl chlorides ([9]).

$$RC=NOH + H_2NNH_2 \longrightarrow R-C=NOH$$

XXIX

$+ H_2NNH-C-SCH_3 \longrightarrow$

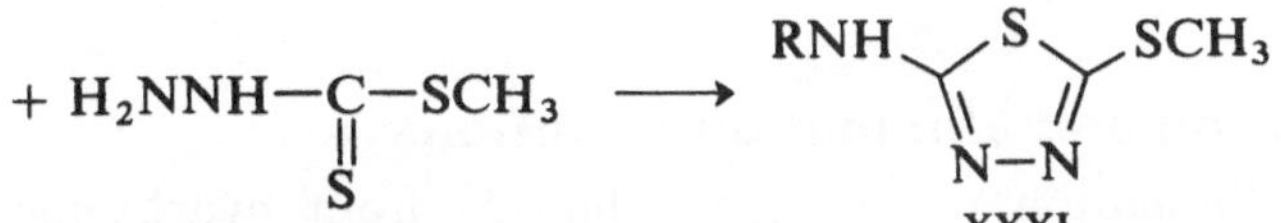

XXX   XXXI

Azide ion attacks hydroxamoyl chloride readily, and the linear reaction product cyclizes instantaneously to yield the isolated tetrazole derivative XXXII ([13]).

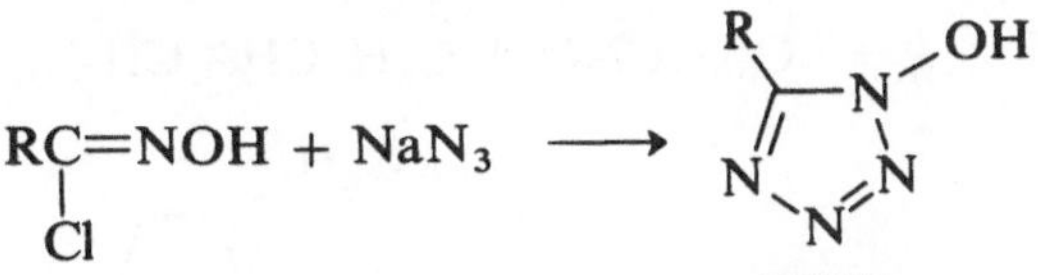

$$RC=NOH + NaN_3 \longrightarrow$$

XXXII

The reaction of hydroxamoyl chlorides with silver nitrate yields nitrolic acids XXXIII ([41]).

$$C_6H_5C=NOH + AgNO_3 \longrightarrow C_6H_5C=NOH$$

XXXIII

D. Reaction with Carbon–Hydrogen Bonds

The displacement of the chloro group in hydroxamoyl chlorides by sodium ethyl cyanoacetate was first demonstrated by Weygand and Bauer in 1927. ([71]). The intermediate linear compound **XXXIV** cyclizes readily to the corresponding isoxazole derivative (**XXXV**) ([49,74]).

$$R-\underset{\underset{Cl}{|}}{C}=NOH + NaCH(CN)COOEt \longrightarrow R-\underset{\underset{CH(CN)COOEt}{|}}{C}=NOH$$

XXXIV

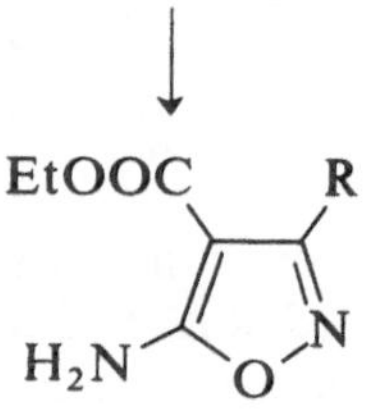

XXXV

This reaction has been used by Quilico and his students ([46–48]) and by Vecchio and his co-workers ([8,62,63,64]) to synthesize a great variety of isoxazole derivatives. The substrates used were cyanoketones ([46]), acetoacetates ([46,47]), β-oxo-esters ([62,64]), and malonic acid amides ([63]).

E. Addition and Elimination Reactions

The elimination of hydrogen chloride from hydroxamoyl chlorides occurs in refluxing toluene, as evidenced by the fact that the 1,3-cycloadducts of benzonitrile oxide with olefins and acetylenes are obtained ([1,12]). For example, heating of benzhydroxamoyl chloride in the presence of styrene affords the cycloadduct **XXXVI** ([12]).

$$C_6H_5CCl=NOH \xrightarrow[-HCl]{\Delta} C_6H_5CNO + C_6H_5CH=CH_2 \longrightarrow$$

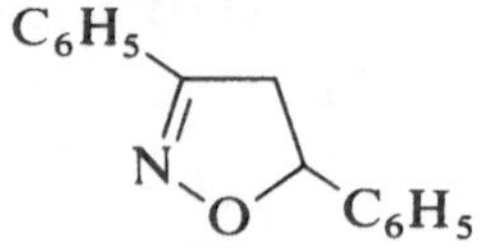

XXXVI

However, the regular procedure of dehydrohalogenation of hydroxamoyl chlorides involves the use of a base as the hydrogen chloride scavenger. While sterically hindered arylhydroxamoyl chlorides afford arylnitrile oxides in high yield, phenylhydroxamoyl chloride adds partially to the generated benzonitrile oxide to afford the cycloadduct XXXVII, which eliminates hydrogen chloride to yield the oxadiazole derivative XXXVIII ([71]).

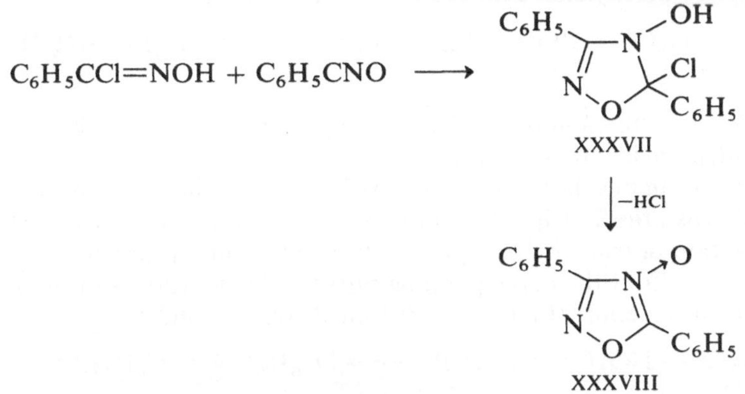

$$C_6H_5CCl{=}NOH + C_6H_5CNO \longrightarrow$$

XXXVII

$$\downarrow -HCl$$

XXXVIII

In the dehydrochlorination of acethydroxamoyl chloride, using aqueous sodium carbonate, interaction of the hydroxamic acid and the generated nitrile oxide with formation of the 1:1 adduct XXXIX was observed ([71]).

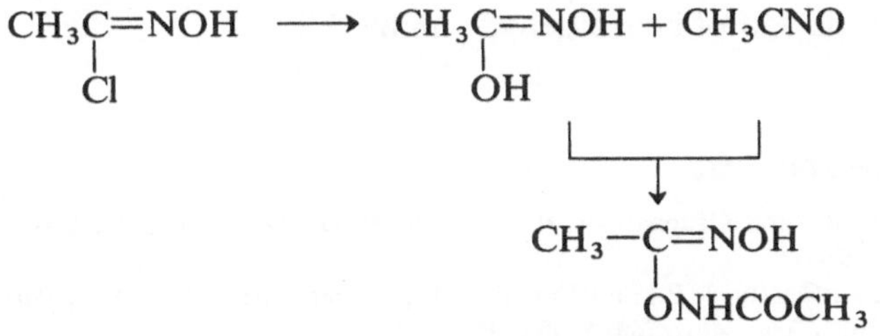

$$CH_3\underset{Cl}{C}{=}NOH \longrightarrow CH_3\underset{OH}{C}{=}NOH + CH_3CNO$$

$$CH_3{-}\underset{ONHCOCH_3}{C}{=}NOH$$

XXXIX

In order to avoid side reactions, the dehydrochlorination of hydroxamoyl chlorides should be conducted at 0°C; using aqueous sodium hydrogen carbonate as the hydrogen chloride acceptor ([71]). In many cases an inert solvent system may offer advantages. For example, Huisgen and his co-workers ([25-27]) conduct the dehydrochlorination in diethyl ether at -20°C, using triethylamine as the hydrogen chloride scavenger. Nevertheless, complications due to the reaction of the nitrile oxide with tertiary amines can occur ([73]).

F. Miscellaneous Reactions

The reaction of hydroxamoyl chlorides with the pyridine-SO_3 complex affords the salts of the hydroxamic acid ester chlorides XL ([11]).

$$RCCl{=}NOH + SO_3 \longrightarrow R{-}CCl{=}NOSO_3K$$

XL

The Grignard reaction of hydroxamoyl chlorides affords ketone oximes. For example, benzhydroxamoyl chloride, on treatment with phenylmagnesium bromide, yields benzophenone oxime (XLI) ([37]).

$$C_6H_5CCl{=}NOH + C_6H_5MgBr \longrightarrow (C_6H_5)_2C{=}NOH$$
$$\text{XLI}$$

However, this and some of the subsequent reactions most likely proceed *via* a nitrile oxide intermediate.

For example, heating of benzhydroxamoyl chloride and triphenylphosphorus affords triphenylphosphine oxide and benzonitrile ([15]). Since the reaction of trimethylphosphite with nitrile oxides proceeds in a similar manner to afford the corresponding nitriles ([18]), dehydrohalogenation and subsequent reaction of the generated nitrile oxide is indicated.

$$C_6H_5CCl{=}NOH + (C_6H_5)_3P \longrightarrow C_6H_5CN + (C_6H_5)_3PO + HCl$$

In the reaction of benzhydroxamoyl chloride and benzoquinone in the presence of sodium ethoxide the 1,3-cycloadduct of benzonitrile oxide was isolated ([48]).

The reduction of benzhydroxamoyl chloride with stannous chloride affords the tin salt of benzaldehyde imine ([52]).

V. REFERENCES

1. Arbasino, M., and Gruenanger, P., *Ric. Sci. Rend., Sez. A* **7**, 561 (1964); *Chem. Abstr.* **63**, 6985 (1965).

2. Armand, J., Guette, J. P., and Valentini, F., *Compt. Rend. Acad. Sci., Paris, Ser. C* **263**, 1388 (1966); *Chem. Abstr.* **66**, 5149 (1967).

3. Avogadro, L., and Tavolo, G., *Gazz. chim. Ital.* **55**, 323 (1925).

4. Benn, M. H., *Can. J. Chem.* **42**, 2393 (1964).

5. Brachwitz, H., *Z. Chem.* **6**, 313 (1966); *Chem. Abstr.* **65**, 16891 (1966).

6. Casnati, G., and Ricca, A., *Tetrahedron Letters* 327 (1967).

7. Claisen, L., and Manasse, O., *Ann.* **274**, 95 (1893).

8. D'Alcontres, G. S., Vecchio, G. L., and Lamonica, G., *Gazz. Chim. Ital.* **91**, 1005 (1961).

9. Dornow, A., and Fischer, K., *Chem. Ber.* **99**, 72 (1966).

10. Dyatkin, B. L., Gevorkyan, A. A., and Knunyants, I. L., *Zh. Obshch. Khim.* **36**, 1326 (1966); *Chem. Abstr.* **65**, 16855 (1966).

11. Ettlinger, M. D., and Dateo, G. P., Jr., United States Pat. 3,146,227 (1964); *Chem. Abstr.* **62**, 6550 (1965).

12. Finzi, P. V., and Arbasino, M., *Ric. Sci., Rend., Sez. A* **8**, 1484 (1965); *Chem. Abstr.* **65**, 7162 (1966).

13. Foster, M. O., *J. Chem. Soc.* **95**, 184 (1909).

14. Fusco, R., and Músante, C., *Gazz. Chim. Ital.* **68**, 665 (1938); *Chem. Abstr.* **33**, 2518 (1939).

15. Gruenanger, P., and Langella, M. R., *Atti Accad. Naz. Lincei. Rend., Classe Sci. Fis., Mat. Nat.* **36**, 387 (1964); *Chem. Abstr.* **62**, 3973 (1965).

16. Grundmann, C., *Angew. Chem. Intern. Ed.* **2**, 260 (1963).
17. Grundmann, C., Mini, V., Dean, J. M., and Frommeld, H. D., *Ann.* **687**, 191 (1965).
18. Grundmann, C., and Frommeld, H. D., *J. Org. Chem.* **30**, 2077 (1965).
19. Grundmann, C., and Dean, J. M., *J. Org. Chem.* **30**, 2809 (1965).
20. Grundmann, C., *Houben-Weyl, Methoden der Organischen Chemie*, Fourth Edition, Vol. 10, Part 3, G. Thieme, Stuttgart (1965), p. 847.
21. Grundmann, C., and Frommeld, H. D., *J. Org. Chem.* **31**, 157 (1966).
22. Grundmann, C., and Frommeld, H. D., *J. Org. Chem.* **31**, 4235 (1966).
23. Houben, H., and Kauffmann, H., *Ber.* **46**, 2823 (1913).
24. Houben, H., and Kauffmann, H., *Ber.* **46**, 3821 (1913).
25. Huisgen, R., and Mack, W., *Tetrahedron Letters* 583 (1961).
26. Huisgen, R., Mack, W., and Anneser, E., *Tetrahedron Letters* 587 (1961).
27. Huisgen, R., and Christi, M., *Angew. Chem.* **79**, 471 (1967).
28. Kinney, C. R., Smith, E. W., Wooley, C., and Willey, A. R., *J. Am. Chem. Soc.* **55**, 3418 (1933).
29. Levin, N., and Hartung, W. H., *J. Org. Chem.* **7**, 408 (1942).
30. Ley, H., *Ber.* **31**, 2127 (1898).
31. Ley, H., and Ulrich, M., *Ber.* **47**, 2938 (1914).
32. Lossen, W., *Ber.* **18**, 1189 (1885).
33. Musante, C., *Gazz. Chim. Ital.* **68**, 331 (1938).
34. Navech, J., Mathis, F., and Mathis-Noel, R., *Compt. Rend.* **244**, 1913 (1957).
35. Nef, J., *Ann.* **280**, 291 (1894).
36. Nenitzescu, C. D., and Isacescu, D. A., *Bull. Soc. Chim. Romania* **14**, 53 (1932); *Chem. Abstr.* **27**, 964 (1933).
37. Palazzo, G., *Gazz. Chim. Ital.* **77**, 214 (1947); *Chem. Abstr.* **42**, 904 (1948).
38. Palazzo, G., *Gazz. Chim. Ital.* **78**, 654 (1948).
39. Paul, R., and Tchelitcheff, S., *Bull. Soc. Chim. France* 2215 (1962).
40. Perold, G. W., Steyn, A. P., and v. Reiche, F. V. K., *J. Am. Chem. Soc.* **79**, 462 (1959).
41. Piloty, O., and Steinbock, H., *Ber.* **35**, 3103 (1902).
42. Ponzio, G., *Gazz. Chim. Ital.* **60**, 886 (1930).
43. Ponzio, G., *Gazz. Chim. Ital.* **60**, 433 (1930).
44. Prandtl, W., and Sennewald, K., *Ber.* **62**, 1754 (1929).
45. Prandtl, W., and Dollfus, W., *Ber.* **65**, 754 (1932).
46. Quilico, A., and Fusco, R., *Rend. Ist. Lombardo Sci. Lettere, A.* **69**, 439 (1936); *Chem. Abstr.* **32**, 7454 (1938).
47. Quilico, A., and Fusco, R., *Gazz. Chim. Ital.* **67**, 589 (1937); *Chem. Abstr.* **32**, 2117 (1938).
48. Quilico, A., and D'Alcontres, G. S., *Gazz. Chim. Ital.* **80**, 140 (1950); *Chem. Abstr.* **45**, 606 (1951).
49. Rajagopalan, P., and Talaty, C. N., *Tetrahedron* **23**, 3541 (1967).
50. Rheinboldt, H., and Schmitz-Dumont, O., *Ann.* **444**, 113 (1925).
51. Rheinboldt, H., *Ann.* **451**, 161 (1927).
52. Sonn, A., and Meyer, W., *Ber.* **58**, 1096 (1925).
53. Souchay, P., and Armand, J., *Compt. Rend.* **256**, 4907 (1963); *Chem. Abstr.* **59**, 5832 (1963).
54. Souchay, P., Armand, J., and Valentini, F., *Compt. Rend.*, Ser. C. **262**, 985 (1966); *Chem. Abstr.* **65**, 3686 (1966).
55. Steinkopf, W., and Bohrmann, L., *Ber.* **41**, 1051 (1908).
56. Steinkopf, W., *J. Prakt. Chem.* [2]**81**, 198 (1910).
57. Steinkopf, W., and Jürgens, B., *J. Prakt. Chem.* [2]**83**, 453 (1911).
58. Steinkopf, W., and Jürgens, B., *J. Prakt. Chem.* [2]**84**, 712 (1911).

59. Sutherland, J. K., and Widdowson, D. A., *J. Chem. Soc.* 4651 (1964).
60. Tiemann, F., and Krüger, P., *Ber.* **18**, 732 (1885).
61. Truce, W. E., and Naik, A. R., *Can. J. Chem.* **44**, 297 (1966).
62. Vecchio, G. L., and Crisafulli, M., *Atti Soc. Peloritana Sci. Fis. Mat. Nat.* **7**, 223 (1961); *Chem. Abstr.* **59**, 12778 (1963).
63. Vecchio, G. L., Cumm, G., and Lamonica, G., *Atti Soc. Peloritana Sci. Fisc. Mat. Nat.* **8**, 351 (1962); *Chem. Abstr.* **60**, 10663 (1964).
64. Vecchio, G. L., Lamonica, G., and Cumm, G., *Gazz. Chim. Ital.* **93**, 15 (1963); *Chem. Abstr.* **59**, 12779 (1963).
65. Werner, A., *Ber.* **27**, 2193, 2846 (1894).
66. Werner, A., and Buss, H., *Ber.* **27**, 2193 (1894).
67. Werner, A., *Ber.* **29**, 1161 (1896).
68. Werner, A., and Skiba, W., *Ber.* **32**, 1654 (1899).
69. Werner, A., *Ber.* **32**, 1775 (1899).
70. Wieland, H., *Ann.* **353**, 65 (1907).
71. Wieland, H., *Ber.* **40**, 1667 (1907).
72. Wieland, H., *Ber.* **42**, 4201 (1909).
73. Wieland, H., and Höchtler, A., *Ann.* **505**, 247 (1937).
74. Weygand, C., and Bauer, E., *Ann.* **459**, 123 (1927).
75. Wiley, R. H., and Wakefield, B. J., *J. Org. Chem.* **25**, 546 (1960).
76. Zinner, G., and Günther, H., *Angew. Chem. Intern. Ed.* **3**, 383 (1964).
77. Zinner, G., and Günther, H., *Chem. Ber.* **98**, 1353 (1965).

Chapter 7

HYDRAZIDOYL HALIDES

I. INTRODUCTION

The chlorides derived from the reaction of aroyl phenylhydrazines with phosphorus pentachloride are recorded in the literature under a variety of names. While *Chemical Abstracts* lists these compounds as the phenyl-hydrazones of the corresponding acid chlorides (for example, benzoyl chloride phenylhydrazone for $C_6H_5C(Cl)=NNHC_6H_5$), Beilstein uses the terms α-chlorobenzyliden phenylhydrazine or α-chlorobenzal phenyl-hydrazine for this compound. In view of the relationship of the chlorides I with the corresponding acid hydrazides (II), their hydrolysis products, the names hydrazide chlorides or hydrazidoyl chlorides are more appropriate. In order to preserve conformity of nomenclature, the latter name is used in this book.

$$
\begin{array}{cc}
R-C=NNHR' & R-C=NNHR' \\
\mid & \mid \\
Cl & OH \\
I & II
\end{array}
$$

N-Phenylbenzhydrazidoyl chloride was first synthesized by v. Pech-mann in 1894. In their reactions hydrazidoyl chlorides bear a close resemb-lance to the other imidoyl chlorides. They have been used by Fusco and his co-workers to synthesize a variety of pyrazole derivatives, and recently Huisgen and his students utilized this class of compounds to generate the highly reactive nitrile imides. The latter compounds react as 1,3-dipoles with a wide variety of substrates to form the corresponding five-membered ring heterocycles, which are often quite difficult to synthesize by other routes.

In addition to the N-arylhydrazidoyl chlorides, numerous N-substituted compounds, such as the N-carboxylic acid derivatives, have been synthesized.

Although the reaction of the corresponding carboxylic acid hydrazides with phosphorus halides can be used to synthesize the compounds under discussion, several different methods have been utilized in the past to synthe-size hydrazidoyl halides. For example, the coupling of benzdiazonium halides with aliphatic α-halomethylene compounds affords hydrazidoyl halides in excellent yields. Likewise, reaction of benzdiazonium halides with diazo compounds gives rise to the formation of hydrazidoyl halides. Huisgen and Koch ([59]) used this method to synthesize N-4-nitrophenylformhydrazidoyl chloride, the only member of this class of compounds having hydrogen attached to the imidoyl carbon atom. Furthermore, direct halogenation of arylhydrazones has been used by Bülow and his students and by Chattaway and his co-workers to synthesize a wide variety of halogenated benzhydrazidoyl halides. This reaction often can not be stopped at the imidoyl halide stage because halogenation of the ring attached to nitrogen becomes competitive.

The reactivity of hydrazidoyl halides in nucleophilic reactions, as well as the ready elimination of hydrogen halides to afford the highly reactive nitrile imines renders this class of compounds quite useful in organic synthesis.

II. SYNTHESIS

A. From Carboxylic Acid Hydrazides and Phosphorous Halides

The reaction of arylcarboxylic acid hydrazides with phosphorus pentachloride is the classical method of synthesis of hydrazidoyl chlorides I ([89,90]).

$$R-\underset{\underset{O}{\|}}{C}-NHNHR' + PCl_5 \longrightarrow R-\underset{\underset{\underset{I}{Cl}}{|}}{C}=NNHR' + POCl_3 + HCl$$

This reaction can be conducted either without a solvent or using inert organic solvents, such as chloroform, carbon tetrachloride or methylene chloride, and the yields are quite good. Sometimes, the polarity of the solvents has some influence on the course of the reaction. For example, in the reaction of 2,4,6-trimethoxybenzoic acid-2,4,6-trinitrophenylhydrazide with phosphorus pentachloride in benzene, chlorination of one of the aryl groups occurs, whereas in methylene chloride the desired 2,4,6-trimethoxy-benz-2,4,6-trinitrophenylhydrazidoyl chloride is obtained in good yield ([69]).

2,4,6-Trimethoxybenz-2,4,6-trinitrophenylhydrazidoyl chloride ([69]). To a suspension of 4 g (0.00915 moles) of 2,4,6-trimethoxybenz-2,4,6-trinitrophenylhydrazide in 100 ml of methylene chloride is added 2.5 g (0.012 mole) of phosphorus pentachloride in 40 ml of methylene chloride over a period of 40 min with ice-cooling and stirring. After stirring for an additional 15 min, the reaction mixture is added to 300 ml of methanol, and, after standing for 13 hours, and crystallization from acetone, 2.75 g (66%) of

2,4,6-trimethoxybenz-2,4,6-trinitrophenylhydrazidoyl chloride, m.p. 194–196°C is obtained.

In the reaction of β-benzoylphenylhydrazone with phosphorus pentachloride Huisgen et al. [64] added phenol to the reaction mixture in order to generate the hydrazidoyl chloride from the phosphorus-containing intermediate $C_6H_5C(Cl)=N-N(POCl_2)C_6H_5$. Recently, Stille and coworkers synthesized the bis-hydrazidoyl chloride III by this method [97].

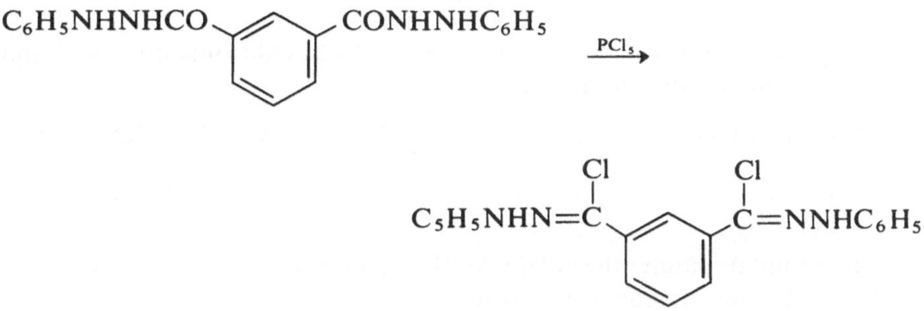

$$C_6H_5NHNHCO \diagdown \diagup CONHNHC_6H_5 \quad \xrightarrow{PCl_5}$$

$$C_5H_5NHN=\overset{\overset{\textstyle Cl}{|}}{C} \diagdown \diagup \overset{\overset{\textstyle Cl}{|}}{C}=NNHC_6H_5$$

III

Hydrazidoyl bromides are obtained on treatment of carboxylic acid hydrazides with a mixture of phosphorus tribromide and phosphorus pentabromide [22].

B. From Diazonium Halides

The coupling of diazonium halides with suitable halogenated activated methylene groups is an excellent method of synthesis of a variety of acyl and aroyl hydrazidoyl halides. This method has been used by Favrel [41–43] and by Dieckmann and Platz [36] to synthesize the acetyl derivative (IV).

$$C_6H_5N_2^{\oplus}Cl^{\ominus} + ClCH_2COCH_3 \longrightarrow C_6H_5NHN=\overset{\overset{\textstyle}{|}}{\underset{\overset{\textstyle}{Cl}}{C}}-COCH_3 + HCl$$

IV

Similarly, carboxyl and carbalkoxy hydrazidoyl chlorides are readily synthesized by this method [5,6–17,24,37–39,40,49–51,77,80]. For example, coupling of 2-nitrobenzdiazonium chloride with ethyl chloroacetoacetate affords the carbethoxyhydrazidoyl chloride V [50].

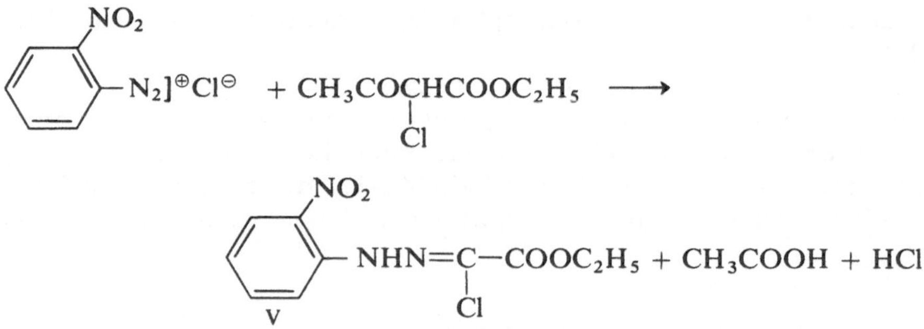

$$\text{[o-NO}_2\text{C}_6\text{H}_4\text{—N}_2\text{]}^\oplus\text{Cl}^\ominus + \text{CH}_3\text{COCHCOOC}_2\text{H}_5 \longrightarrow$$

with the Cl substituent on the CH:

$$\underset{\text{V}}{\text{o-NO}_2\text{C}_6\text{H}_4\text{—NHN=C—COOC}_2\text{H}_5} + \text{CH}_3\text{COOH} + \text{HCl}$$

with Cl on the =C.

The free carboxylic acid VI is obtained from chloromalonic acid and the corresponding diazonium halide ([77,80]).

$$\text{RN}_2^\oplus\text{Cl}^\ominus + \text{ClCH(COOH)}_2 \longrightarrow \underset{\underset{\text{VI}}{\text{Cl}}}{\text{RNHN=C—COOH}} + \text{CO}_2 + \text{HCl}$$

In a similar manner the aldehyde VII is obtained from chloromalonaldehyde and benzenediazonium chloride ([36]).

$$\text{RN}_2^\oplus\text{Cl}^\ominus + \text{ClCH(CHO)}_2 \longrightarrow \underset{\underset{\text{VII}}{\text{Cl}}}{\text{RNHN=C—CHO}}$$

The reaction of 4-nitrobenzdiazonium chloride (VIII) with diazomethane yields the formhydrazidoyl chloride (IX) in 81 % yield ([59]).

$$\underset{\text{VIII}}{\text{4-O}_2\text{NC}_6\text{H}_4\text{N}_2^\oplus\text{Cl}^\ominus} + \text{CH}_2\text{N}_2 \longrightarrow \underset{\underset{\text{IX}}{\text{Cl}}}{\text{O}_2\text{NC}_6\text{H}_4\text{NHN=C—H}}$$

Similarly, reaction of ethyl diazoacetate with diazonium chlorides yields carboxyalkyl hydrazidoyl chlorides X ([59]).

$$\text{RN}_2^\oplus\text{Cl}^\ominus + \text{N}_2\text{CHCOOEt} \longrightarrow \underset{\underset{\text{X}}{\text{Cl}}}{\text{RNHN=C—COOEt}}$$

C. Halogenation of Hydrazones

The halogenation of the arylhydrazones of arylaldehydes is an excellent method of synthesis of hydrazidoyl halides. However, halogenation of the ring attached to nitrogen can not be avoided. For example, reaction of benzaldehyde phenylhydrazone (XI) with bromine in glacial acetic acid yields benz-2,4-dibromophenylhydrazidoyl bromide (XII) ([19,20–23]). Upon further bromination p-bromobenz-2,4,6-tribromophenylhydrazidoyl bromide (XIII) is obtained ([19]).

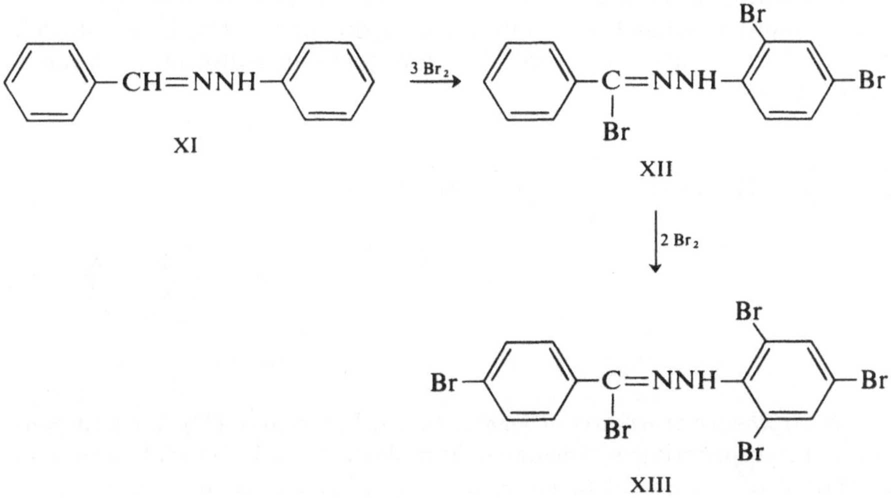

The mechanism of the bromination of arylhydrazones has been recently investigated by Hegarty and Scott ([58]).

Likewise, chlorination of benzaldehyde o-tolylhydrazone yields the hydrazidoyl chloride XIV and finally the N-chloro derivative XV, which rearranges upon heating in glacial acetic acid to afford the hydrazidoyl chloride XVI ([19,28,54]).

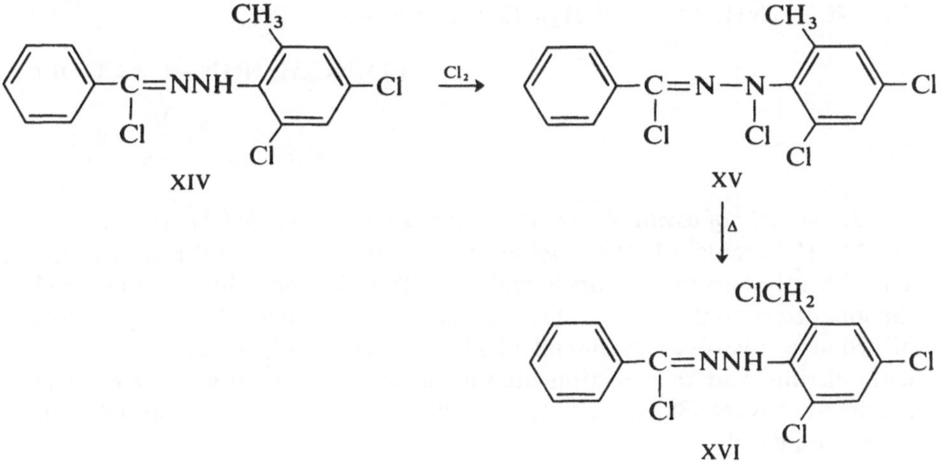

The Chattaway-Adamson rearrangement ([28]) of the N-chloro derivatives was recently investigated by Gibson ([54]).

If the phenyl ring attached to nitrogen is deactivated, bromination to
the hydrazidoyl bromides without ring bromination can be achieved ([18,95]).
For example, benzaldehyde 4-nitrophenylhydrazone (XVII) is brominated
in glacial acetic acid to afford the corresponding hydrazidoyl bromide
(XVIII) ([95]).

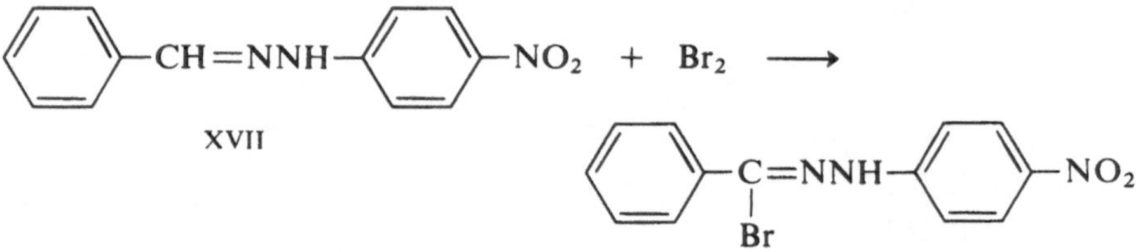

$$C_6H_5-CH=NNH-C_6H_4-NO_2 \ + \ Br_2 \ \longrightarrow$$

XVII

$$C_6H_5-\underset{\underset{Br}{|}}{C}=NNH-C_6H_4-NO_2$$

XVIII

N-Arylbenzhydrazidoyl bromides. General procedure ([95]). To a suspension of the appropriate hydrazone (0.04 mole) in 100 ml of glacial acetic acid
a solution of bromine (0.08 mole) in 20 ml of glacial acetic acid is added
with vigorous stirring. The solid reaction products are collected by filtration.
The reported yields are about 60%.

The halogenation of arylhydrazones of arylaldehyde has been extended
to the arylhydrazones of ethyl acetoacetate ([5-8,10,13,17,30,31,96]). For
example, reaction of XIX with bromine in glacial acetic acid/acetic anhydride,
in the presence of sodium acetate, affords the hydrazidoyl bromide XX in
excellent yield ([96]).

$$4\text{-}O_2NC_6H_4NHN=C(COCH_3)COOEt \ \longrightarrow$$
XIX

$$4\text{-}O_2NC_6H_4NHN=\underset{\underset{Br}{|}}{C}-COOEt$$
XX

Ethyl ester of oxalic 4-nitrophenylhydrazidoyl bromide ([96]). The amount
of 344 g (1.23 mole) of XIX is added to a mixture of 1650 ml glacial acetic
acid, 900 ml of acetic anhydride and 246 g (3 moles) of sodium acetate, and
the mixture is cooled to 0°C. Over a period of two hours 199 g (1.23 mole)
of bromine, dissolved in 300 ml of glacial acetic acid, is added dropwise
with stirring, and the reaction mixture is added to 7 liters of water. The
precipitated ester (382 g, 97.5%, m.p. 201–203°C) is filtered, washed with
water, and dried.

The chlorination of the carboxylic acid derivative XXI yields the acylhydrazidoyl chloride XXII ([7]).

$$C_6H_5NHN{=}C(COCH_3)COOH \xrightarrow{Cl_2} 2,4\text{-}Cl_2C_6H_3NHN{=}\underset{\underset{\text{XXII}}{\overset{|}{Cl}}}{C}{-}COCH_3$$

XXI

If the bromination of the ester XXIII is conducted under anhydrous conditions, the acetyl group is attacked, and the hydrazidoyl bromide XXIV is formed. In the presence of water and sodium acetate, the expected hydrazidoyl bromide XXV is obtained ([30,32]).

$$2,4\text{-}Br_2C_6H_3NHN{=}\underset{\underset{\text{XXV}}{\overset{|}{Br}}}{C}{-}COOEt$$

$$\overset{Br_2}{\nearrow}$$

C$_6$H$_5$NHN=C(COCH$_3$)COOEt

XXIII $\qquad \underset{\text{anhydrous}}{\overset{Br_2}{\searrow}}$

$$2,4\text{-}Br_2C_6H_3NHN{=}\underset{\underset{\text{XXIV}}{\overset{|}{Br}}}{C}{-}COCHBr_2$$

Acetophenone phenylhydrazone (XXVI) yields the hydrazidoyl bromide XXVII upon bromination in glacial acetic acid ([19]).

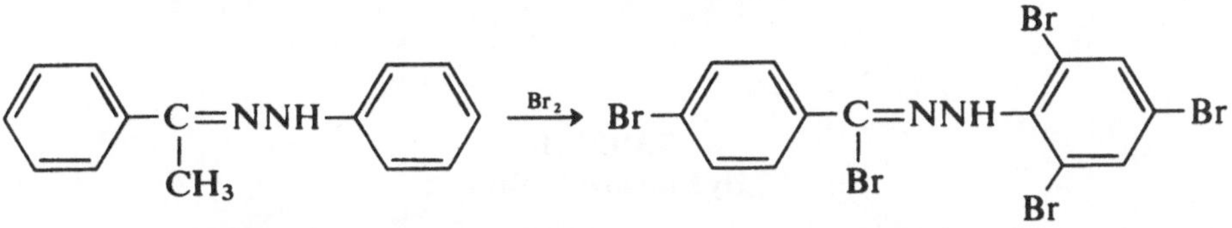

XXVI $\qquad$ XXVII

D. Halogenation of Azo Compounds and Azines

In the reaction of chloral with phenylhydrazine the azo ethylene compound XXVIII is obtained. Further chlorination yields the hydrazidoyl chloride XXIX ([24]).

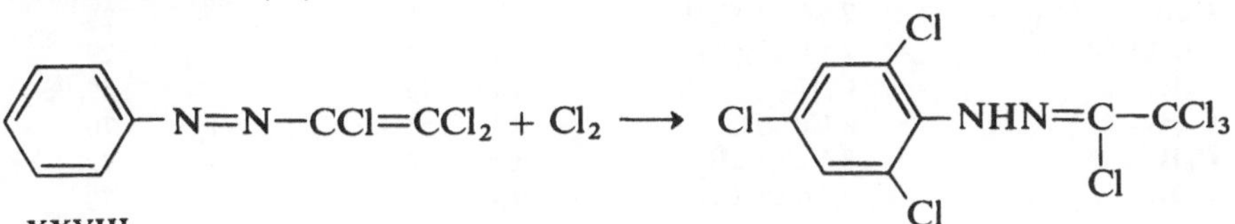

XXVIII $\qquad$ XXIX

Furthermore, chlorine is readily added to azines, and the corresponding bis-hydrazidoyl chlorides are obtained in good yield ([26,73,98]). For example, from benzalazine (XXX) and chlorine the bis-benzhydrazidoyl chloride XXI is obtained ([73,98]).

$$C_6H_5CH{=}N{-}N{=}CHC_6H_5 + 2\,Cl_2 \longrightarrow C_6H_5\underset{\underset{Cl}{|}}{C}{=}N{-}N{=}\underset{\underset{Cl}{|}}{C}C_6H_5$$

$$\text{XXX} \qquad\qquad\qquad\qquad\qquad\qquad\qquad \text{XXXI}$$

This reaction resembles the direct chlorination of aldoximes, which yields hydroxamoyl chlorides (see Chapter 6).

E. Miscellaneous Methods

The reaction of the salt XXXII with benzdiazonium sulfate gives rise to the formation of the aroylhydrazidoyl bromide XXXIII ([76]).

$$[C_6H_5COCH_2\overset{\overset{\textstyle CH_3}{|}}{\underset{\oplus}{S}}C_6H_5]Br^{\ominus} + C_6H_5N_2^{\oplus}HSO_4^{\ominus} \longrightarrow$$

$$\text{XXXII}$$

$$C_6H_5CO\underset{\underset{Br}{|}}{C}{=}NNHC_6H_5 + C_6H_5SCH_3$$

$$\text{XXXIII}$$

The hitherto-synthesized hydrazidoyl halides are listed in Tables I and II.

TABLE I
Hydrazidoyl Halides

$$R{-}\underset{\underset{X}{|}}{C}{=}NNHR'$$

R	R'	X	M.p., °C	Yield, %	Reference
H	4-$O_2NC_6H_4$	Cl	141–142	81	59
CCl_3	2,4,6-$Cl_3C_6H_2$	Cl	104	—	24
$CH_3CCl_2CCl_2$	2,4,6-$Cl_3C_6H_2$	Cl	84–85	—	29
C_6H_5	C_6H_5	Cl	131	—	89, 90
C_6H_5	2-ClC_6H_4	Cl	85–86	71	70
C_6H_5	2,4-$Cl_2C_6H_3$	Cl	90	—	72
C_6H_5	2,4,6-$Cl_3C_6H_2$	Cl	98	—	72

Table I—*continued*

R	R′	X	M.p., °C	Yield, %	Reference
C_6H_5	$4\text{-}O_2N(2\text{-}Br)C_6H_3$	Br	172	—	18, 95
C_6H_5	$2,4\text{-}Br_2C_6H_3$	Br	116–117	—	18, 20
C_6H_5	$2,4,6\text{-}Br_3C_6H_2$	Br	113	—	71
C_6H_5	$4\text{-}O_2NC_6H_4$	Cl	189–191	85	66a
C_6H_5	$4\text{-}O_2NC_6H_4$	Br	190	—*	95
C_6H_5	$2,4\text{-}(O_2N)_2C_6H_3$	Br	224–225	—	18
C_6H_5	$2\text{-}C_5H_4N$†	Br	114–115	—	56
C_6H_5	(pyrazol-3-yl, 5-C_6H_5, NH)	Br	168–170	—	95
C_6H_5	(triazol, NH)	Br	176	72	91
$4\text{-}CH_3C_6H_4$	$4\text{-}O_2NC_6H_4$	Br	192	—	95
$4\text{-}(CH_3)_2CHC_6H_4$	$4\text{-}O_2NC_6H_4$	Br	167	—	95
$4\text{-}ClC_6H_4$	$4\text{-}O_2NC_6H_4$	Br	222–223	—	95
$4\text{-}BrC_6H_4$	$4\text{-}O_2NC_6H_4$	Br	236–237	—	95
$4\text{-}BrC_6H_4$	$2\text{-}CH_3\text{-}4,6\text{-}Br_2C_6H_2$	Br	143–144	—	21
$4\text{-}BrC_6H_4$	$2,4,6\text{-}Br_3C_6H_2$	Br	180–181	—	19
$4\text{-}CH_3OC_6H_4$	$2,4\text{-}Cl_2C_6H_3$	Cl	111	—	99
$4\text{-}CH_3OC_6H_4$	$2,4\text{-}Br_2C_6H_3$	Br	135	—	34
$2,4\text{-}(CH_3O)_2C_6H_3$	$2,4\text{-}(O_2N)_2C_6H_3$	Cl	184–186	53	69
$2,4,6\text{-}(CH_3O)_3C_6H_2$	$2,4,6\text{-}(O_2N)_3C_6H_2$	Cl	194–196	66	69
$2,4,6\text{-}(CH_3O)_3\text{-}3\text{-}ClC_6H$	$2,4\text{-}(O_2N)_2C_6H_3$	Cl	184–186	67	69
$2\text{-}O_2NC_6H_4$	$2,4\text{-}Cl_2C_6H_3$	Cl	132	—	22
$2\text{-}O_2NC_6H_4$	$2,4\text{-}Cl_2C_6H_3$	Br	122	—	22
$2\text{-}O_2NC_6H_4$	$2,4\text{-}Br_2C_6H_3$	Br	110	—	22
$2\text{-}O_2NC_6H_4$	$2,4,6\text{-}Cl_3C_6H_2$	Cl	107	—	22
$2\text{-}O_2NC_6H_4$	$2,4,6\text{-}Cl_3C_6H_2$	Br	115–116	—	22
$3\text{-}O_2NC_6H_4$	$2,4\text{-}Cl_2C_6H_3$	Cl	158	—	21
$3\text{-}O_2NC_6H_4$	$2,4\text{-}Br_2C_6H_3$	Br	175–176	—	21
$3\text{-}O_2NC_6H_4$	$2\text{-}CH_3\text{-}4,6\text{-}Cl_2C_6H_2$	Cl	140–141	—	20, 21
$3\text{-}O_2NC_6H_4$	$2,4,6\text{-}Cl_3C_6H_2$	Cl	157	—	21
$3\text{-}O_2NC_6H_4$	$2,4,6\text{-}Br_3C_6H_2$	Br	172–174	—	20, 21
$4\text{-}O_2NC_6H_4$	$2,4\text{-}Cl_2C_6H_3$	Cl	199	—	21
$4\text{-}O_2NC_6H_4$	$2,4\text{-}Br_2C_6H_3$	Br	214	—	21
$4\text{-}O_2NC_6H_4$	$2,4,6\text{-}Cl_3C_6H_2$	Cl	164	—	21
$3\text{-}C_6H_5NHN{=}C(Cl)$	C_6H_5	Cl	165–166	58	97

* Reference 95 reports yields of approximately 60 %.
† C_5H_4N = Pyridyl.

TABLE II
Acyl and Aroylhydrazidoyl Halides

$$RCOC=NNHR'$$
$$\vert$$
$$X$$

R	R'	X	M.p., °C	Yield, %	Reference
H	C_6H_5	Cl	141	100	36
CH_3	C_6H_5	Cl	136.5	100	36, 42, 43
CH_3	$2,4-Cl_2C_6H_3$	Cl	125	—	7
C_6H_5	$4-CH_3C_6H_4$	Cl	142–143	74	37, 39
C_6H_5	C_6H_5	Br	—	—	76
EtO	C_6H_5	Cl	80–81	—	5, 41
EtO	C_6H_5	Br	83–84	—	5
EtO	$2-CHOC_6H_4$	Cl	88–89	75	40
EtO	$2,4-Cl_2C_6H_3$	Cl	98	—	6
EtO	$2,4,6-Cl_3C_6H_2$	Cl	73.5	—	24
EtO	$2,4,6-Cl_3C_6H_2$	Br	75	—	24
EtO	$2,4,y-Br_3C_6H_2$	Cl	108.5	—	24
EtO	$2,4,6-Br_3C_6H_2$	Br	102.5	—	24
EtO	$2-O_2NC_6H_4$	Cl	120	100	50
EtO	$4-O_2NC_6H_4$	Cl	191–193	64	59
EtO	$4-O_2NC_6H_4$	Br	201–203	97.5	96
EtO	$2-O_2N-\alpha-C_{10}H_6$	Cl	146–147	60	51
HO	$3-O_2NC_6H_4$	Cl	204	50	77
HO	$2-CH_3-4-O_2NC_6H_3$	Cl	194	30	80
HO	$2,4,6-Cl_3C_6H_2$	Cl	151.5	—	24
H_2N	$2,4-Cl_2C_6H_3$	Cl	232	—	7, 8
C_6H_5NH	C_6H_5	Cl	161	80	16
$2-ClC_6H_4NH$	$2,4-Cl_2C_6H_3$	Cl	199.5	—	16

III. PHYSICOCHEMICAL PROPERTIES

Hydrazidoyl halides are solid compounds which contain a reactive halo group. The halo group participates readily in nucleophilic reactions (see Section IV) and perhaps elimination of hydrogen halide occurs prior to substitution. The elimination of hydrogen halide affords the highly reactive nitrile imides XXXIV, which Huisgen and his co-workers ([35,60–70]) have utilized in a wide variety of 1,3-cycloaddition reactions.

$$RC=NNHR' \xrightarrow{-HCl} RC\equiv\overset{\oplus}{N}-\overset{\ominus}{N}R' \longleftrightarrow \overset{\oplus}{RC}=\overset{\ominus}{N}-\overset{\ominus}{N}R' \longleftrightarrow$$
$$\vert$$
$$Cl \qquad\qquad\qquad\qquad\qquad XXXIV$$

$$\overset{\ominus}{RC}=\overset{\oplus}{N}=NR'$$

Scott and co-workers ([92-95]), who investigated the hydrolysis of hydrazidoyl bromides in aqueous dioxane, prefer a stepwise nucleophilic reaction sequence, involving a carbonium ion-like intermediate.

The hydrolysis of N-heterocyclic hydrazidoyl bromides proceeds at a faster rate than the hydrolysis of the N-aryl compounds ([2]).

The characteristic spectral feature of the hydrazidoyl halides is their C=N double bond absorption in the infrared, which appears at 1580–1610 cm^{-1} ([69,81,84]).

IV. CHEMICAL BEHAVIOR

A. Reaction with Oxygen–Hydrogen Bonds

The reaction of hydrazidoyl halides with water is rather sluggish, most likely due to their insolubility in water ([90]), because in dioxane rapid reaction to the corresponding carboxylic acid hydrazides (XXXV) has been observed ([95]).

$$
\begin{array}{c}
\text{RC}{=}\text{NNHR}' + \text{H}_2\text{O} \longrightarrow \text{RCNHNHR}' \\
| \qquad\qquad\qquad\qquad\quad || \\
\text{Cl} \qquad\qquad\qquad\qquad\;\; \text{O} \\
\text{XXXV}
\end{array}
$$

If the hydrolysis is conducted in the presence of aqueous potassium hydrogencarbonate a cyclic transition state may account for the product formation ([2]).

$$
\text{R}{-}\underset{\underset{\text{Br}}{|}}{\text{C}}{=}\text{NNHR}' \longrightarrow
\left[
\begin{array}{c}
\text{NHR}' \\
| \\
\text{R}{-}\text{C}{\overset{\displaystyle\nearrow}{\diagup}}\text{N} \\
|\qquad\qquad \\
\text{O}\diagdown\;\;\diagup\text{O}{-}\text{H} \\
\text{C} \\
|| \\
\text{O}
\end{array}
\right]
\longrightarrow \text{RCONHNHR}' + \text{CO}_2
$$

The hydrolysis of hydrazidoyl halides in hydrochloric acid gives rise to the formation of the corresponding arylhydrazines ([11,12,14]). Reduction with zinc in ethanol yields arylhydrazones ([72]).

Hydrolysis in 95% ethanol, buffered with sodium acetate, yields the corresponding ethers (XXXVI) ([95]).

$$
\begin{array}{c}
\text{R}{-}\text{C}{=}\text{NNHR}' + \text{R}''\text{OH} \longrightarrow \text{R}{-}\text{C}{=}\text{NNHR}' \\
| \qquad\qquad\qquad\qquad\qquad\qquad | \\
\text{Cl} \qquad\qquad\qquad\qquad\qquad\;\; \text{OR}'' \\
\qquad\qquad\qquad\qquad\qquad\qquad\qquad \text{XXXVI}
\end{array}
$$

Hydrazidoyl halides, upon treatment with sodium acetate or sodium benzoate, yield the N-substituted rearrangement products XXXVII ([19,27,28]).

$$RC{=}NNHR' + NaOOCR'' \longrightarrow RCONNHR'$$

$$\underset{Br}{|} \qquad\qquad\qquad\qquad \underset{\underset{XXXVII}{COR''}}{|}$$

B. Reaction with Sulfur–Hydrogen Bonds

Although the corresponding thio compounds should be readily available from hydrazidoyl halides and hydrogen sulfide or mercaptans, respectively, these reactions are not reported in the literature. However, Fusco and Musante ([46]) investigated the reaction of thiocarbonic acid O-esters with hydrazidoyl bromides.

Recently, Barnish and Gibson ([3]) treated hydrazidoyl halides, substituted in the o-position of the phenyl ring attached to nitrogen by fluoro and bromo groups, with potassium thioacetate, and the authors obtained 1,3,4-benzothiadiazine derivatives XXXVIII.

X = F or Br   XXXVIII

C. Reactions with Nitrogen–Hydrogen Bonds

The reaction of hydrazidoyl halides with ammonia, primary and secondary amines, and hydrazines occurs as expected to produce the corresponding substitution products XXXIX ([5–10,14,17,18,20,21,22,27,30,31,33, 38,50,78]).

$$RC{=}NNHR' + R_2''NH \longrightarrow RC{=}NNHR'$$

$$\underset{Cl}{|} \qquad\qquad\qquad\qquad \underset{\underset{XXXIX}{NR_2''}}{|}$$

In the case of the carboxylic acid chloride XL, selective reaction on the halide attached to the carbonyl group has been observed, indicating enhanced reactivity over the halo group attached to the C=N double bond which is, of course, deactivated by the neighboring carbonyl group ([84,85]).

$$Cl{-}\underset{\underset{O}{\|}}{C}{-}\underset{\underset{Cl}{|}}{C}{=}NNHR + R'NH_2 \longrightarrow R'NHCO\underset{\underset{Cl}{|}}{C}{=}NNHR$$

$$\qquad XL \qquad\qquad\qquad\qquad\qquad XLI$$

In contrast, XL yields the bis-substitution products upon treatment with aliphatic amines ([83]).

The reaction of XL with arylhydrazines proceeds similarly to afford substitution on the carboxylic acid chloride function ([84]).

The carbonyl chlorides XL, which are obtained from the corresponding N-carboxylic acids and thionyl chloride ([85]), are reactive compounds; Lozinskii and his co-workers have used these intermediates to synthesize a huge variety of N-carboxylic acid derivatives ([77-87]). For example, reaction of XL with sodium thiocyanate gives rise to the formation of the corresponding isothiocyanates, which were treated with arylamines to yield the corresponding thiourea derivatives ([81,83]). In the reaction of XL with sodium azide the corresponding azides are formed, which upon heating in toluene generate the isocyanates. However, only the cyclic trimers could be isolated in this reaction ([87]). Reaction of XL with o-phenylenediamine, o-aminophenol, and o-aminothiophenol gave mixtures of the linear and heterocyclic products XLII and XLIII ([82,86]).

$$
\underset{\text{XL}}{\text{RNHN}=\underset{\underset{\text{Cl}}{|}}{\text{C}}-\text{COCl}} + \underset{\text{HX}}{\overset{\text{H}_2\text{N}}{\bigcirc}} \longrightarrow \underset{\text{XLII}}{\underset{\text{HX}}{\overset{\text{RNHN}=\underset{\underset{\text{Cl}}{|}}{\text{C}}-\text{CONH}}{\bigcirc}}}
$$

$$
+ \quad \underset{\text{XLIII}}{\text{RNHN}=\underset{\underset{\text{Cl}}{|}}{\text{C}}-\text{C}\underset{\text{X}}{\overset{\text{N}}{\bigcirc}}}
$$

In the reaction of hydrazidoyl halides XLIV with secondary amines decarboxylation occurs, and the substitution products XLV are isolated ([78]).

$$
\underset{\underset{\text{XLIV}}{}}{\text{RNHN}=\underset{\underset{\text{Cl}}{|}}{\text{C}}-\text{COOH}} + \bigcirc\hspace{-1.2em}\text{NH} \longrightarrow \underset{\text{XLV}}{\text{RNHN}=\text{CH}-\text{N}\hspace{-0.3em}\bigcirc}
$$

Tertiary amines afford either amine salts or ylides upon mixing with hydrazidoyl halides ([15,76,88]).

Similarly, reaction of hydrazidoyl halides with sodium cyanate ([96]), potassium selenocyanate ([46]), and sodium azide ([78]) has been reported. For example, decarboxylation occurs when the carboxylic acid derivative XLIV is treated with sodium azide, and the tetrazole derivative XLVI is isolated ([78,94]).

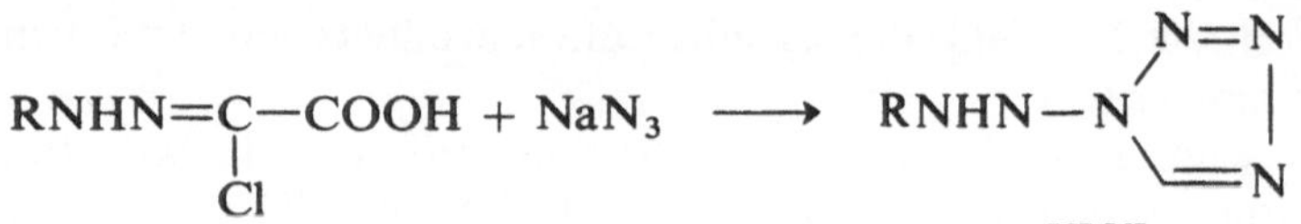

$$\underset{\underset{\textstyle\text{XLIV}}{}}{\text{RNHN}=\underset{\underset{\textstyle\text{Cl}}{|}}{\text{C}}-\text{COOH}} + \text{NaN}_3 \longrightarrow \underset{\underset{\textstyle\text{XLVI}}{}}{\text{RNHN}-\text{N}\begin{array}{c}\text{N}=\text{N}\\|\\=\text{N}\end{array}}$$

D. Reactions with Carbon–Hydrogen Bonds

The reaction of hydrazidoyl halides with potassium cyanide yields the nitriles XLVII ([7]).

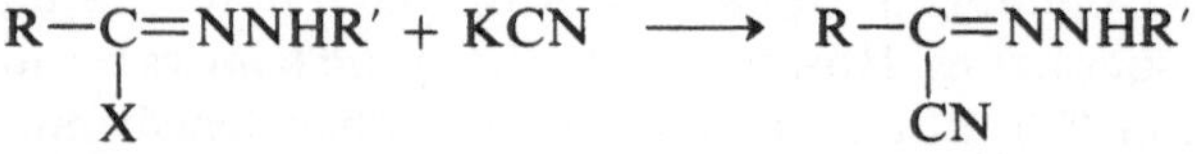

$$\underset{\underset{\textstyle\text{X}}{|}}{\text{R}-\text{C}=\text{NNHR}'} + \text{KCN} \longrightarrow \underset{\underset{\textstyle\underset{\text{XLVII}}{\text{CN}}}{|}}{\text{R}-\text{C}=\text{NNHR}'}$$

The reactions of hydrazidoyl halides with a variety of activated methylene compounds, such as ethyl cyanoacetate, diethyl malonate, malononitrile, etc., have been used to synthesize a huge variety of pyrazole derivatives ([4,44–48,52,74–75]).

E. Addition and Elimination Reactions

The dehydrochlorination of hydrazidoyl chlorides was first investigated by Fusco and Romani ([49]), but Huisgen and his co-workers ([35,60–70]) recognized the chemistry of the generated species XLVIII.

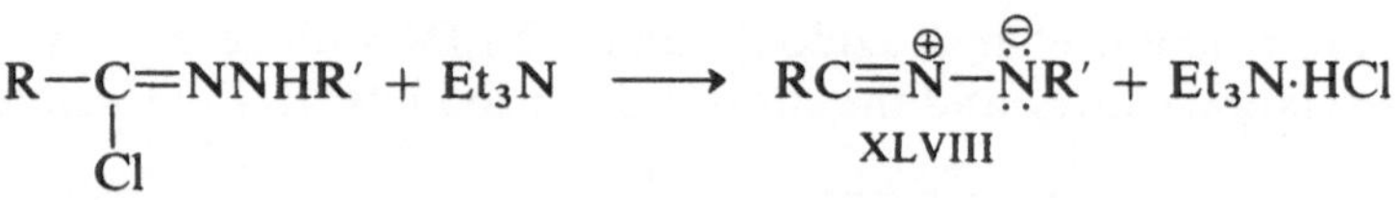

$$\underset{\underset{\textstyle\text{Cl}}{|}}{\text{R}-\text{C}=\text{NNHR}'} + \text{Et}_3\text{N} \longrightarrow \underset{\text{XLVIII}}{\text{RC}\equiv\overset{\oplus}{\text{N}}-\overset{\ominus}{\text{N}}\text{R}'} + \text{Et}_3\text{N}\cdot\text{HCl}$$

In the absence of a substrate the generated XLVIII dimerizes to yield the 1,4-dihydro-1,2,4,5-tetrazine derivative XLIX ([66,91]).

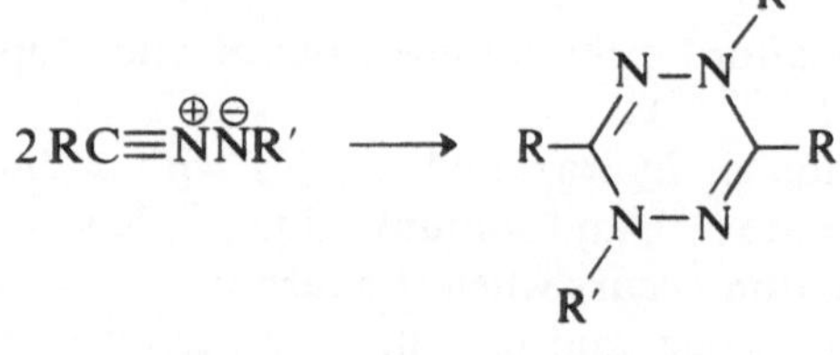

$$2\,\text{RC}\equiv\overset{\oplus}{\text{N}}\overset{\ominus}{\text{N}}\text{R}' \longrightarrow \underset{\underset{\textstyle\text{R}'}{}}{\overset{\overset{\textstyle\text{R}'}{}}{\text{R}-\underset{\text{N}-\text{N}}{\overset{\text{N}-\text{N}}{\bigvee\bigwedge}}-\text{R}}}$$

XLIX

If N-acylnitrile imides are generated by this procedure, cyclization with formation of 1,3,4-oxadiazoles L occurs ([61]).

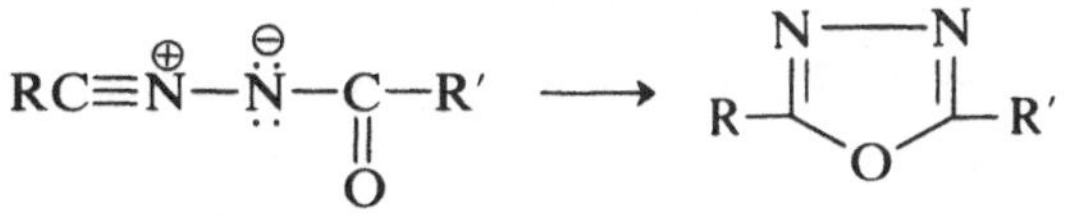

L

In a similar manner N-imidoylnitrile imides afford 1,2,4-triazoles ([62]), and N-thioacylnitrile imides cyclize to yield 1,3,4-thiadiazoles ([63]).

The cyclization of nitrile imides, having a nitro group attached to the *o*-position of the phenyl ring attached to carbon was already observed by Chattaway and Walker ([22,23]). The structure of the isolated products LI was established recently by Gibson ([1,55]).

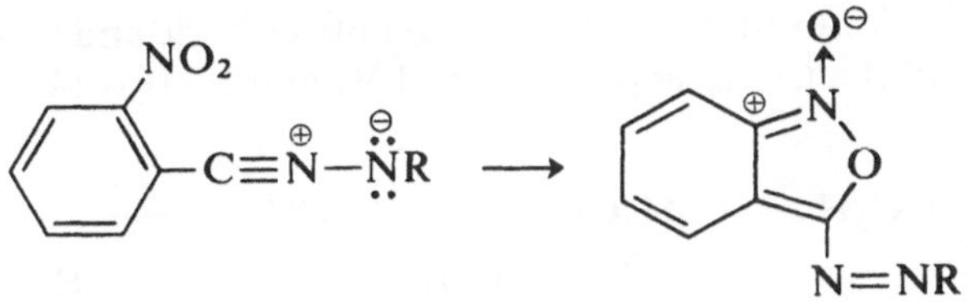

LI

The heterocyclic hydrazidoyl bromide LII also undergoes ready cyclization to yield the bicyclic heterocycle LIII ([56]).

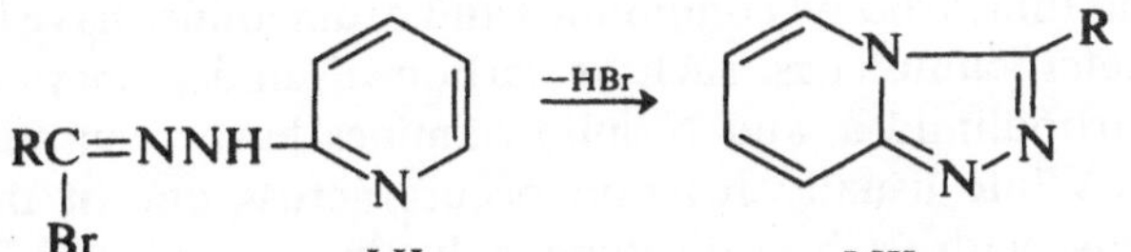

Of course, this cyclization occurs *via* the intermediate nitrile imide. The hydrazidoyl bromide, obtained from benzaldehyde 4-quinazolyl-hydrazone and bromine, could not be isolated, and only the cyclization product has been obtained in good yield ([1,56]).

Similarly, the heterocyclic-substituted hydrazidoyl bromide LIV eliminates bromide ion in 95% ethanol at 25°C, and the major product is 5-azido-3-phenyl-1,2,4-triazole (LV), resulting from tetrazolyl anchimerism *via* a cyclic transition state ([92]).

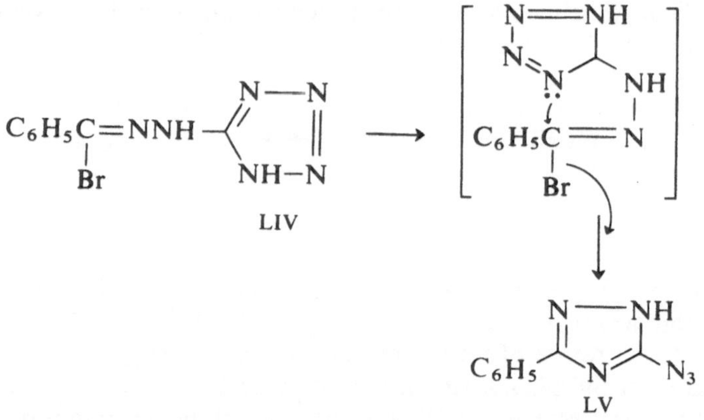

$$\text{C}_6\text{H}_5\underset{\underset{\text{Br}}{|}}{\text{C}}=\text{NNH}-\left\langle\begin{array}{c}\text{N}-\text{N}\\ \|\\ \text{NH}-\text{N}\end{array}\right\rangle \longrightarrow \left[\text{C}_6\text{H}_5\underset{\underset{\text{Br}}{|}}{\text{C}}=\text{N}\cdots\left\langle\begin{array}{c}\text{N}=\text{NH}\\ \text{NH}\\ |\end{array}\right\rangle\right]$$

LIV

$$\text{C}_6\text{H}_5\left\langle\begin{array}{c}\text{N}-\text{NH}\\ \text{N}\end{array}\right\rangle\text{N}_3$$

LV

If nitrile imides are generated in the presence of a nucleophilic olefin, the corresponding 1,3-cycloaddition products are isolated, usually in good yield ([53,60–70]). For example, from diphenylnitrile imide and methyl cinnamate a mixture of the isomeric pyrazolines LVI and LVII is obtained in 95% yield ([35]).

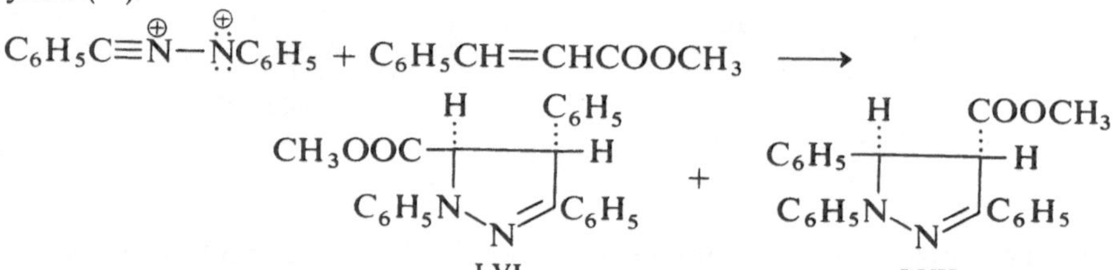

$$\text{C}_6\text{H}_5\text{C}\overset{\oplus}{\equiv}\overset{\oplus}{\text{N}}-\overset{\cdots}{\text{N}}\text{C}_6\text{H}_5 + \text{C}_6\text{H}_5\text{CH}=\text{CHCOOCH}_3 \longrightarrow$$

LVI LVII

In addition to activated olefins, other double-bond substrates, such as carbonyl and thiocarbonyl compounds and azomethines have been used ([66]). Likewise, heterocumulenes, such as carbon disulfide, isocyanates, isothiocyanates, carbodiimides, and N-sulfinylamines have been added to nitrile imides ([66]). While usually reaction occurs across one of the cumulative double bonds, with carbon disulfide only the bis-adduct LVIII could be isolated ([65]).

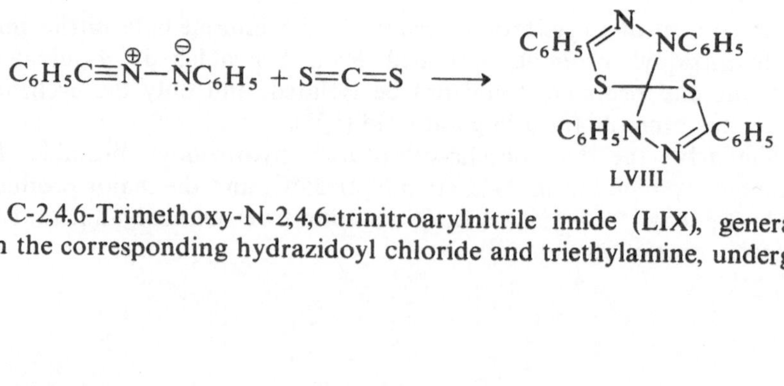

$$\text{C}_6\text{H}_5\text{C}\overset{\oplus}{\equiv}\overset{\ominus}{\text{N}}-\overset{\cdots}{\text{N}}\text{C}_6\text{H}_5 + \text{S}=\text{C}=\text{S} \longrightarrow$$

LVIII

C-2,4,6-Trimethoxy-N-2,4,6-trinitroarylnitrile imide (LIX), generated from the corresponding hydrazidoyl chloride and triethylamine, undergoes

an intramolecular redox reaction to yield the benztriazole derivative
LX ([1,69]).

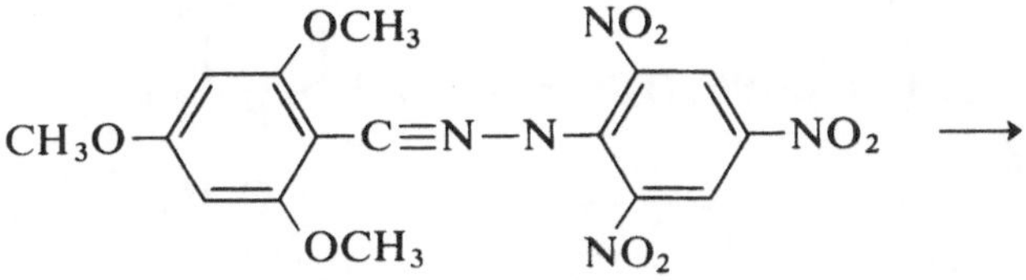

Grünanger and Langella ([57]) and Huisgen and his co-workers ([64]) have
synthesized pyrazole derivatives in the reaction of nitrile imides with
acetylene derivatives.

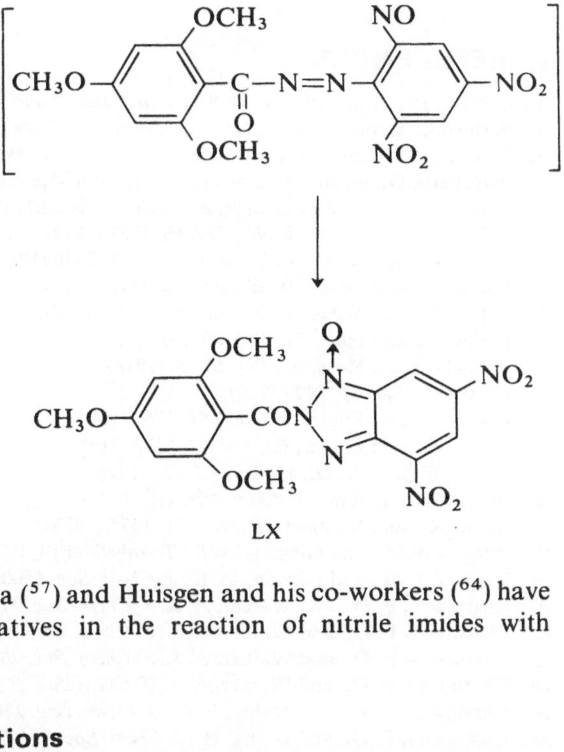

F. Miscellaneous Reactions

The bis-hydrazidoyl chloride LXI, on treatment with the alkali phos-
phides LXII yields the bis-phosphine LXIII ([73]).

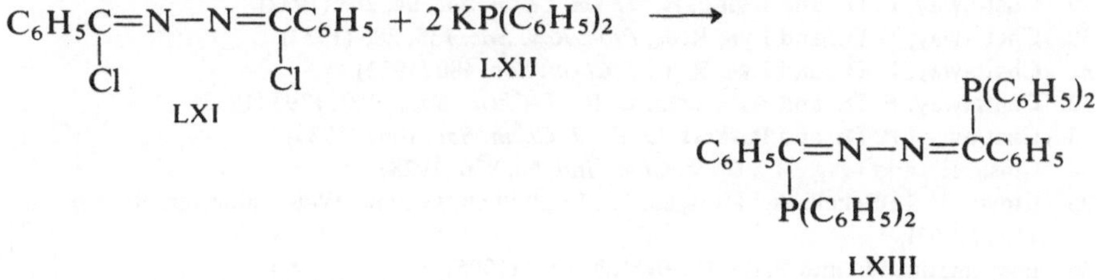

Likewise, reaction of LXI with $Li_2PC_6H_5$ yields the P-heterocycle LXIV ([73]).

$$C_6H_5\underset{\underset{Cl}{|}}{C}{=}N{-}N{=}\underset{\underset{Cl}{|}}{C}C_6H_5 + Li_2PC_6H_5 \longrightarrow$$

LXI

LXIV

V. REFERENCES

1. Barnish, I. T., and Gibson, M. S., *Chem & Ind.* (*London*) 1699 (1965).
2. Barnish, I. T., and Gibson, M. S., *J. Chem. Soc.* 2999 (1965).
3. Barnish, I. T., and Gibson, M. S., *Chem. Commun.* 1001 (1967).
4. Bianchetti, G., Pocar, D., Dalla Croce, P., and Vigevani, A., *Ber.* **98**, 2715 (1965).
5. Bowack, D. A., and Lapworth, A., *J. Chem. Soc.* **87**, 1854 (1905).
6. Bülow, C., and Neber, P. W., *Ber.* **45**, 3732 (1912).
7. Bülow, C., and Neber, P. W., *Ber.* **46**, 2032, 2370 (1913).
8. Bülow, C., and Neber, P. W., *Ber.* **46**, 2370 (1913).
9. Bülow, C., and Neber, P. W., *Ber.* **49**, 2179 (1916).
10. Bülow, C., and Huss, R., *Ber.* **50**, 1480 (1917).
11. Bülow, C., and Huss, R., *Ber.* **51**, 28 (1918).
12. Bülow, C., *Ber.* **51**, 399 (1918).
13. Bülow, C., and Engler, R., *Ber.* **51**, 1264 (1918).
14. Bülow, C., and Engler, R., *Ber.* **52**, 632 (1919).
15. Bülow, C., and Seidel, F., *Ber.* **57**, 629 (1924).
16. Bülow, C., and King, E., *Ann.* **439**, 216 (1924).
17. Bülow, C., and Spengler, W., *Ber.* **58**, 1375 (1925).
18. Burgess, J. M., and Gibson, M. S., *Tetrahedron* **18**, 1001 (1962).
19. Burgess, J. M., and Gibson, M. S., *J. Chem. Soc.* 1500 (1964).
20. Chattaway, F. D., and Walker, A. J., *J. Chem. Soc.* 975 (1925).
21. Chattaway, F. D., and Walker, A. J., *J. Chem. Soc.* 1687 (1925).
22. Chattaway, F. D., and Walker, A. J., *J. Chem. Soc.* 2407 (1925).
23. Chattaway, F. D., and Walker, A. J., *J. Chem. Soc.* 323 (1927).
24. Chattaway, F. D., and Daldy, F. G., *J. Chem. Soc.* 2760 (1928).
25. Chattaway, F. D., and Irving, H., *J. Chem. Soc.* 91 (1930).
26. Chattaway, F. D., and Farinholt, L. H., *J. Chem. Soc.* 95 (1930).
27. Chattaway, F. D., and Adamson, A. B., *J. Chem. Soc.* 158 (1930).
28. Chattaway, F. D., and Adamson, A. B., *J. Chem. Soc.* 843 (1930).
29. Chattaway, F. D., and Irving, H., *J. Am. Chem. Soc.* **54**, 263 (1932).
30. Chattaway, F. D., and Lye, R. J., *Proc. Roy. Soc.* **135**, 286 (1932).
31. Chattaway, F. D., and Lye, R. J., *J. Chem. Soc.* 480 (1933).
32. Chattaway, F. D., and Ashworth, D. R., *J. Chem. Soc.* 1390, 1393 (1933).
33. Chattaway, F. D., and Parkers, G. D., *J. Chem. Soc.* 1007 (1935).
34. Ciusa, R., and Mega, P., *Gazz. Chim. Ital.* **58**, 836 (1928).
35. Clovis, J. S., Eckell, A., Huisgen, R., Wallbillich, G., and Weberndoerfer, B., *Ber.* **100**, 1593 (1967).
36. Dieckmann, W., and Platz, L., *Ber.* **38**, 2987 (1905).

37. Dubenko, R. G., and Pel'kis, P. S., *Zh. Obshch. Khim.* **33**, 3917 (1963).
38. Dubenko, R. G., and Pel'kis, P. S., *Zh. Obshch. Khim.* **34**, 679 (1964).
39. Dubenko, R. G., and Pel'kis, P. S., *Zh. Obshch Khim.* **34**, 3481 (1964).
40. Evdokimov, U., *Gazz. Chim. Ital.* **87**, 1191 (1957).
41. Favrel, G., *Compt. Rend.* **134**, 1312 (1902).
42. Favrel, G., *Bull. Soc. Chim. France* [4]**41**, 1494 (1927).
43. Favrel, G., and Chrz, J., *Bull. Soc. Chim. France* [4]**41**, 1603 (1927).
44. Fusco, R., and Justoni, R., *Gazz. Chim. Ital.* **67**, 3 (1937).
45. Fusco, R., and Musante, C., *Gazz. Chim. Ital.* **68**, 147, 154 (1938).
46. Fusco, R., and Musante, C., *Gazz. Chim. Ital.* **68**, 665, 675, 677 (1938).
47. Fusco, R., *Gazz. Chim. Ital.* **69**, 344, 348, 353, 359 (1939).
48. Fusco, R., and Pizzotti, R., *Gazz. Chim. Ital.* **72**, 411 (1942).
49. Fusco, R., and Romani, R., *Gazz. Chim. Ital.* **78**, 342 (1948).
50. Fusco, R., and Rossi, S., *Gazz. Chim. Ital.* **86**, 491 (1956).
51. Fusco, R., and Bianchetti, G., *Gazz. Chim. Ital.* **87**, 441 (1957).
52. Fusco, R., Bianchetti, G., and Pocar, D., *Gazz. Chim. Ital.* **91**, 1233 (1961).
53. Gaudiano, G., Umani-Ronchi, A., Bravo, P., and Acampora, M., *Tetrahedron Letters* 107 (1967).
54. Gibson, M. S., *J. Chem. Soc.* 2270 (1962).
55. Gibson, M. S., *Tetrahedron* **18**, 1377 (1962).
56. Gibson, M. S., *Tetrahedron* **19**, 1587 (1963).
57. Grünanger, P., and Langella, M. R., *Gazz. Chim. Ital.* **90**, 229, 234 (1960).
58. Hegarty, A. F., and Scott, F. L., *Tetrahedron Letters* 3801 (1965).
59. Huisgen, R., and Koch, H. J., *Ann.* **591**, 218, 226 (1955).
60. Huisgen, R., Seidel, M., Sauer, J., McFarland, J. W., and Wallbillich, G., *J. Org. Chem.* **24**, 892 (1959).
61. Huisgen, R., Sauer, J., Sturm, H. J., and Markgraf, J. H., *Ber.* **93**, 2106 (1960).
62. Huisgen, R., Sauer, J., and Seidel, M., *Ber.* **93**, 2885 (1960).
63. Huisgen, R., Strum, H. J., and Seidel, M., *Ber.* **94**, 1555 (1961).
64. Huisgen, R., Seidel, M., Wallbillich, G., and Knupfer, H., *Tetrahedron* **17**, 3, 18 (1962).
65. Huisgen, R., Grashey, R., Seidel, M., Knupfer, H., and Schmidt, R., *Ann.* **685**, 169 (1962).
66. Huisgen, R., *Angew. Chem. Intern. Ed.* **2**, 565 (1963).
66a. Huisgen, R., Grashey, R., Knupfer, H., Kunz, R., and Seidel, M., *Ber.* **97**, 1085 (1964).
67. Huisgen, R., Grashey, R., Aufderhaar, E., and Kunz, R., *Ber.* **98**, 642 (1965).
68. Huisgen, R., Aufderhaar, E., and Wallbillich, G., *Ber.* **98**, 1476 (1965).
69. Huisgen, R., and Weberndörder, V., *Ber.* **100**, 71 (1967).
70. Huisgen, R., Sustmann, R., and Wallbillich, G., *Ber.* **100**, 1786 (1967).
71. Humphries, J. E., Bloom, E., and Evans, R., *J. Chem. Soc.* **123**, 1766 (1923).
72. Humphries, J. E., Humble, H., and Evans, R., *J. Chem. Soc.* 1304 (1925).
73. Issleib, K., and Balszuweit, A., *Ber.* **99**, 1316 (1966).
74. Justoni, R., *Gazz. Chim. Ital.* **68**, 49 (1938).
75. Justoni, R., and Fusco, R., *Gazz. Chim. Ital.* **68**, 59 (1938).
76. Krollpfeiffer, F., and Hartmann, H., *Ber.* **83**, 90, 92 (1950).
77. Lozinskii, M. O., and Pel'kis, P. S., *Zh. Obshch. Khim.* **31**, 1621 (1961).
78. Lozinskii, M. O., and Pel'kis, P. S., *Zh. Obshch. Khim.* **32**, 526 (1962).
79. Lozinskii, M. O., Pel'kis, P. S., and Sanova, S. N., *Zh. Obshch. Khim.* **33**, 2231 (1963).
80. Lozinskii, M. O., *Ukrain. Chem. J.* **30**, 68 (1964).
81. Lozinskii, M. O., and Pel'kis, P. S., *Zh. Organ. Khim.* **1**, 1415 (1965); *Chem. Abstr.* **64**, 621 (1966).

82. Lozinskii, M. O., and Pel'kis, P. S., *Zh. Organ. Khim.* **1**, 1793 (1965); *Chem. Abstr.* **64**, 5063 (1966).

83. Lozinskii, M. O., Pel'kis, P. S., and Sanova, S. N., *Zh. Organ. Khim.* **1**, 1800 (1965); *Chem. Abstr.* **64**, 4974 (1966).

84. Lozinskii, M. O., Lopatinskaya, N. A., and Pel'kis, P. S., *Zh. Organ. Khim.* **1**, 1877 (1965); *Chem. Abstr.* **64**, 3411 (1966).

85. Lozinskii, M. O., and Pel'kis, P. S., *Zh. Organ. Khim.* **1**, 1976 (1965); *Chem. Abstr.* **64**, 9619 (1966).

86. Lozinskii, M. O., and Pel'kis, P. S., *Zh. Organ. Khim.* **2**, 692 (1966); *Chem. Abstr.* **65**, 8814 (1966).

87. Lozinskii, M. O., Sanova, S. N., and Pel'kis, P. S., *Zh. Organ. Khim.* **2**, 1716 (1966); *Chem. Abstr.* **66**, 85766 (1967).

88. Neber, P. W., and Wörner, H., *Ann.* **526**, 173 (1936).

89. Pechmann, v. H., *Ber.* **27**, 320 (1894).

90. Pechmann, v. H., and Seeberger, L., *Ber.* **27**, 2122 (1894).

91. Scott, F. L., Morrish, W. N., and Reilly, J., *J. Org. Chem.* **32**, 692 (1957).

92. Scott, F. L., and Holland, N. M., *Proc. Chem. Soc.* 106 (1962).

93. Scott, F. L., and Cronin, D. A., *Tetrahedron Letters* 715 (1963).

94. Scott, F. L., and Cronin, D. A., *Chem. & Ind.* (*London*) 1757 (1964).

95. Scott, F. L., and Aylward, J. B., *Tetrahedron Letters* 841 (1965).

96. Sharp, D. B., and Hamilton, C. S., *J. Am. Chem. Soc.* **68**, 588 (1946).

97. Stille, J. K., Harris, F. W., and Bedford, M. A., *J. Heteroc. Chem.* **3**, 155 (1966).

98. Stolle, R., *J. Prakt. Chem.* **85**, 386 (1911).

99. Vanghelovici, M., *Bl. Soc. Chim. Romania* **8**, 20 (1926); *Chem. Abstr.* **20**, 1341 (1928); *Bl. Soc. Chim. Romania* **9**, 59 (1927).

Chapter 8

CYCLIC IMIDOYL HALIDES

I. INTRODUCTION

In the cyclic series imidoyl halides are reported only for five-, six-, and seven-membered ring systems. However, it was demonstrated recently that 2,2-dichloroaziridines, which are obtained by addition of dichloro-carbene to C=N double bond compounds, react as iminium chlorides. The synthetic methods for the cyclic imidoyl halides are similar to those used for the preparation of the linear species. Therefore, emphasis is focused on the reactions of cyclic imidoyl halides rather than on their synthesis. The organization of this chapter also differs from that of the previous chapters because it is based on ring size rather than on the difference attributable to substituents attached to the C=N double bond system.

Several of the cyclic imidoyl halides reported in the old literature are crystallized from aqueous systems. Therefore, the structural assignment of some of the reported imidoyl halides is doubtful. From all available evidence it is indicated that true cyclic imidoyl chlorides are quite reactive.

For example, the imidoyl chlorides I–III have either been isolated or their intermediacy in certain reactions has been demonstrated.

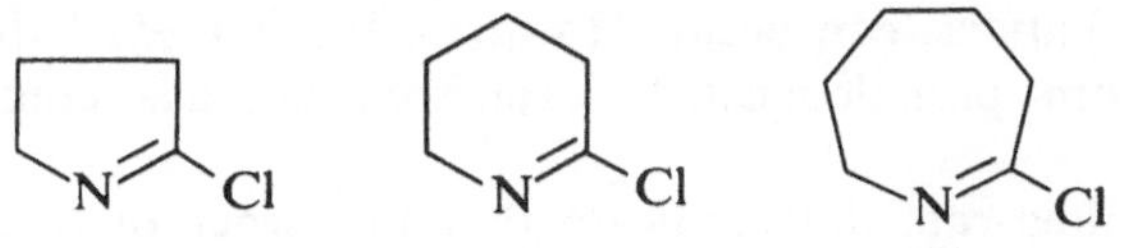

From the available data it is indicated that the six-membered ring compounds are more reactive than the five-membered ring homologs. Another group of compounds with imidoyl halide character are geminate halides adjacent to a nitrogen atom. The degree of imidoyl halide character would, of course, depend on the basicity of the adjacent nitrogens. For example, compounds IV and VI are predominantly in the nonpolar form, whereas V is a true iminium chloride.

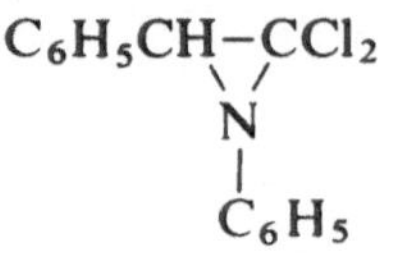

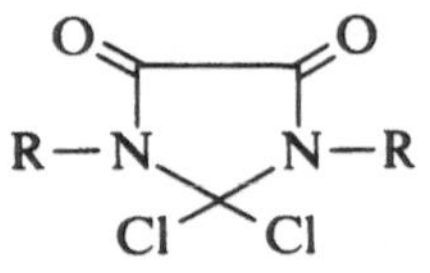

$$C_6H_5CH-CCl_2$$
$$\underset{\underset{C_6H_5}{|}}{N}$$

IV

V

VI

R = Alkyl

Since numerous halogen derivatives of nitrogen-containing heterocyclic ring systems resemble imidoyl halides, it is not possible to discuss these compounds within the framework of this monograph.

However, data from rate studies related to the nucleophilic displacement of halogens attached to N-heterocycles allow a limited generalization. In heterocyclic ring systems, such as pyridine, the aromatic character dominates the chemistry of 2-halopyridines VII; however, in the corresponding benzoderivatives, such as 2-haloquinolines IX, the imidoyl halide character is increased. Furthermore, an increase in hetero atoms as ring building blocks lowers the aromaticity and therefore increases the imidoyl halide character. For example, cyanuric halides VIII react considerably faster with nucleophiles than VII and IX, and the following order of reactivity is indicated:

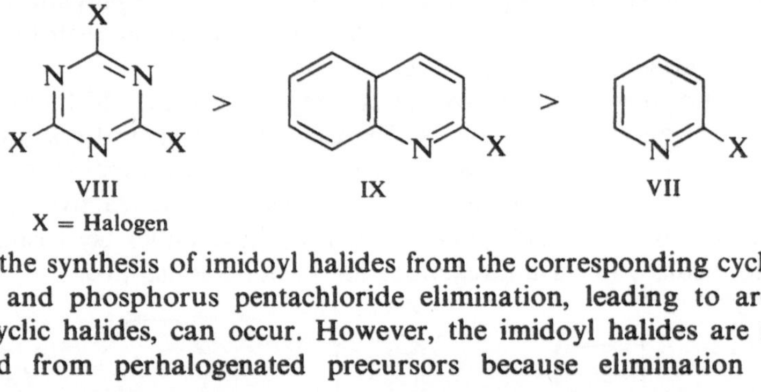

VIII   >   IX   >   VII

X = Halogen

In the synthesis of imidoyl halides from the corresponding cyclic precursors and phosphorus pentachloride elimination, leading to aromatic heterocyclic halides, can occur. However, the imidoyl halides are readily obtained from perhalogenated precursors because elimination is not possible.

The limited-rate data indicate that the order of reactivity of cyclic imidoyl halides with regard to nucleophilic substitution reactions is the following: $F > Cl > Br > I$. As expected, elimination reactions are of no significance in the cyclic series.

II. THREE-MEMBERED RING IMIDOYL HALIDES

Three-membered ring imidoyl chlorides, such as X, have not been synthesized. However, 2,2-dichloroaziridines XI, which are in equilibrium with the corresponding iminium chlorides (XII), are well known.

$$R_2'C \overset{\displaystyle\diagdown\diagup}{\underset{\overset{|}{R}}{N}} C-Cl \qquad R_2'C \overset{\displaystyle\diagdown\diagup}{\underset{\overset{|}{R}}{N}} CCl_2 \longleftrightarrow R_2'C \overset{\displaystyle\diagdown\overset{\oplus}{N}\diagup}{\underset{\overset{|}{R}}{}} CCl]Cl^{\ominus}$$

$$\text{X} \qquad\qquad\qquad \text{XI} \qquad\qquad\qquad\qquad \text{XII}$$

2,2'-Dichloroaziridines are synthesized from the corresponding C=N double-bond compounds and dichlorocarbene [7,10,11,12,19,20,21,24,25]. For example, 1,3-diphenyl-2,2-dichloroaziridine (XIII), the first member of this class of compounds, was synthesized in 1959 by Fields and Sandri [12].

$$C_6H_5CH{=}NC_6H_5 + :CCl_2 \longrightarrow C_6H_5CH \overset{\displaystyle\diagdown\diagup}{\underset{\overset{|}{C_6H_5}}{N}} CCl_2$$

$$\text{XIII}$$

The dihalocarbenes can be generated from chloroform [7,12,19,21], hexachloroacetone [20], and trihaloacetic acids [24,25], and in addition to azomethines, benzophenone anils [10,11] and ketimines [25] have been used as substrates. Similarly, monochlorocarbene reacts with azomethines and anils to yield 2-chloroaziridines [10,11]. The monochlorocarbene was generated from methylene chloride using *n*-butyllithium [11].

1-p-Ethoxyphenyl-3-phenyl-2,2-dichloroaziridine [7]. To a stirred mixture of 22.5 g (0.1 mole) of benzylidene-*p*-chloroaniline and 44.9 g (0.4 mole) of potassium *t*-butoxide in 350 ml of hexane, 47.8 g (0.4 mole) of chloroform is added. After stirring for 18 hours at room temperature, and short re-fluxing and filtration, the solvent is removed *in vacuo* leaving 28.2 g (91%) of 1-*p*-ethoxyphenyl-3-phenyl-2,2-dichloroaziridine, m.p. 76.5–77.5°C after crystallization from hexane.

Heating with excess water results in the elimination of hydrogen chloride and formation of α-chloro-α-phenyl-*p*-ethoxyacetanilide, m.p. 145–146.5°C. The yield is quantitative.

The thermolysis of XIII (refluxing in toluene for 24 hr) yields N-phenyl-α-phenyl-α-chloroacetimidoyl chloride (XIV) [16].

$$C_6H_5CH \overset{\displaystyle\diagdown\diagup}{\underset{\overset{|}{C_6H_5}}{N}} CCl_2 \overset{\Delta}{\longrightarrow} C_6H_5CH\underset{\overset{|}{Cl}}{-}C\underset{\overset{|}{Cl}}{=}NC_6H_5$$

$$\text{XIII} \qquad\qquad\qquad\qquad\qquad \text{XIV}$$

A similar rearrangement could account for the formation of the oxindol derivative XV obtained by Klamann *et al.* [21] in the reaction of dichlorocarbene (generated from chloroform, ethylene oxide, and tetraethyl-ammonium bromide) with benzophenone anil.

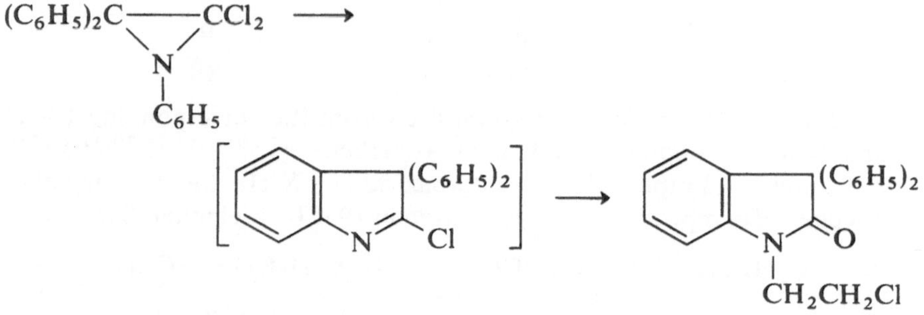

$$(C_6H_5)_2C\text{———}CCl_2 \longrightarrow$$

$$\left[\begin{array}{c} (C_6H_5)_2 \\ N\text{===}Cl \end{array}\right] \longrightarrow \begin{array}{c} (C_6H_5)_2 \\ N\text{—}O \\ | \\ CH_2CH_2Cl \end{array}$$

XV

The nucleophilic reactions of N-phenyl-2,2-dichloroaziridines are best explained by the assumption that the iminium chloride is the intermediate leading to the linear reaction products.

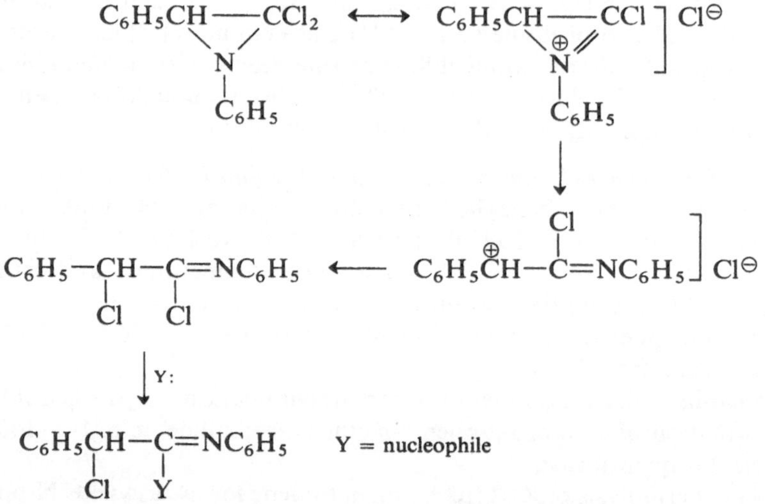

$$C_6H_5CH\text{———}CCl_2 \longleftrightarrow \left[C_6H_5CH\text{———}CCl \right] Cl^{\ominus}$$

$$C_6H_5\text{—}CH\text{—}C\text{=}NC_6H_5 \longleftarrow \left[C_6H_5\overset{\oplus}{C}H\text{—}C\text{=}NC_6H_5 \right] Cl^{\ominus}$$

$$C_6H_5CH\text{—}C\text{=}NC_6H_5 \qquad Y = \text{nucleophile}$$

For example, reaction of 1,3-diphenyl-2,2-dichloroaziridine (XIII) with aliphatic primary amines yields the amidines XVI ([19]).

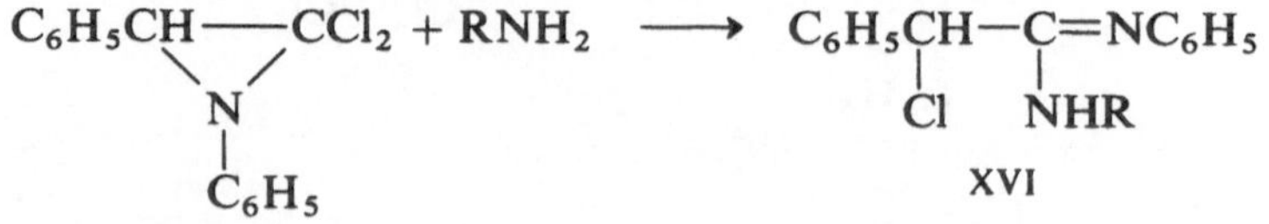

$$C_6H_5CH\text{———}CCl_2 + RNH_2 \longrightarrow C_6H_5CH\text{—}C\text{=}NC_6H_5$$

XIII   XVI

In the reaction of XIII with water (^{7,16}) and carboxylic acids (^{19}), the corresponding α-chloro-α-phenylacetanilide is obtained in high yield.

The reaction of XIII with ethanol affords aniline hydrochloride and ethyl α-chlorophenylacetate (XVII) (^{19}).

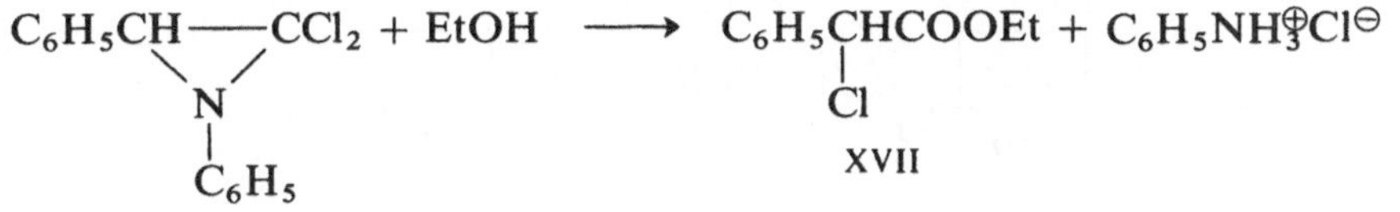

$$C_6H_5CH\!-\!\!-\!\!CCl_2 + EtOH \longrightarrow C_6H_5\underset{Cl}{\overset{}{C}}HCOOEt + C_6H_5NH_3^{\oplus}Cl^{\ominus}$$

$$\underset{\underset{\displaystyle C_6H_5}{|}}{N}$$

XIII   XVII

Heating of 1,3,3-triphenyl-2,2-dichloroaziridine (XVIII) with sodium iodide affords the corresponding ketenimine (XIX) in almost quantitative yield (^{19}).

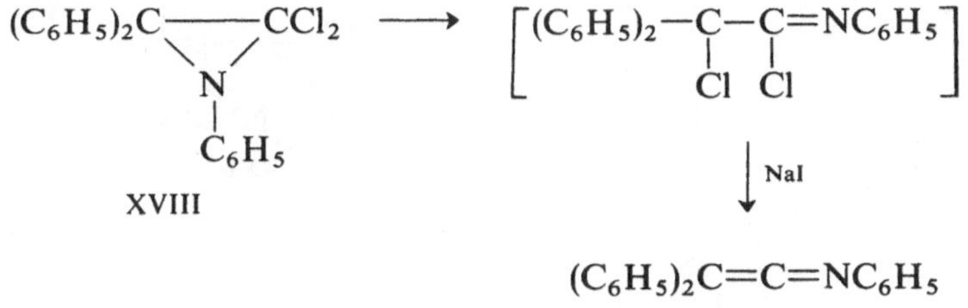

$$(C_6H_5)_2C\!-\!\!-\!\!CCl_2 \longrightarrow \left[(C_6H_5)_2\!-\!\underset{Cl}{\overset{}{C}}\!-\!\underset{Cl}{\overset{}{C}}\!=\!NC_6H_5\right]$$

$$\underset{\underset{\displaystyle C_6H_5}{|}}{N}$$

XVIII

$$\downarrow \text{NaI}$$

$$(C_6H_5)_2C\!=\!C\!=\!NC_6H_5$$

XIX

Interestingly, nucleophilic substitution reactions of 2-chloroaziridines do not proceed *via* the linear imidoyl chloride, as evidenced by the retention of the cyclic structure (^{11}).

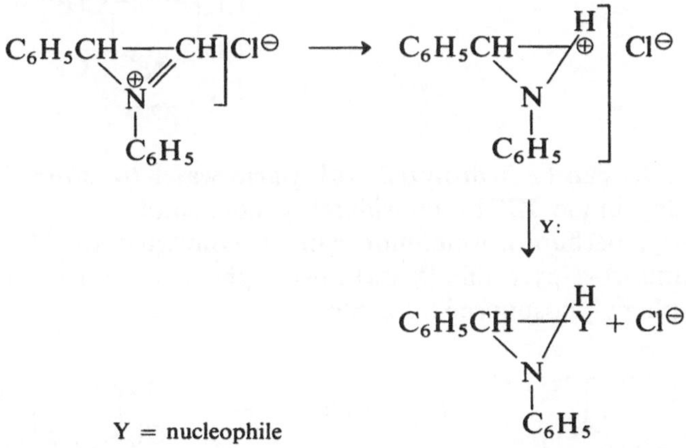

$$C_6H_5CH\!-\!\!-\!\!CH\big]Cl^{\ominus} \longrightarrow \left[C_6H_5CH\!-\!\!\!\overset{H}{\underset{}{{}_{\oplus}}}\right]Cl^{\ominus}$$

Y = nucleophile

$$\downarrow \text{Y:}$$

$$C_6H_5CH\!-\!\!\!\overset{H}{Y} + Cl^{\ominus}$$

The reaction of 2-chloroaziridines with nucleophiles proceeds rapidly, with inversion of the stereochemistry (^{11}).

III. FIVE-MEMBERED RING IMIDOYL HALIDES

In 1907 Tafel and Wassmuth ([41]) treated pyrrolidone with phosphorus pentachloride and isolated a compound, m.p. 50–51°C, for which the cyclic imidoyl chloride structure XX has been postulated.

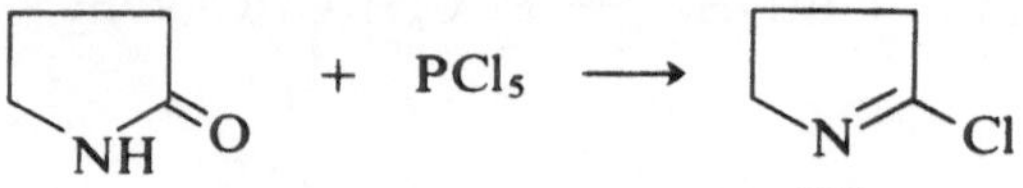

The compound was crystallized from aqueous acetone, indicating either a rather low reactivity, or a different structure.

The perchlorinated derivatives XXI and XXII are obtained upon reaction of dichloromaleic imide with phosphorus pentachloride ([1,6]).

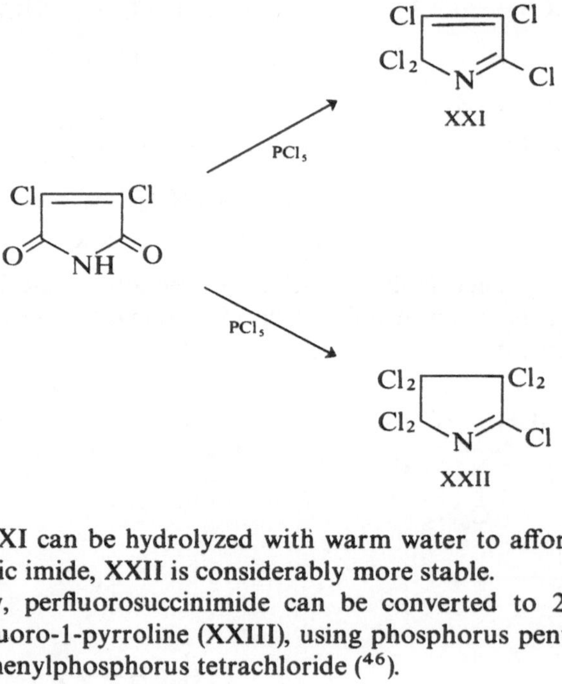

While XXI can be hydrolyzed with warm water to afford the starting dichloromaleic imide, XXII is considerably more stable.

Similarly, perfluorosuccinimide can be converted to 2,5,5-trichloro-3,3,4,4-tetrafluoro-1-pyrroline (XXIII), using phosphorus pentachloride, or, preferably, phenylphosphorus tetrachloride ([46]).

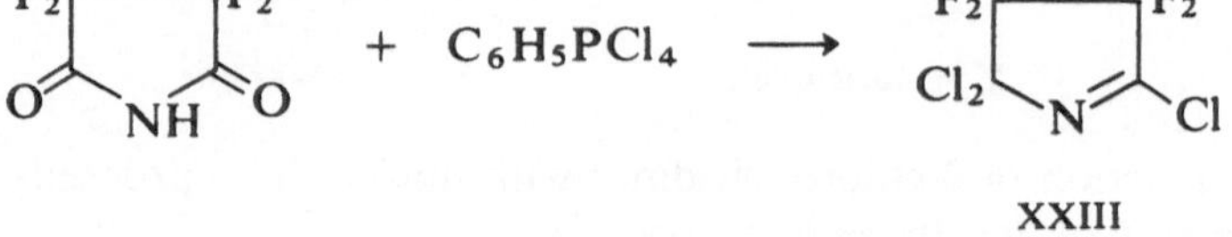

2,5,5-Trichloro-3,3,4,4-tetrafluoro-1-pyrroline ([46]). A mixture of 85.5 g (0.5 mole) of perfluorosuccinimide and 300 g (1.2 moles) of phenylphosphorous tetrachloride is refluxed until the evolution of hydrogen chloride ceases (*ca.* 5 hours). Rectification affords 110.3 g (91 %) of 2,5,5-trichloro-3,3,4,4-tetrafluoro-1-pyrroline, b.p. 112–114°C.

In contrast to XXII, the tetrafluoro homolog XXIII is considerably more reactive toward nucleophiles, as evidenced by the rapid conversion to the starting imide in the presence of water ([46]).

The reaction of XXIII with ammonia and aniline produces 2-amino-5-imino-3,3,4,4-tetrafluoropyrrolines (XXIV) ([46]).

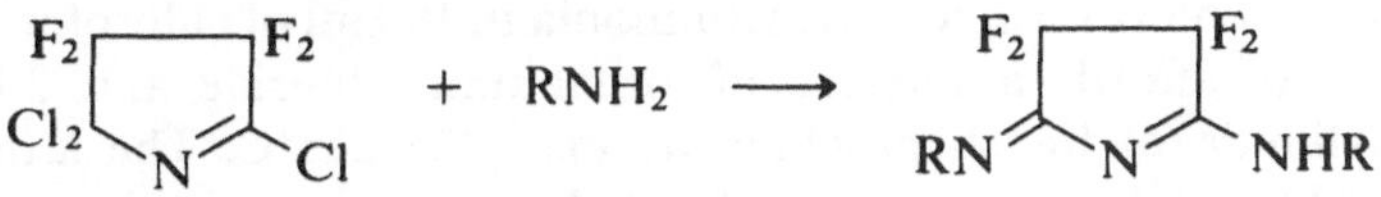

$$R = H, C_6H_5$$

The cyclic analogs of chloroformamidines are virtually unknown. Suzuki ([40]) mentioned the synthesis of 2-chloroimidazolidine but reported no physical constants. The cyclic chloroformamidinium chloride XXV can be obtained from 1,3-dimethylimidazolidine-2-one (XXVI) and carbonyl chloride, but this reaction is considerably slower than the phosgenation of the tetrasubstituted linear urea ([49]).

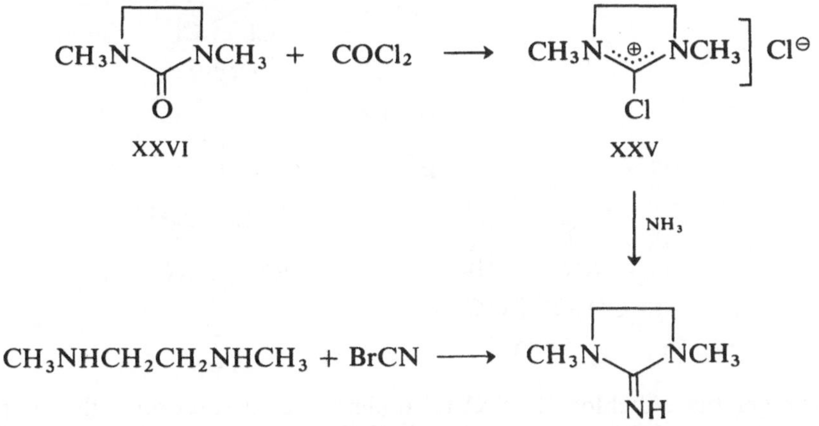

The polar structure XXV of the hygroscopic solid obtained was indicated by its physical properties, as well as by the C=N double bond absorption in the infrared spectrum (1626 cm^{-1}). Further verification of the postulated structure is obtained by the facile conversion of XXV to XXVII, which was synthesized independently from N,N'-dimethylethylene-diamine and cyanogen bromide ([49]).

2-Chloro-1,3-dimethylimidazolidinium chloride ([49]). To 20.5 g (0.18 mole) of 1,3-dimethylimidazolidine-2-one in 200 ml of ethylene dichloride excess carbonyl chloride is added at reflux temperature over a period of 60 min. The excess carbonyl chloride is removed in a stream of nitrogen and after evaporation of the solvent 30 g (98.5%) of the very hygroscopic 2-chloro-1,3-dimethylimidazolidinium chloride, m.p. 108–119°C, is obtained.

Addition of 21.5 g (0.13 mole) of the imidazolidinium chloride dissolved in 100 ml of chloroform to excess ammonia in 100 ml of chloroform at room temperature affords a mixture of ammonium chloride and 2-imino-1,3-dimethylimidazolidine hydrochloride, m.p. 225–228°C. The latter can be separated from the ammonium chloride by extraction with hot chloroform. The free base XXVII, b.p. 50–51°C/0.1 mm, n_D^{24} 1.5010, can be generated with sodium hydroxide.

The heterocyclic chlorides XXVIII, obtained by the addition of oxalyl chloride to carbodiimide ([37,47,48]), have the nonpolar geminate dichloride structure, as evidenced by the absence of C=N absorption in their infrared spectra.

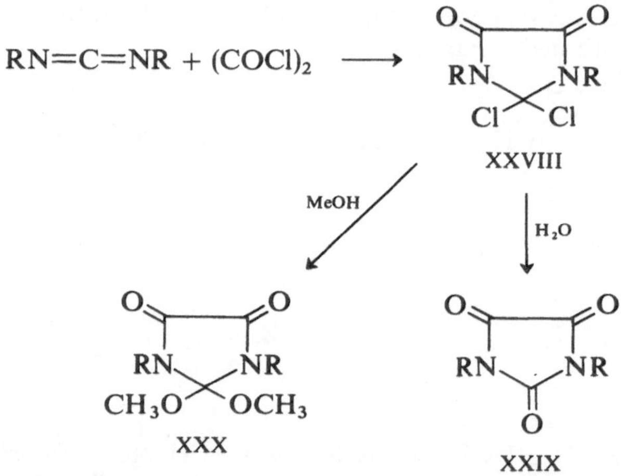

$$RN{=}C{=}NR + (COCl)_2 \longrightarrow$$

XXVIII

MeOH

H$_2$O

XXX

XXIX

The geminate dichlorides XXVIII undergo rapid reaction with water to yield the parabanic acids XXIX ([37,47,48]), and heating in methanol produces the parabanic acid dimethyl acetals XXX ([48]).

Apparently the lowering of the basicity of the nitrogen due to the adjacent carbonyl groups is sufficient to stabilize XXVIII in the nonpolar geminate dichloride structure.

The fully aromatic heterocyclic five-membered ring imidoyl halides have only a limited degree of imidoyl reactivity. However, their benzo derivatives are often considerably more reactive. For example, 2-chlorobenzthiazole (XXXI) reacts considerably faster than 2-chlorothiazole (XXXII) in nucleophilic substitution reactions ([50]).

XXXI   XXXIII   XXXIV

XXXII   XXXV

The 2-halobenzthiazole system is very well investigated, and the rate of reaction of 2-halobenzthiazoles with nucleophiles, such as $CH_3O^{\ominus}$ and $PhS^{\ominus}$, has been measured. For example, the following order of reactivity of 2-halobenzthiazoles with $MeO^{\ominus}$ has been established: $F \gg Cl > Br > I$ ([31,43]). Likewise, the activation energies and entropies for the reaction of halobenzthiazoles and $MeO^{\ominus}$ and $PhS^{\ominus}$ have been measured ([32]).

The effect of substituents in the benzene ring has also been investigated and it was generally found that electron-withdrawing substituents increase and electron-donating groups retard the rate of nucleophilic attack ([33,44,45]).

The rate of the reaction of 2-chlorobenzoxazole (XXXIII) with $PhS^{\ominus}$ is faster than the rate of $PhS^{\ominus}$ with XXXI under comparable conditions ([13]). Likewise, electron-withdrawing substituents enhance the rate of reaction of 2-chlorobenzoxazoles with nucleophiles ([5]). 2-Chlorobenzimidazole (XXXIV) has been synthesized from benzimidazolone and phosphorus pentachloride ([26]); however, its reaction with nucleophiles has not been investigated.

2-Chlorothiazoline (XXXV), the cyclic homolog of chlorothioformimidate, has been synthesized from phenyl isothiocyanate and phosphorus pentachloride ([17]), but again nothing is known about its reactivity with nucleophiles.

Apparently, an increase of heteroatoms as ring building blocks increases the imidoyl character of the corresponding halide. For example, 1-phenyl-5-chlorotetrazole (XXXVI) undergoes rapid reaction with phenolate ion to produce the corresponding phenyltetrazolyl ether (XXXVII), which can be hydrogenated under mild conditions to form 1-phenyl-5-tetrazolone (XXXVIII) and the corresponding hydrocarbon ([27]).

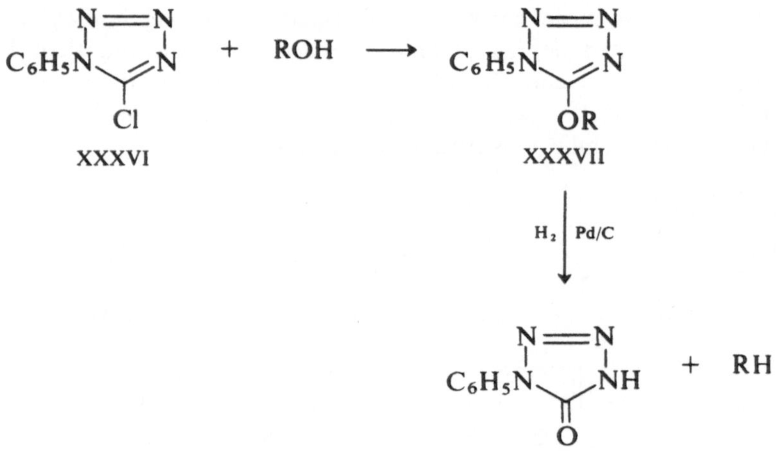

$$\text{XXXVI} + \text{ROH} \longrightarrow \text{XXXVII}$$

$$\xrightarrow{\text{H}_2 \;|\; \text{Pd/C}} \text{XXXVIII} + \text{RH}$$

This reaction sequence provides a facile method to replace phenolic hydroxyl groups by hydrogen, and in addition to **XXXVI**, 2-chlorobenzthiazole (**XXXI**) and 2-chlorobenzoxazole (**XXXIII**) can be used with equally good results ([27]).

IV. SIX-MEMBERED RING IMIDOYL HALIDES

The parent compound of this series, 2-chloro-3,4,5,6-tetrahydropyridine (**XXXIX**), was obtained as an intermediate in the Beckmann rearrangement of cyclopentanone oxime ([39]). Thus, heating of cyclopentanone oxime with thionyl chloride affords about equal amounts of the lactam hydrochloride **XL** and the imidoyl chloride **XXXIX** ([39]).

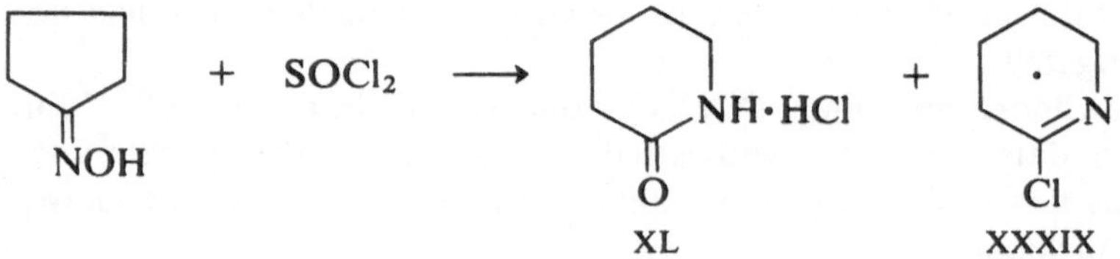

$$+ \text{SOCl}_2 \longrightarrow \underset{\text{XL}}{\text{NH·HCl}} + \underset{\text{XXXIX}}{\text{Cl}}$$

However, the imidoyl chloride **XXXIX** could not be isolated, and it was therefore reacted *in situ* with methyl anthranilate to afford 6,7,8,9-tetrahydropyrido-[2,1-*b*]-quinazole-11-one (**XLI**). Since this condensation occurs in high yield ([23]), the amount of **XXXIX** present in the original reaction mixture can be established in this manner.

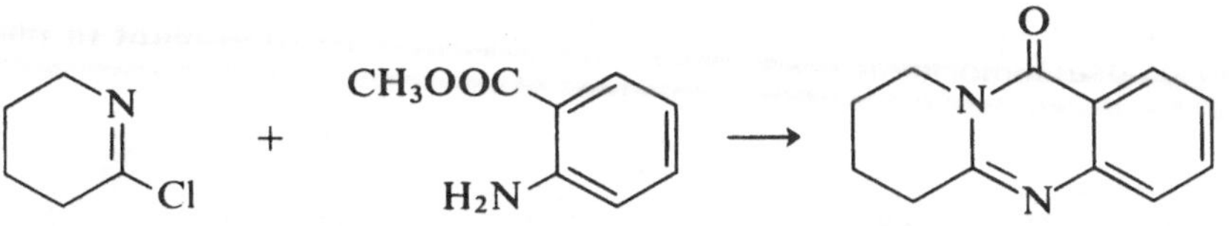

XXXIX   XLI

In contrast, J. v. Braun and his co-workers ([4]) have isolated 2,3,3-trichloro-3,4,5,6-tetrahydropyridine (XLII), m.p. 21°C in the reaction of piperidinone with excess phosphorus pentachloride. The structure of XLII was confirmed by hydrolysis to 3,3-dichloropiperidinone (XLIII) ([4]).

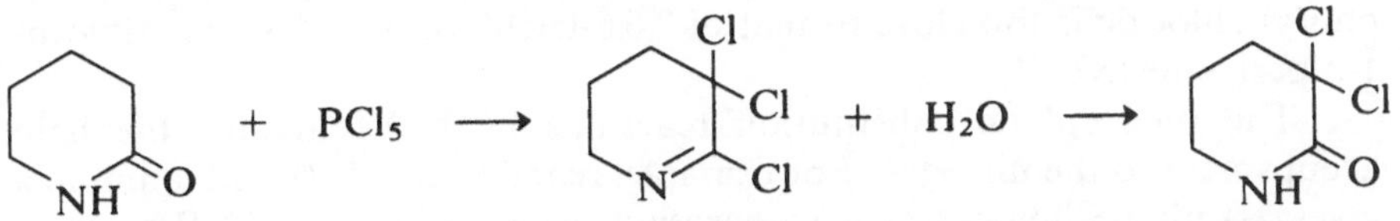

XLII   XLIII

Similarly, reaction of cyclic imides with phosphorus pentahalides should afford the corresponding dihaloimidoyl halides. In 1882 Bernheimer ([3]) treated glutarimide with phosphorus pentachloride, and he assigned the imidoyl chloride structure XLIV to the reaction product. However, it was later shown that the compound was 2,3,6-trichloropyridine (XLV) ([8]).

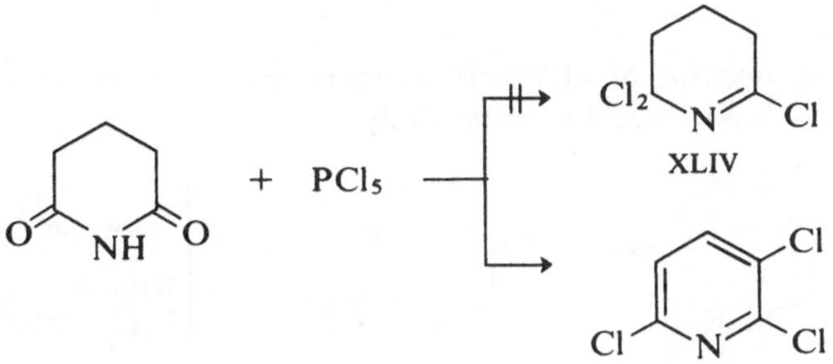

XLIV

XLV

This facile synthesis of halopyridines could be extended to α-methyl- and α,α'-dimethylglutarimide ([8]). The reaction of glutarimide with phosphorus pentabromide proceeds similarly to afford a mixture of 2,3,6-tribromopyridine and 2,3,5,6-tetrabromopyridine ([9]).

If the elimination of hydrogen halide is prevented by halogen substitution, the corresponding imidoyl halides are obtained. For example, the reaction of hexafluoroglutarimide with phosphorus pentachloride produces the imidoyl chloride XLVI in high yield. The reaction proceeds *via* the intermediate XLVII, which can be isolated if a deficient amount of phosphorus pentachloride is used ([46]).

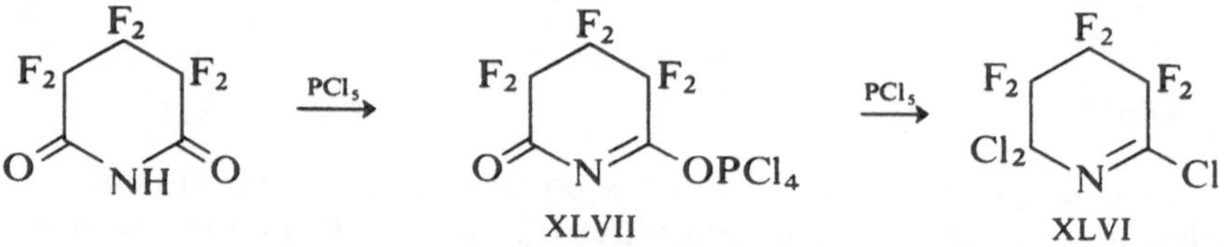

XLVII   XLVI

However, it is advantageous to use phenylphosphorus tetrachloride rather than phosphorus pentachloride, because the boiling point of phosphoryl chloride is too close to that of 2,6,6-trichloro-3,3,4,4,5,5-hexafluoro-1-piperideine (XLVI).

The nucleophilic substitution reactions in XLVI involve the halo groups next to the nitrogen. For example, reaction of XLVI with ammonia gives 2-amino-6-imino-3,3,4,4,5,5-hexafluoro-1-piperideine (XLVIII), which can be homopolymerized upon heating ([46]).

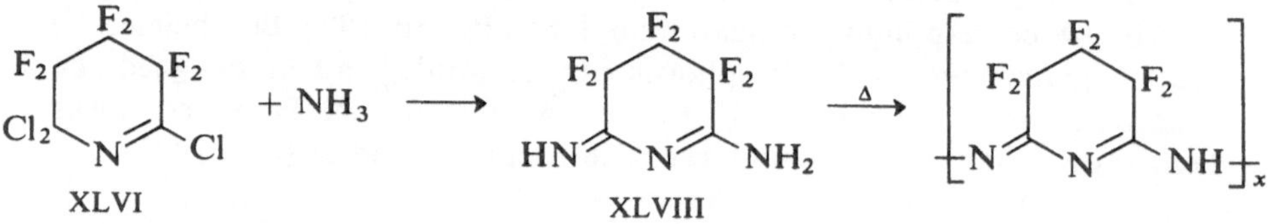

XLVI   XLVIII

Similarly, reaction of XLVI with primary and secondary amines produces compounds XLIX and L, respectively.

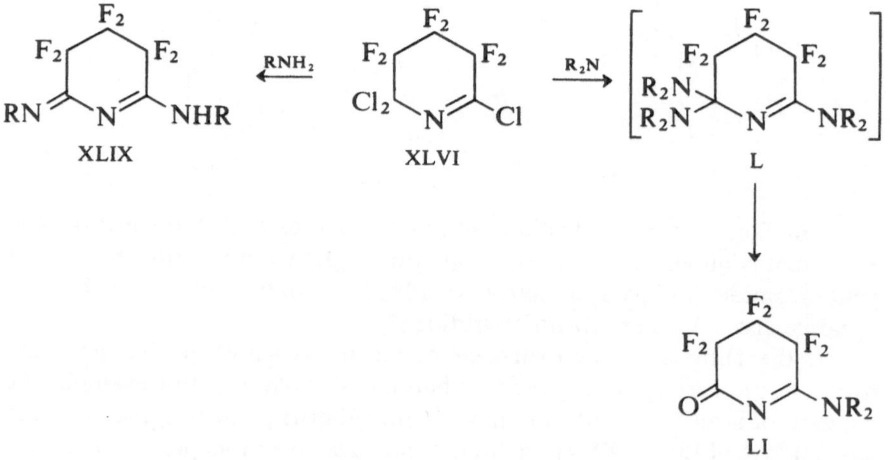

XLIX   XLVI   L

LI

The reaction products with secondary amines could not be isolated in pure form, but since they are aminoacetals of the corresponding carbonyl compound, hydrolysis with water to LI is easily accomplished ([46]).

The chloro groups in XLVI can also be displaced by fluoride ion under mild conditions, to afford the perfluorinated imidoyl fluoride LII ([46]).

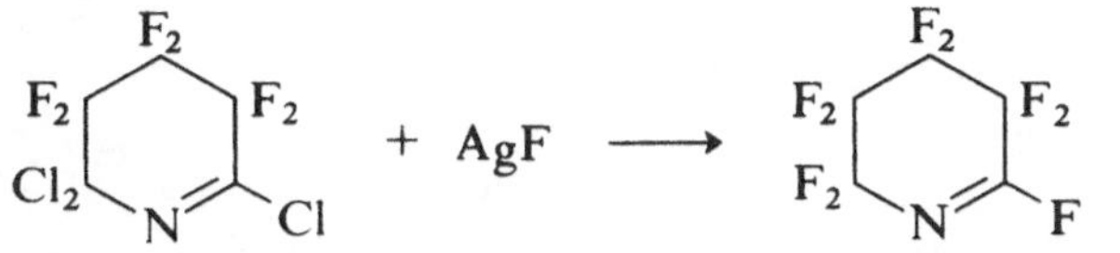

2,2,6-*Trichloro-3,3,4,4,5,5-hexafluoro-1-piperideine* ([46]). A mixture of 44.2 g (0.2 mole) of hexafluoroglutarimide and 120 g (0.48 mole) of phenylphosphorus tetrachloride is refluxed until the evolution of hydrogen chloride ceases (approximately 15 hours). Fractional distillation gives 55.9 g (95%) of 2,2,6-trichloro-3,3,4,4,5,5-hexafluoro-1-piperideine, b.p. 135–136.5 (760 mm).

Perfluoro-1-piperideine ([46]). The amount of 35 g of 2,2,6-trichloro-3,3,4,4,5,5-hexafluoro-1-piperideine is added in small portions to 100 g finely powdered silver fluoride. Upon each addition a vigorous reaction occurs. Fractional distillation from the silver salts affords 20.7 g (71%) of perfluoro-1-piperideine, b.p. 43°C.

Perhalogenated imidoyl chlorides are also obtained by the direct high-temperature chlorination of carbamoyl chlorides. For example, chlorination of morpholine N-carbonyl chloride (LIII) at 200°C yields the perchloro derivative LIV ([18]).

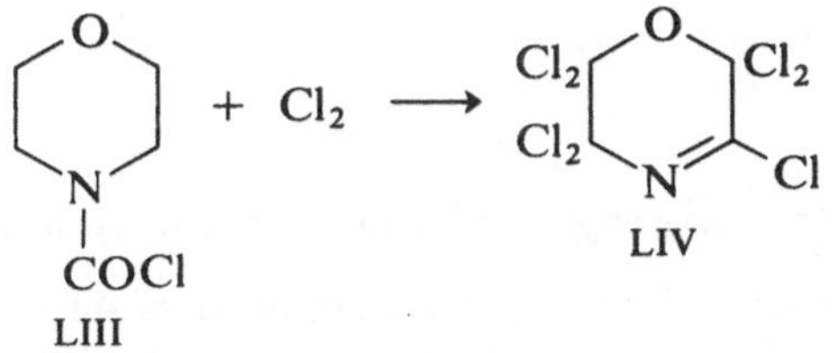

Again one might expect that the benzoderivatives of "aromatic" six-membered ring imidoyl halides have reactive halo groups. Recently, an interesting route to six-membered ring benzimidoyl chlorides has been reported ([35]). Reaction of o-cyanomethylbenzoic acid (LV) with 2 moles of phosphorus pentachloride produces 1,3-dichloroisoquinoline (LVI).

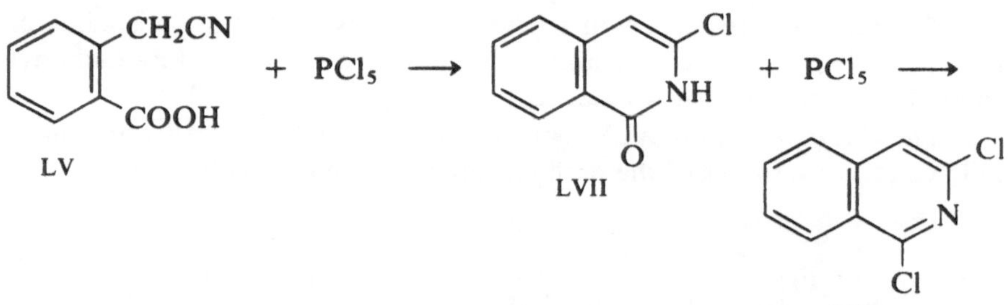

If only one mole of phosphorus pentachloride is used, the intermediate LVII is obtained ([35]). However, LVI does not undergo rapid reaction with nucleophiles as evidenced by the fact that the compound can be recrystallized from ethanol ([35]).

In a similar manner, reaction of *o*-thiocyanatobenzoic acid (LVIII) with phosphorus pentachloride produces *o*-thiocyanatobenzoyl chloride (LIX), which in the presence of hydrogen chloride cyclizes to 2-chloro-1,3-benzothiazine-4-one (LX) ([35]).

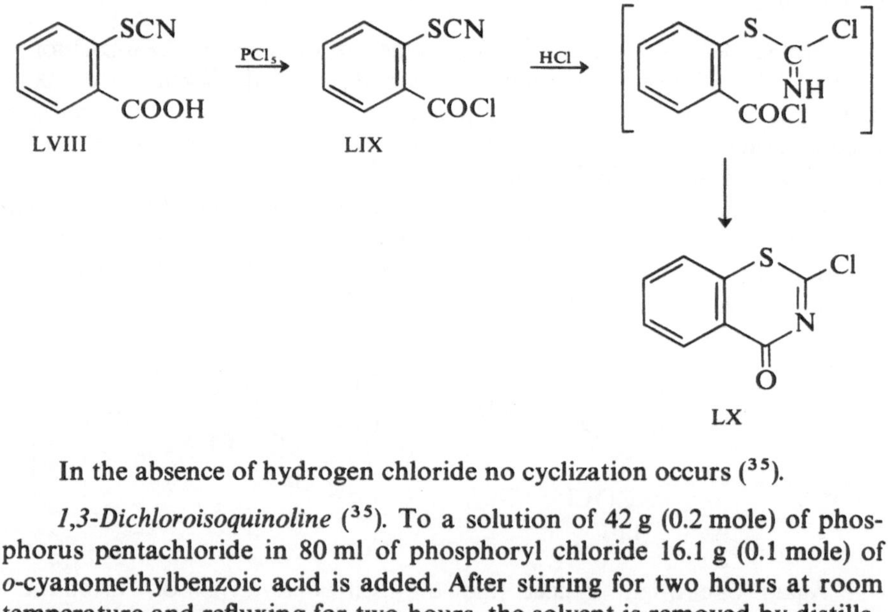

In the absence of hydrogen chloride no cyclization occurs ([35]).

1,3-Dichloroisoquinoline ([35]). To a solution of 42 g (0.2 mole) of phosphorus pentachloride in 80 ml of phosphoryl chloride 16.1 g (0.1 mole) of *o*-cyanomethylbenzoic acid is added. After stirring for two hours at room temperature and refluxing for two hours, the solvent is removed by distillation and the residue is crystallized from ethanol to afford 18.5 g (94%) of 1,3-dichloroisoquinoline, m.p. 120°C.

The heterocyclic six-membered ring halogen derivatives with more than one heteroatom in the ring have a higher degree of imidoyl halide character. For example, the three chloro groups in trichloro-1,3,5-triazine (cyanuric chloride) (LXI) are quite reactive, and can be replaced stepwise with a wide variety of nucleophiles ([36]). Similarly, trifluoro-1,3,5-triazine (LXII) reacts rapidly with nucleophiles ([14,22]).

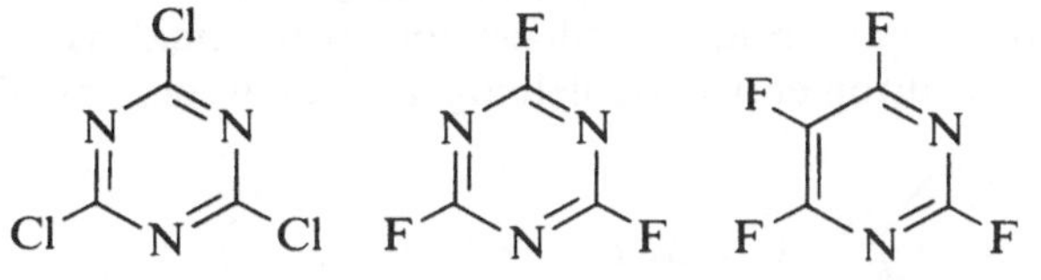

LXI   LXII   LXIII

Similarly, reaction of tetrafluoropyrimidine (LXIII) with nucleophiles occurs under mild conditions ([34]).

In order to establish the degree of imidoyl halide character in N-heterocycles far more has to be known about their rate of reaction with nucleophiles, and the indicated compounds are selected because we have some knowledge of their reactivity.

V. SEVEN-MEMBERED RING IMIDOYL HALIDES

In the reaction of caprolactam with phosgene, the corresponding seven-membered ring imidoyl chloride LXIV is a logical intermediate. However, the product of reaction with a second mole of phosgene, 1-chlorocarbonyl-2-chloro-4,5,6,7-tetrahydro-1H-azepine (LXV), has only been isolated ([28]). The mechanism of this reaction is discussed in detail in Chapter 3.

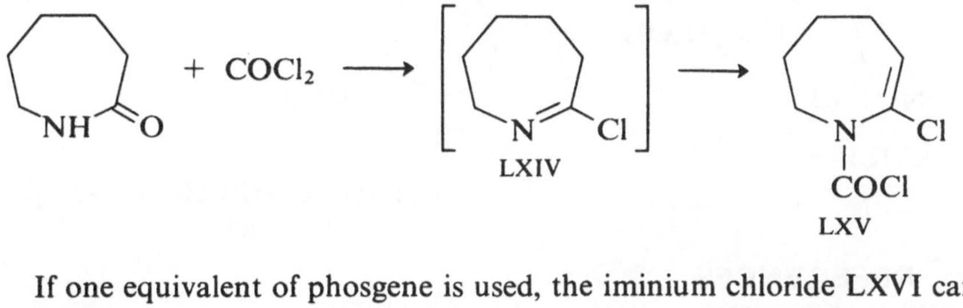

If one equivalent of phosgene is used, the iminium chloride LXVI can be obtained. The latter affords the cyclic amidine LXVII upon reaction with aniline ([38]).

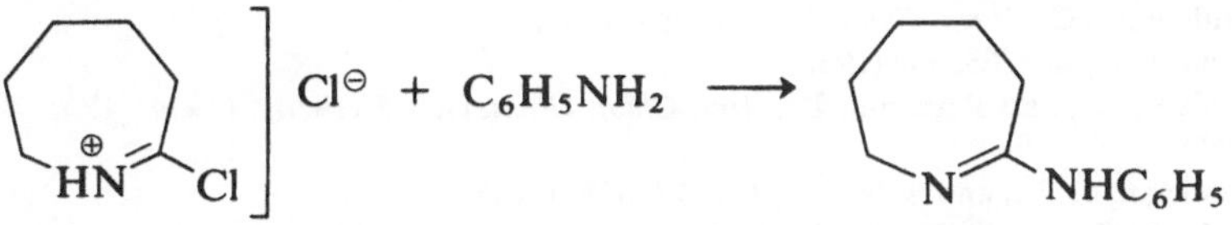

LXVI   LXVII

Nitration ([15]) and chlorination ([29]) of LXV gives rise to substitution in the α-position, which renders compound LXV useful for the synthesis of d,l-lysin. Likewise, dichlorination *via* the enamine derived from LXIV is possible ([2]).

If caprolactam is treated with phosphoryl chloride or one equivalent of phosphorus pentachloride, condensation of the imidoyl chloride LXIV with its enamine tautomer occurs, yielding the azepine derivative LXVIII ([30]).

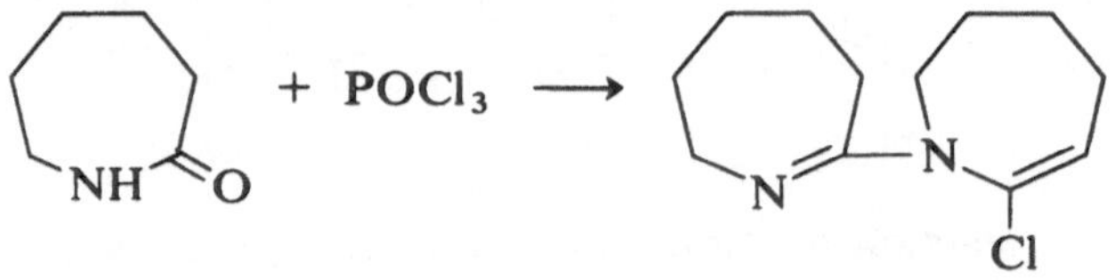

LXVIII

1-Chlorocarbonyl-2-chloro-4,5,6,7-tetrahydro-1H-azepine ([28]). To 400 g of phosgene dissolved in 750 ml of chloroform a solution of 113 g (1 mole) of caprolactam in 250 ml of chloroform is added dropwise at 3–4°C. After stirring for 2 hours the reaction mixture is gradually heated to reflux and the excess phosgene is removed with nitrogen. Removal of the solvent and vacuum distillation affords 187.5 g (96%) of 1-chlorocarbonyl-2-chloro-4,5,6,7-tetrahydro-1H-azepine.

The solvolysis of LXV with ethanol was investigated recently and mixtures of 3,4,5,6-tetrahydro-7-ethoxy-2H-azepine (LXIX) and ethyl N-carbethoxy-ε-aminocaproate (LXX) were obtained ([42]).

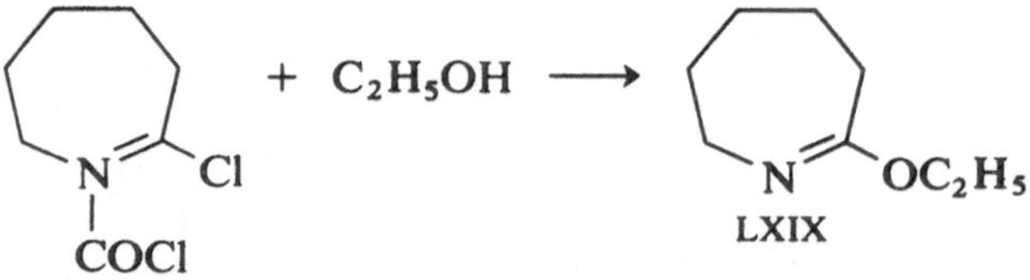

LXIX

LXV

$+ C_2H_5OOCNH(CH_2)_5COOC_2H_5$

LXX

VI. REFERENCES

1. Anschütz, R., and Schroeter, G., *Ann.* **295**, 82 (1896).
2. Beer, L., and Metzger, H., French Pat. 1,396,153 (1965); *Chem. Abstr.* **63**, 1709 (1965).
3. Bernheimer, O., *Gazz. Chim. Ital.* **12**, 283 (1882).
4. Braun, J. v., *Ber.* **63**, 502 (1930).
5. Cerniani, A., and Passerini, R., *Ann. Chim.* (*Rome*) **44**, 3 (1954); *Chem. Abstr.* **49**, 4626 (1955).
6. Ciamician, G. L., and Silber, P., *Ber.* **17**, 553 (1884).
7. Cook, A. G., and Fields, E. K., *J. Org. Chem.* **27**, 3686 (1962).

8. Crouch, W. W., and Lochte, H. L., *J. Am. Chem. Soc.* **65**, 270 (1943).
9. Crouch, W. W., and Lochte, H. L., *Rec. Trav. Chim.* **67**, 380 (1948).
10. Deyrup, J. A., and Greenwald, R. B., *Tetrahedron Letters* 321 (1965).
11. Deyrup, J. A., and Greenwald, R. B., *J. Am. Chem. Soc.* **87**, 4538 (1965).
12. Fields, E. K., and Sandri, J. M., *Chem & Ind.* 1216 (1959).
13. Foa, M., Ricci, A., Todesco, P. E., and Vivarelli, P., *Bull. Sci. Chim. Ind. Bologna* **23**, 89 (1965); *Chem. Abstr.* **63**, 13236 (1965).
14. Grisley, D. W., Gluesenkamp, E. W., and Heininger, S. A., *J. Org. Chem.* **23**, 1802 (1958).
15. Haan, J. de, and Hoff, J. P. H. van der, Belgian Pat. 614,233 (1962); *Chem. Abstr.* **58**, 1356 (1963).
16. Heine, H. W., and Smith A. B. III, *Angew. Chem. Intern. Ed.* **2**, 400 (1963).
17. Hofmann, A. W., *Ber.* **12**, 1127 (1879).
18. Holtschmidt, H., *Angew. Chem. Intern. Ed.* **1**, 632 (1962).
19. Ichimura, K., and Ohta, M., *Tetrahedron Letters* 807 (1966).
20. Kadaba, P. K., and Edwards, J. O., *J. Org. Chem.* **25**, 1431 (1960).
21. Klamann, D., Wache, H., Ulm, K., and Nerdel, F., *Chem. Ber.* **100**, 1870 (1967).
22. Kober, E., Schroeder, H., Rätz, R. F. W., Ulrich, H., and Grundemann, C., *J. Org. Chem.* **27**, 2577 (1962).
23. Levy, P. R., and Stephen, H., *J. Chem. Soc.* 985 (1956).
24. Lukasiewicz, A., *Tetrahedron* **20**, 1113 (1964).
25. Lukasiewicz, A., and Lesinka, J., *Tetrahedron* **21**, 3247 (1965).
26. Manuelli and Recchi, *Atti della Reale Acad. dei Lincei*, **9**, 271 (1900).
27. Musliner, W. J., and Gates, J. W. Jr., *J. Am. Chem. Soc.* **88**, 4271 (1966).
28. Ottenheym, J. H., and Garritsen, J. W., British Pat. 901,169 (1962); *Chem. Abstr.* **58**, 6810 (1963).
29. Ottenheym, J. H., and Garritsen, J. W., German Pat. 1,154,118 (1963); *Chem. Abstr.* **60**, 2789 (1964).
30. Prajsnar, B., *Roczniki Chem.* **36**, 1449 (1962); *Chem. Abstr.* **59**, 5136 (1963).
31. Ricci, A., Todesco, P. E., and Vivarelli, P., *Gazz. Chim. Ital.* **95**, 101 (1965); *Chem. Abstr.* **63**, 1676 (1965).
32. Ricci, A., Foa, M., Todesco, P. E., and Vivarelli, P., *Tetrahedron Letters* 1935 (1965).
33. Ricci, A., Todesco, P. E., and Vivarelli, P., *Gazz. Chim. Ital.* **95**, 490 (1965); *Chem. Abstr.* **63**, 8171 (1965).
34. Schroeder, H., Kober, E., Ulrich, H., Rätz, R., Agahigian, H., and Grundmann, C., *J. Org. Chem.* **27**, 2580 (1962).
35. Simchen, G., *Angew. Chem. Intern. Ed.* **5**, 663 (1966).
36. Smolin, E. M., and Rapoport, L., *S-Triazines and Derivatives*, Interscience, New York, 1959, p. 55.
37. Stachel, H. D., *Angew. Chem.* **71**, 246 (1959).
38. Stamicarbon, N. V., Belgian Pat. 609,822 (1962); *Chem. Abstr.* **57**, 16505 (1962).
39. Stephen, T., and Stephen H., *J. Chem. Soc.* 4694 (1956).
40. Suzuki, K., Japanese Pat. 1528 (1952); *Chem. Abstr.* **48**, 2120 (1954).
41. Tafel, J., and Wassmuth, O., *Ber.* **40**, 2831 (1907).
42. Tetenbaum, M. T., *J. Org. Chem.* **31**, 4298 (1966).
43. Todesco, P. E., and Vivarelli, P., *Bull. Sci. Fac. Chim. Ind. Bologna* **20**, 143 (1962); *Chem. Abstr.* **59**, 403 (1963).
44. Todesco, P. E., and Vivarelli, P., *Gazz. Chim. Ital.* **92**, 1221 (1962); *Chem. Abstr.* **59**, 396 (1963).
45. Todesco, P. E., and Vivarelli, P., *Gazz. Chim. Ital.* **94**, 372 (1964); *Chem. Abstr.* **61**, 6902 (1964).

46. Ulrich, H., Kober, E., Schroeder, H., Rätz, R., and Grundmann, C., *J. Org. Chem.* **27**, 2585 (1962).
47. Ulrich, H., and Sayigh, A. A. R., *J. Org. Chem.* **30**, 2781 (1965).
48. Ulrich, H., Tucker, B., and Sayigh, A. A. R., *Tetrahedron* **22**, 1565 (1966).
49. Ulrich, H., and Sayigh, A. A. R., unpublished results.
50. Young, T. E., and Amstutz, E. D., *J. Am. Chem. Soc.* **73**, 4773 (1951).

Appendix

This appendix is a brief review of recent literature, and covers approximately eight months (the last half of 1967 and the first two months of 1968). The large number of additional references compiled over so short a period of time reflects the amount of work currently conducted in this area.

A. CARBONIMIDOYL DIHALIDES

The direct chlorination of chlorosulfonyl isocyanate (I) to chlorosulfonylcarbonimidoyl dichloride (II) by means of phosphorus pentachloride has been reported [42]. However, severe conditions are required and the yield is low. Selective fluorination of II with antimony trifluoride produces the sulfonyl fluoride III [42].

$$ClSO_2NCO + PCl_5 \longrightarrow ClSO_2N=CCl_2 + SbF_3 \longrightarrow$$
$$\qquad I \qquad\qquad\qquad\qquad II$$

$$FSO_2N=CCl_2$$
$$III$$

In a recent publication by Holtschmidt and his co-workers the mechanism of the high-temperature chlorination of N-methylpyrrolidone was discussed [43]. In this reaction 1-chlorocarbonylhexachloropropyl-3-carbonimidoyl dichloride is formed as a major product (see Chapter 2). Further work related to the nucleophilic substitution reactions of carbonimidoyl dichlorides with water [48], hydrogen sulfide [48], alcohols [31], mercaptans [31], and amines [31] has been reported, and a step-wise displacement of the chloro groups was demonstrated by Neidlein and his students [31,32]. For example, reaction of arenesulfonylcarbonimidoyl dichlorides IV with methylmercaptan yields the 1-chlorothioformimidates V, which, upon further reaction with primary and secondary amines, yield the S,N-acetals VI [31].

211

$$RSO_2N=CCl_2 + CH_3SH \longrightarrow RSO_2N=\underset{\underset{Cl}{|}}{C}-SCH_3$$

IV
V

$$RSO_2N=\underset{\underset{Cl}{|}}{C}-SCH_3 + R'_2NH \longrightarrow RSO_2N=\underset{\underset{NR'_2}{|}}{C}-SCH_3$$

VI

With 1,2-diols and dithiols the corresponding cyclic acetals are obtained ([31]). Likewise, aminoalcohols, aminothiols, and diamines yield cyclic acetals upon reaction with acyl carbonimidoyl dichlorides ([6]). The reaction of IV with acylhydrazides produces N-sulfonylamino-1,3,4-oxadiazoles VII ([30]).

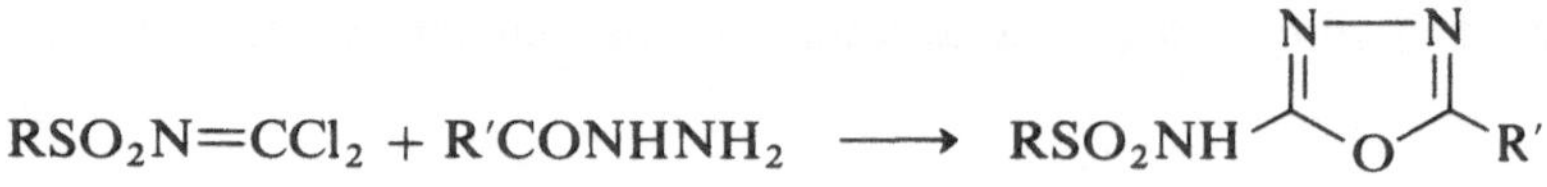

$$RSO_2N=CCl_2 + R'CONHNH_2 \longrightarrow RSO_2NH\underset{O}{\overset{N---N}{\diagup \diagdown}}R'$$

VII

The reaction of amide oximes with carbonimidoyl dichlorides has been reported as well ([13]).

Several recent articles are concerned with the synthesis and reactions of perfluorinated carbonimidoyl difluorides. Glemser and his students ([3]) reported the formation of trifluoromethylcarbonimidoyl difluoride in the reaction of cyanogen chloride and nitrogen trifluoride at 450–500°C. Further work related to the chemistry of perfluoro-bis-carbonimidoyl difluorides has been presented by Ogden and Mitsch ([35–37]). For example, photolysis of difluoromethane-1,1-bis-carbonimidoyl difluoride (VIII) generates the interesting radical species $CF_2=N\cdot$, which can be trapped by a variety of radical, carbene, and olefin substrates ([35]). In the presence of fluoride ion VIII rearranges to bis(trifluoromethyl)carbodiimide (IX) ([36]). Hydrolysis of VIII yields 1,3-bis (trifluoromethyl)urea (X) ([37]).

$$F_3CN=C=NCF_3$$

IX

$$F^{\ominus}\nearrow$$

$$F_2C=N-CF_2-N=CF_2$$

VIII

$$\searrow H_2O$$

$$F_3CNHCONHCF_3$$

X

B. IMIDOYL HALIDES

A novel method of synthesis of imidoyl chlorides has been reported by Kresze and Wucherpfennig ([23]). The authors treated N-sulfinylsulfonamides XI with aliphatic and aromatic carboxylic acid chlorides and obtained N-arenesulfonylimidoyl chlorides XII in excellent yield.

$$RSO_2N{=}S{=}O + R'COCl \longrightarrow RSO_2N{=}CR' + SO_2$$
$$\quad XI \hspace{8cm} | $$
$$\hspace{9.5cm} Cl$$
$$\hspace{9.5cm} XII$$

The formation of imidoyl bromides (XIII) in the reaction of 2-bromo-acetophenone oxime (XIV) with triphenylphosphorus has been reported by Masaki et al. ([28]).

$$C_6H_5C{-}CH_2Br + P(C_6H_5)_3 \longrightarrow C_6H_5N{=}C{-}CH_3 + OP(C_6H_5)_3$$
$$\quad \| \hspace{11cm} |$$
$$\quad NOH \hspace{9.5cm} Br$$
$$\quad XIV \hspace{9.7cm} XIII$$

Imidoyl bromides are also obtained when oximes are treated with the phosphonium salt XV. For example, reaction of XV with the oxime XVI produces the imidoyl bromide XVII, which can be dehydrohalogenated to the ketenimine XVIII by means of triethylamine ([28]).

$$C_6H_5{-}C{-}CH_2C_6H_5 + Br\overset{\oplus}{P}(C_6H_5)_3Br^{\ominus} \longrightarrow C_6H_5N{=}C{-}CH_2C_6H_5$$
$$\quad \| \hspace{10cm} |$$
$$\quad NOH \hspace{4cm} XV \hspace{6cm} Br$$
$$\quad XVI \hspace{11cm} XVII$$

$$\Big\downarrow Et_3N$$

$$C_6H_5N{=}C{=}CHC_6H_5$$
$$XVIII$$

The significance of imidoyl halides as intermediates in the Beckmann rearrangement has been pointed out earlier (see Chapter 3). In the Chapmann rearrangement imidoyl chlorides are used to prepare imidates, which undergo rearrangement to yield N-aroyl derivatives of diarylamines ([4]). For example, reaction of the sodium salt of dimethyl 4,6-dihydroxyisophthalate (XIX) with the benzimidoyl chloride XX yields the bis-imidate XXI, which can be thermally rearranged and hydrolyzed to produce the diamines XXII ([4]).

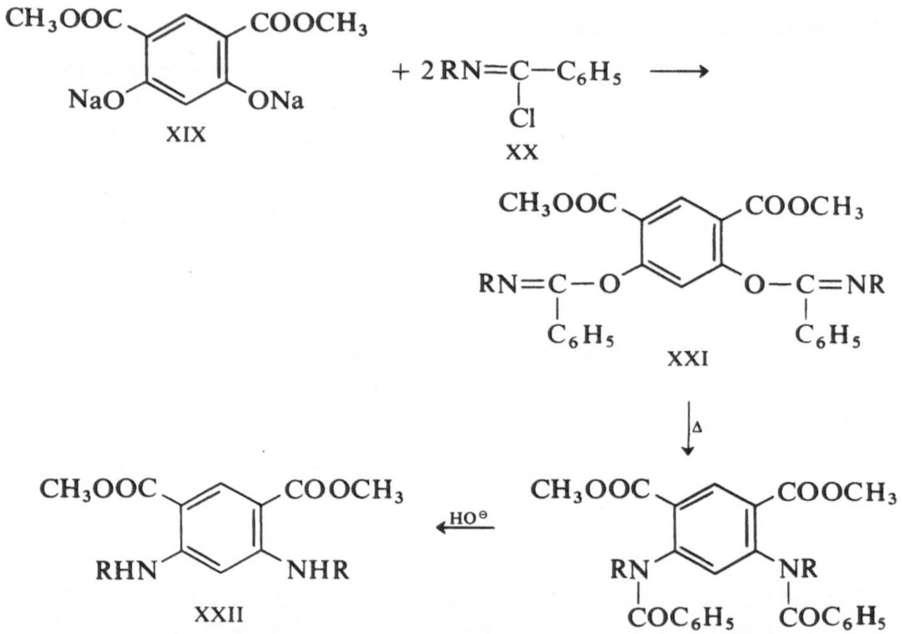

CH$_3$OOC — COOCH$_3$, NaO — ONa

XIX

$+ 2 RN=\overset{\underset{|}{Cl}}{C}-C_6H_5 \longrightarrow$

XX

CH$_3$OOC — COOCH$_3$

$RN=\overset{\underset{|}{C_6H_5}}{C}-O$ — $O-\overset{\underset{|}{C_6H_5}}{C}=NR$

XXI

$\downarrow \Delta$

CH$_3$OOC — COOCH$_3$, RHN — NHR

XXII

$\xleftarrow{\text{HO}^\ominus}$

CH$_3$OOC — COOCH$_3$

RN — NR

$\underset{COC_6H_5}{|}$ $\underset{COC_6H_5}{|}$

This reaction is quite useful to convert an aryl hydroxy compound to the corresponding amino derivative.

The hexachloroantimonate salts of unsubstituted imidoyl halides have been converted to the corresponding azides ([44]). Boron trichloride complexes of N-substituted benzimidoyl chlorides are reported by Hall and co-workers ([14]).

Two recent review articles related to the Vilsmeier–Haack reaction have appeared ([17,34]), and the formylation of 4H-quinolizine-4-ones by means of the DMF–POCl$_3$ complex has been reported ([47]).

The reaction of chlorodimethylformiminium chloride (XXII) with diazoketones and ethyl diazoacetate gives rise to the formation of α-diazoaldehydes XXIII ([46]). In contrast, diazomethane undergoes reaction with XXII to yield 2-dimethylamino-1,3-dichloropropane ([46]).

$RCOCH_2N_2 + (CH_3)_2\overset{\oplus}{N}\!\!\!=\!\!\!CHCl]Cl^\ominus \longrightarrow$

XXII

$[(CH_3)_2\overset{\oplus}{N}\!\!\!=\!\!\!CH-CN_2COR]Cl^\ominus$

$\downarrow H_2O$

$OCHCN_2COR$

XXIII

The formylation of polyene aldehydes ([33]) and α,β-unsubstituted ketones ([24]) with the Vilsmeier reagent (DMF–POCl$_3$) has been reported recently. For example, reaction of the ketones XXIV with the Vilsmeier reagent produces the iminium salts XXV, which are hydrolyzed to the aldehydes XXVI. The latter compounds form the conjugated ene-yne derivatives XXVII upon addition of their dioxane solution to dilute sodium hydroxide at 80–90°C ([24]).

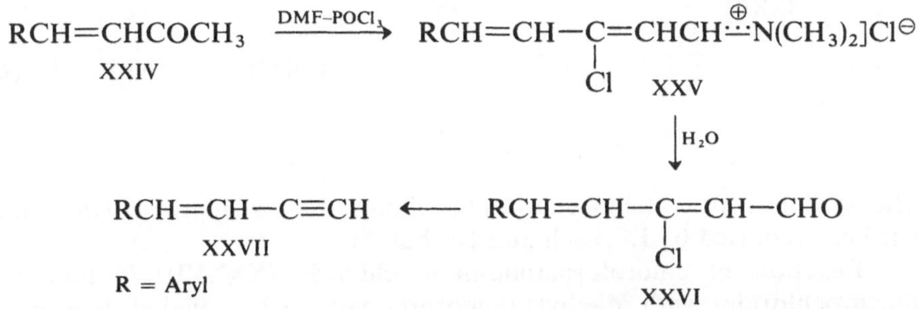

$$RCH{=}CHCOCH_3 \xrightarrow{\text{DMF–POCl}_3} RCH{=}CH{-}\underset{\underset{Cl}{|}}{C}{=}CHCH{\overset{\oplus}{\doteq}}N(CH_3)_2]Cl^{\ominus}$$

XXIV   XXV

$$\downarrow {\scriptstyle H_2O}$$

$$RCH{=}CH{-}C{\equiv}CH \longleftarrow RCH{=}CH{-}\underset{\underset{Cl}{|}}{C}{=}CH{-}CHO$$

XXVII   XXVI

R = Aryl

Arylchloroformates XXVIII react with dimethylformamide to liberate carbon dioxide and form the iminium salts XXIX. Solvolysis of XXIX with methanol produces the corresponding phenol, N,N-dimethylformamide and methyl chloride ([41]).

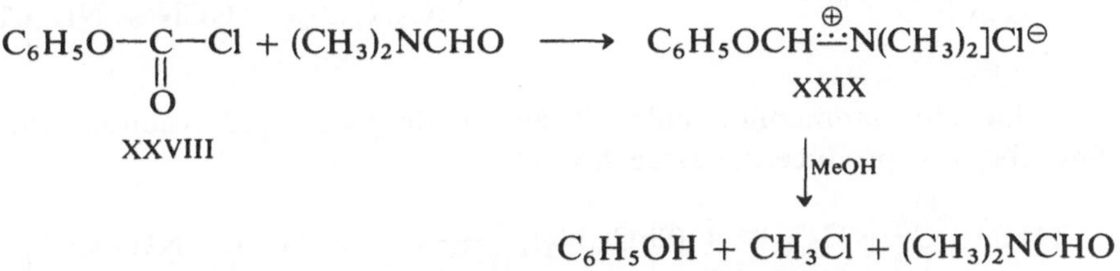

$$C_6H_5O{-}\underset{\underset{O}{\|}}{C}{-}Cl + (CH_3)_2NCHO \longrightarrow C_6H_5OCH{\overset{\oplus}{\doteq}}N(CH_3)_2]Cl^{\ominus}$$

XXVIII   XXIX

$$\downarrow {\scriptstyle MeOH}$$

$$C_6H_5OH + CH_3Cl + (CH_3)_2NCHO$$

The reaction of sulfur dichloride with N,N-dimethylformamide and thioformamide has been investigated by Hasserodt ([16]).

C. HALOFORMAMIDINES

The reaction of N-acylthioureas XXX with carbonyl chloride yields the chloroformamidines XXXI, which are readily dehydrochlorinated to N-acylcarbodiimides XXXII ([29]).

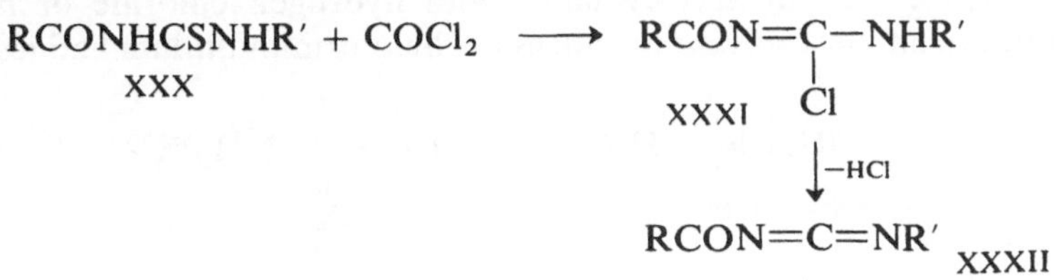

$$RCONHCSNHR' + COCl_2 \longrightarrow RCON{=}\underset{\underset{Cl}{|}}{C}{-}NHR'$$

XXX   XXXI

$$\downarrow {\scriptstyle -HCl}$$

$$RCON{=}C{=}NR'$$ XXXII

N-Acetylchloroformamidines XXXIII undergo reaction with methylene-bis-thioamides XXXIV to produce mixtures of 4H-1,3,5-thiadiazines XXXV and N-acylthioureas XXXVI ([15]).

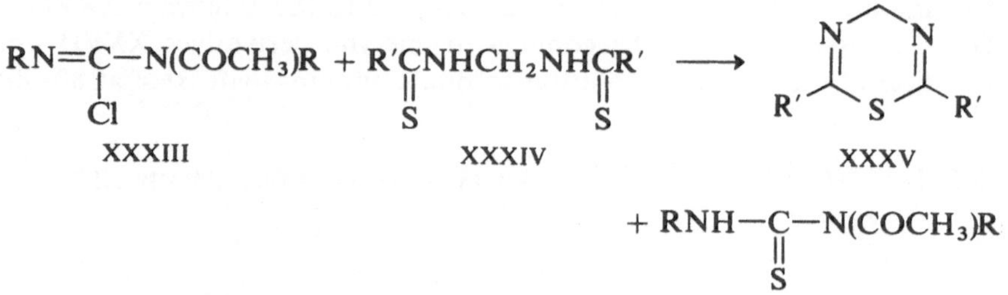

$$RN=\underset{\underset{Cl}{|}}{C}-N(COCH_3)R + R'\underset{\underset{S}{||}}{C}NHCH_2NH\underset{\underset{S}{||}}{C}R' \longrightarrow$$

XXXIII   XXXIV   XXXV

$$+ RNH-\underset{\underset{S}{||}}{C}-N(COCH_3)R$$

XXXVI

The synthesis of N-(chlorofluorophosphonyl)-N'-arylchloroformamidines has been reported by Derkach and Narbut ([9]).

Reaction of chloroformamidinium chloride (XXXVII) (cyanamide dihydrochloride) with diethyl iminocarbonate yields diethyl N-cyano-imidocarbonate (XXXVIII) ([1]).

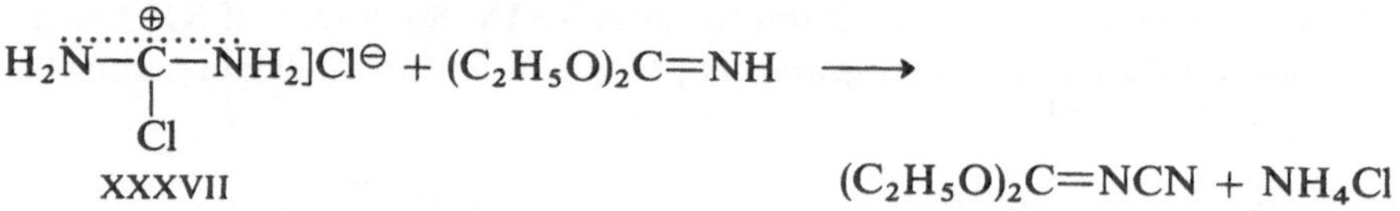

$$H_2\overset{\oplus}{N}-\underset{\underset{Cl}{|}}{C}-\overset{..}{N}H_2]Cl^{\ominus} + (C_2H_5O)_2C=NH \longrightarrow$$

XXXVII

$$(C_2H_5O)_2C=NCN + NH_4Cl$$

XXXVIII

Chloroformamidinium chloride also undergoes rapid reaction with $(SbCl_4N_3)_2$ to produce the azide XXXIX ([45]).

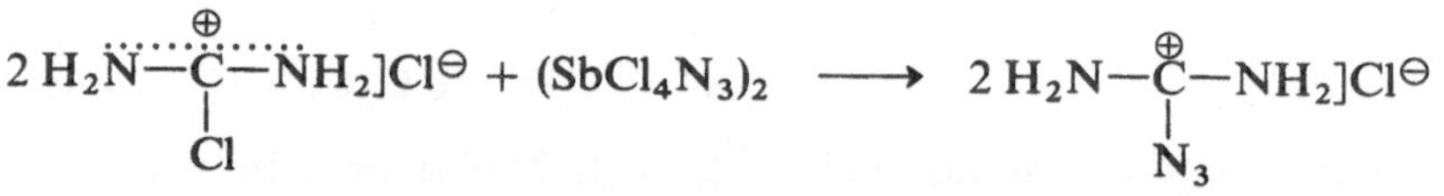

$$2\ H_2\overset{\oplus}{N}-\underset{\underset{Cl}{|}}{C}-\overset{..}{N}H_2]Cl^{\ominus} + (SbCl_4N_3)_2 \longrightarrow 2\ H_2N-\underset{\underset{N_3}{|}}{\overset{\oplus}{C}}-NH_2]Cl^{\ominus}$$

XXXIX

The synthesis of perfluorinated fluoroformamidines has been reported by Koshar and co-workers ([22]).

D. HALOFORMIMIDATES AND HALOTHIOFORMIMIDATES

The reaction of aryl cyanates with hydrogen chloride or hydrogen bromide in an inert solvent produces the haloformimidinium halides XL ([26]).

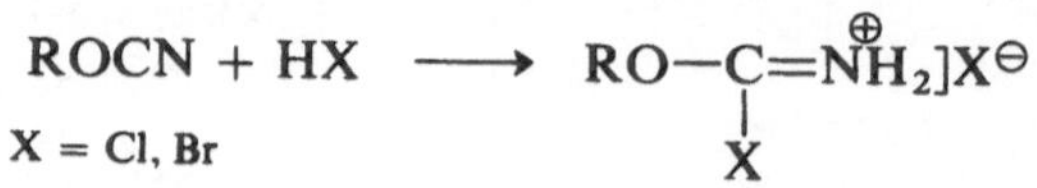

$$ROCN + HX \longrightarrow RO-\underset{\underset{X}{|}}{C}=\overset{\oplus}{N}H_2]X^{\ominus}$$

X = Cl, Br

XL

In a similar manner antimony pentachloride and aluminum trichloride form 1:1 adducts with aryl cyanates ([27]).

A comprehensive article related to the formation of aryl-1-(chloro-thio)formimidoyl chlorides XLI appeared recently ([40]), and the reaction of XLI with a variety of amines has been reported ([38]). The enormous reactivity of XLI was further evidenced by the formation of 1:1 adducts with alde-hydes and vinyl ethers ([39]). For example, reaction of XLI (R = C_6H_5) with vinylethyl ether yields 3-phenyl-1,3-thiazolin-2-one (XLII) ([39]).

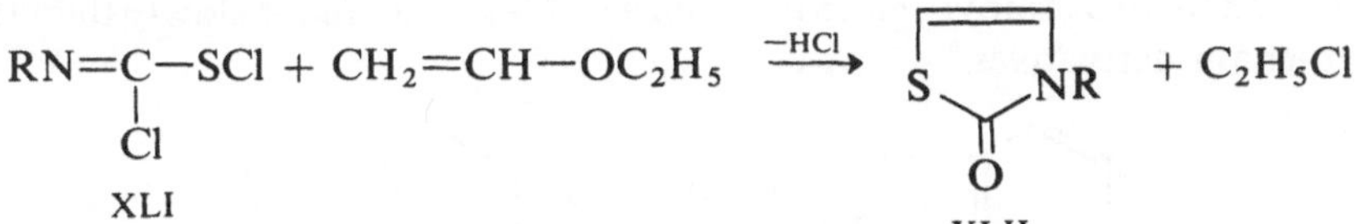

$$RN=C-SCl + CH_2=CH-OC_2H_5 \xrightarrow{-HCl} \quad \underset{\underset{XLII}{O}}{\overset{S \quad NR}{\diagdown}} \quad + C_2H_5Cl$$

$$\underset{XLI}{\overset{|}{Cl}}$$

E. HYDRAZIDOYL HALIDES

Further work related to the cyclization of the nitrile imides, generated from N-o-nitroarylhydrazidoyl bromides, has appeared ([2]), and the reaction of hydrazidoyl halides with azide ion ([19]) and phenylhydrazine ([18]) has been studied. Lozinskii et al. ([25]) have shown that chlorooxalylarylhydrazidoyl chlorides (see Chapter 7) react with azide ion preferentially on the halo group attached to the carbonyl group.

Dehydrohalogenation of hydrazidoyl halides in the presence of furan gave rise to the formation of the 1,3-cycloadducts XLIII of furan and the corresponding nitrile imides ([8]).

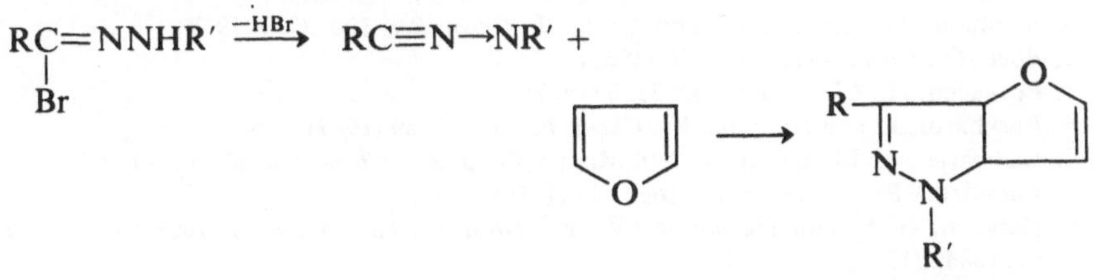

$$\underset{\underset{Br}{|}}{RC=NNHR'} \xrightarrow{-HBr} RC\equiv N{\to}NR' +$$

XLIII

F. CYCLIC IMIDOYL HALIDES

Further work related to the reaction of succinonitrile and glutaro-nitrile and some of its derivatives with hydrogen halide has been reported and the cyclic amidinium structures XLIV and XLV were verified by NMR studies ([10,11]).

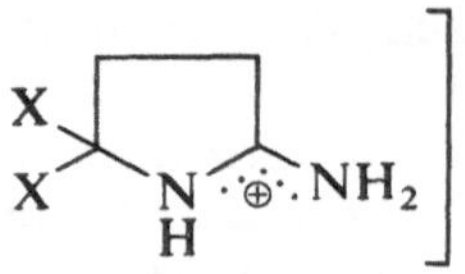

XLIV 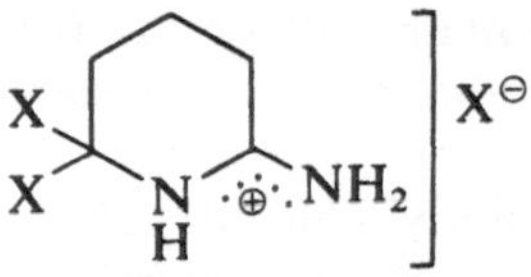   XLV

The nucleophilic substitution reactions of 3-chloro-1,2-benzisoxazole (XLVI) have been investigated by Böshagen ([5]), and the author noted that cyclic benzhydroxamoyl chlorides, such as XLVI, react more slowly than the open-chain derivatives.

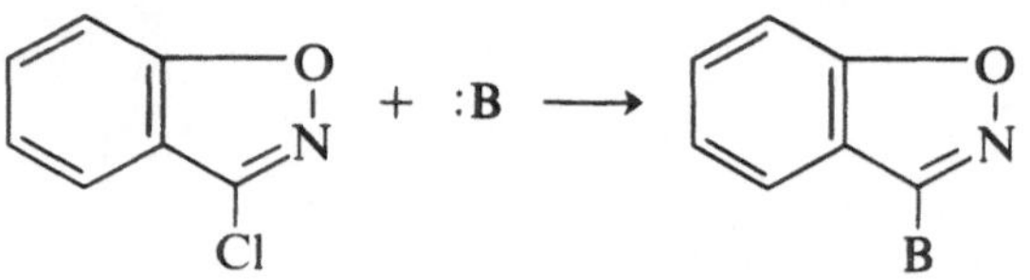

XLVI

$$B = OH^-, OR^-, R_2N^-, N_3^-$$

Kinetic investigations related to nucleophilic substitution reactions of halo-quinolines have been reported recently ([7,12,20,21]) and again the increased reactivity of the halo group adjacent to the ring nitrogen was verified.

REFERENCES

1. Allenstein, E., and Fuchs, R., *Chem. Ber.* **100**, 2604 (1967).
2. Barnish, I. T., and Gibson, M. S., *J. Chem. Soc.* (C), 8 (1968).
3. Biermann, U., Glemser, O., and Knaak, J., *Chem. Ber.* **100**, 3789 (1967).
4. Bock, G., *Chem. Ber.* **100**, 2870 (1967).
5. Böshagen, H., *Chem. Ber.* **100**, 3326 (1967).
6. Burkhardt, J., and Hamann, K., *Chem. Ber.* **100**, 2569 (1967).
7. Calligaris, M., Illuminati, G., and Manno, G., *J. Am. Chem. Soc.* **89**, 3518 (1967).
8. Caramella, P., *Tetrahedron Letters* 743 (1968).
9. Derkach, G. I., and Narbut, A. V., *Zh. Obshch. Khim.* **37**, 1364 (1967); *Chem. Abstr.* **67**, 10840 (1967).
10. Duquette, L. G., and Johnson, F., *Tetrahedron* **23**, 4517 (1967).
11. Duquette, L. G., and Johnson, F., *Tetrahedron* **23**, 4539 (1967).
12. Genel, F., Illuminati, G., and Marino, G., *J. Am. Chem. Soc.* **89**, 3516 (1967).
13. Grundmann, C., and Richter, R., *Tetrahedron Letters* 963 (1968).
14. Hall, D., Ummat, P. K., and Wade, K., *J. Chem. Soc.* (A), 1612 (1967).
15. Hartke, K., *Arch. Pharm.* **300**, 766 (1967).
16. Hasserodt, U., *Chem. Ber.* **101**, 113 (1968).
17. Hazebroucq, G., *Ann. Pharm. Franc.* **24**, 793 (1966); *Chem. Abstr.* **67**, 10854 (1967).
18. Hegarty, A. F., and Scott, F. L., *J. Chem. Soc.* (C), 2507 (1967).
19. Hegarty, A. F., Aylward, J. B., and Scott, F. L., *J. Chem. Soc.* (C), 2587 (1967).

20. Illuminati, G., Marino, G., and Sleiter, G., *J. Am. Chem. Soc.* **89**, 3510 (1967).
21. Illuminati, G., Linda, P., and Marino, G., *J. Am. Chem. Soc.* **89**, 3521 (1967).
22. Koshar, R. J., Husted, D. R., and Wright, C. D., *J. Org. Chem.* **32**, 3859 (1967).
23. Kresze, G., and Wucherpfennig, W., *Chem. Ber.* **101**, 365 (1968).
24. Lötzbeyer, J., and Bodendorf, K., *Chem. Ber.* **100**, 2620 (1967).
25. Lozinskii, M. D., Kukota, S. N., and Pel'kis, P. S., *Probl. Poluch. Poluprod. Prom. Organ. Sin.*, *Akad. Nauk SSSR* 193 (1967); *Chem. Abstr.* **68**, 12957 (1968).
26. Martin, D., and Weise, A., *Chem. Ber.* **100**, 3736 (1967).
27. Martin, D., and Weise, A., *Chem. Ber.* **100**, 3747 (1967).
28. Masaki, M., Fukui, K., and Ohta, M., *J. Org. Chem.* **32**, 3564 (1967).
29. Neidlein, R., and Heukelbach, E., *Arch. Pharm.* **299**, 709 (1966).
30. Neidlein, R., and Haussmann, W., *Arch. Pharm.* **300**, 180 (1967).
31. Neidlein, R., and Haussmann, W., *Arch. Pharm.* **300**, 553 (1967).
32. Neidlein, R., and Heukelbach, E., *Arch. Pharm.* **300**, 567 (1967).
33. Nikolajewski, H. E., Dähne, S., and Hirsch, B., *Chem. Ber.* **100**, 2616 (1967).
34. Ochiai, M., *Kagaku No Ryoiki* **19**, 900 (1965); *Chem. Abstr.* **66**, 32076 (1967).
35. Ogden, P. H., and Mitsch, R. A., *J. Am. Chem. Soc.* **89**, 3868 (1967).
36. Ogden, P. H., and Mitsch, R. A., *J. Am. Chem. Soc.* **89**, 5007 (1967).
37. Ogden, P. H., *J. Chem. Soc.* (C), 2302 (1967).
38. Ottmann, G., and Hooks, H., *Angew. Chem.* **79**, 1062 (1967).
39. Ottmann, G., Hoberecht, H., and Hooks, H., *Angew. Chem.* **79**, 1063 (1967).
40. Ottmann, G., and Hooks, H., *J. Heterocycl. Chem.* **4**, 365 (1967).
41. Pattison, V. A., Colson, J. G., and Carr, R. L. K., *J. Org. Chem.* **33**, 1084 (1968).
42. Roesky, H. W., and Biermann, U., *Angew. Chem.* **79**, 904 (1967).
43. Schmelzer, H. G., Degener, E., Holtschmidt, H., and Heitzer, H., *Tetrahedron Letters* 2801 (1967).
44. Schmidt, A., *Chem. Ber.* **100**, 3319 (1967).
45. Schmidt, A., *Chem. Ber.* **100**, 3725 (1967).
46. Stojanovic, F. M., and Arnold, Z., *Collection Czech. Chem. Commun.* **32**, 2155 (1967).
47. Thyagarajan, B. S., and Gopalakrishnan, P. V., *Tetrahedron* **23**, 3851 (1967).
48. Tullock, C. W., United States Pat. 3,347,644 (1967).

AUTHOR INDEX

Numbers in parentheses are reference numbers and indicate that an author's work is referred to although his name is not cited in the text. Numbers in italics show the page on which the complete reference is listed.

SUBJECT INDEX

Headings and subheadings in italics refer to working examples.